Methane and Methanol Utilizers

BIOTECHNOLOGY HANDBOOKS

Series Editors: Tony Atkinson and Roger F. Sherwood
                PHLS Centre for Applied Microbiology and Research
                Division of Biotechnology
                Salisbury, Wiltshire, England

Volume 1 *PENICILLIUM* AND *ACREMONIUM*
             Edited by John F. Peberdy

Volume 2 *BACILLUS*
             Edited by Colin R. Harwood

Volume 3 CLOSTRIDIA
             Edited by Nigel P. Minton and David J. Clarke

Volume 4 *SACCHAROMYCES*
             Edited by Michael F. Tuite and Stephen G. Oliver

Volume 5 METHANE AND METHANOL UTILIZERS
             Edited by J. Colin Murrell and Howard Dalton

Volume 6 PHOTOSYNTHETIC PROKARYOTES
             Edited by Nicholas H. Mann and Noel G. Carr

A Continuation Order Plan is available for this series. A continuation order will bring delivery of each new volume immediately upon publication. Volumes are billed only upon actual shipment. For further information please contact the publisher.

Methane and Methanol Utilizers

Edited by
J. Colin Murrell and
Howard Dalton
University of Warwick
Coventry, England

Plenum Press • New York and London

Library of Congress Cataloging-in-Publication Data

Methane and methanol utilizers / edited by J. Colin Murrell and Howard
Dalton.
       p.   cm. -- (Biotechnology handbooks ; v. 5)
    Includes bibliographical references and index.
    ISBN 0-306-43878-X
    1. Methylotrophic bacteria. 2. Methylotrophic microorganisms.
 I. Murrell, J. C. (J. Colin) II. Dalton, Howard. III. Series.
 QR92.M47M46 1992
 589.9--dc20   91-42281
                                                                 CIP

ISBN 0-306-43878-X

© 1992 Plenum Press, New York
A Division of Plenum Publishing Corporation
233 Spring Street, New York, N.Y. 10013

All rights reserved

No part of this book may be reproduced, stored in a retrieval system, or transmitted in any form or by any means, electronic, mechanical, photocopying, microfilming, recording, or otherwise, without written permission from the Publisher

Printed in the United States of America

Contributors

Howard Dalton • Department of Biological Sciences, University of Warwick, Coventry CV4 7AL, England

W. de Koning • Department of Microbiology, University of Groningen, 9751 NN Haren, The Netherlands

G. E. de Vries • Department of Microbiology, University of Groningen, 9751 NN Haren, The Netherlands

L. Dijkhuizen • Department of Microbiology, University of Groningen, 9751 NN Haren, The Netherlands

Peter N. Green • NCIMB Ltd., Aberdeen AB2 1RY, Scotland

Richard S. Hanson • Gray Freshwater Biological Institute, University of Minnesota, Navarre, Minnesota 55392

W. Harder • TNO-Institute of Environmental Sciences, 2600 AE Delft, The Netherlands

David J. Leak • Centre for Biotechnology, Imperial College of Science, Technology and Medicine, London SW7 2AZ, England

P. R. Levering • Microbiological R&D Laboratories, Organon International BV, 5340 BH Oss, The Netherlands

Mary E. Lidstrom • Environmental Engineering Science, California Institute of Technology, Pasadena, California 91125

J. Colin Murrell • Department of Biological Sciences, University of Warwick, Coventry CV4 7AL, England

Preface

Methane and its oxidation product, methanol, have occupied an important position in the chemical industry for many years: the former as a feedstock, the latter as a primary chemical from which many products are produced. More recently, the role played by methane as a potent "greenhouse" gas has aroused considerable attention from environmentalists and climatologists alike. This role for C_1 compounds has, of course, been quite incidental to the myriad of microorganisms on this planet that have adapted their life-styles to take advantage of these readily available ambient sources. Methane, a renewable energy source that will always be with us, is actually a difficult molecule to activate; so any microorganism that can effect this may point the way to catalytic chemists looking for controllable methane oxidation. Methanol, formed as a breakdown product of plant material, is also ubiquitous and has also encouraged the growth of prokaryotes and eukaryotes alike.

In an attempt to give a balanced view of how microorganisms have been able to exploit these simple carbon sources, we have asked a number of leading scientists (modesty forbids our own inclusion here) to contribute chapters on their specialist areas of the subject.

This collection is not a comprehensive treatise, but a selective series of articles aimed at providing the reader with a broad overview from which can be distilled the essence of the subject. Each article is rich in primary references to facilitate the acquisition of more detailed information. Following an introductory chapter and a chapter on the general and nomenclatural characteristics of methylotrophs, there are specialized chapters on the physiology, biochemistry, and molecular biology of the principal characters—the methane and methanol utilizers. We can, of course, read and understand each chapter in its own right as an individual authoritative reference, but to understand and appreciate the breadth and depth of our current understanding, we should assimilate the whole volume. Such an eclectic approach is necessary to fully appreciate the industrial potential afforded by methylotrophs, and this is essentially encapsulated in the final chapter, which is on their biotechnological applications.

J. Colin Murrell
Howard Dalton

Coventry, England

Contents

Chapter 1

Introduction 1

Richard S. Hanson

1. Distribution in Nature of Reduced One-Carbon Compounds and Microbes that Utilize Them 1
   Isolation and Characteristics of Methylotrophic Bacteria... 3
2. Ecology of Methylotrophs................................. 7
3. Methylotrophs and Biotechnology 9
4. Genetics of Methylotrophs 11
5. Conclusions... 14
   References... 15

Chapter 2

Taxonomy of Methylotrophic Bacteria 23

Peter N. Green

1. Introduction.. 23
2. Obligate Methylotrophs 25
   2.1. Methane-Utilizing Bacteria 25
   2.2. Taxonomic Structure within the Methanotrophs...... 45
   2.3. Obligate Methanol and Methylated Amine Utilizers ... 53
3. Facultative Methylotrophs 62
   3.1. The Genus *Methylobacterium*....................... 63
   3.2. *Pseudomonas aminovorans* and Related Strains 67
   3.3. Other Non-Pink-Pigmented Motile Rods 69
   3.4. Nonmotile Rods and Coccobacilli................... 70
   3.5. The Hyphomicrobia................................ 71
   3.6. Gram-Positive Facultative Methylotrophs 72

3.7.	Facultative Autotrophs and Phototrophs	72
3.8.	Facultatively Methylotrophic Marine Bacteria	74
3.9.	Anaerobic Methanol-Utilizing Bacteria	74
3.10.	Methylotrophic Halophilic Methanogens	74
4.	Phylogeny of Methylotrophs	75
	References	77

Chapter 3

Methane Oxidation by Methanotrophs: Physiological and Mechanistic Implications 85

Howard Dalton

1.	Introduction	85
2.	Chemical Versus Biological Catalysis	86
3.	Particulate Methane Monooxygenase	88
4.	Soluble Methane Monooxygenase	90
	4.1. The Three Protein Components	90
	4.2. The Hydroxylase	93
5.	Kinetic Mechanism of the MMO Reaction	98
6.	Substrate Specificity—Mechanistic Implications	99
	6.1. Oxygen or Hydrocarbon Activation?	99
	6.2. Substrate Oxidation	101
	6.3. Mechanism of Action of MMO	103
7.	The Physical Nature of the Active Site	108
	References	111

Chapter 4

The Genetics and Molecular Biology of Obligate Methane-Oxidizing Bacteria 115

J. Colin Murrell

1.	Introduction	115
2.	Mutagenesis	116
	2.1. Conventional Mutagenesis	116
	2.2. Dichloromethane Mutagenesis	117

3. Gene Transfer Systems...... 118
   3.1. Plasmids in Methanotrophs...... 118
   3.2. Bacteriophages of Methanotrophs...... 120
   3.3. Conjugation Systems...... 120
   3.4. Transformation Studies and Electroporation...... 123
4. Molecular Biology of Methanotrophs...... 123
   4.1. Nitrogen Metabolism Genes...... 124
   4.2. C_1-Specific Genes...... 130
5. Future Developments...... 142
   References...... 144

Chapter 5

The Physiology and Biochemistry of Aerobic Methanol-Utilizing Gram-Negative and Gram-Positive Bacteria...... 149

L. Dijkhuizen, P. R. Levering, and G. E. de Vries

1. Introduction...... 149
2. The Ribulose Monophosphate Cycle of Formaldehyde Fixation...... 151
   2.1. Discovery and General Properties...... 151
   2.2. Occurrence of the RuMP Pathway of Formaldehyde Fixation...... 154
   2.3. Distribution of Variants of the RuMP Pathway...... 156
3. Biochemistry of Enzymes Involved in Methanol Oxidation... 157
   3.1. Methanol Dehydrogenase in Gram-Negative Bacteria...... 157
   3.2. Methanol Dehydrogenase in Gram-Positive Bacteria... 158
   3.3. Oxidation of Formaldehyde and Formate...... 160
   3.4. Oxidation of Formaldehyde in Methylotrophs Using the RuMP Cycle...... 162
   3.5. Hexulosephosphate Synthase and Hexulosephosphate Isomerase...... 166
4. Metabolic Regulation...... 168
   4.1. Regulation of Methanol Dehydrogenase Synthesis.... 168
   4.2. Regulation of the RuMP Cycle of Formaldehyde Fixation...... 169
   References...... 173

Chapter 6

The Genetics and Molecular Biology of Methanol-Utilizing Bacteria 183

Mary E. Lidstrom

1.	Introduction	183
2.	Plasmids in Methanol-Utilizing Bacteria	186
3.	Mutagenesis in *Methylobacillus flagellatum*	187
4.	Transposon Mutagenesis	187
5.	Chromosomal Mapping	188
	5.1. Auxotrophy Markers in *Methylophilus* Strains	189
	5.2. R-Prime Mapping in *Methylobacterium extorquens* AM1	189
	5.3. Hfr-Like Donors in *M. flagellatum*	190
6.	Expression of Foreign Genes in Methanol-Utilizing Bacteria	191
7.	Methanol Oxidation Genes	191
	7.1. Localization and Identification of Mox Genes in *Methylobacterium* Strains	192
	7.2. Phenotypic Characterization of Mox Mutants in *Methylobacterium* Strains	194
	7.3. Characterization of Mox Genes	196
8.	PQQ Synthesis Genes	199
9.	Genes for Assimilation of C_1 Units	200
	9.1. Serine Cycle Genes	200
	9.2. Ribulose Monophosphate Cycle Genes	202
10.	Summary and Future Prospects	202
	References	203

Chapter 7

Methanol-Utilizing Yeasts 207

W. de Koning and W. Harder

1.	Introduction	207
2.	Isolation and Properties	209
	2.1. Enrichment and Isolation	209
	2.2. Taxonomy	210

3.	Physiology and Biochemistry.............................	213
	3.1. Enzymology and Compartmentation of Methanol Metabolism..	213
	3.2. Regulation of the Synthesis and Activity of Enzymes Involved in Methanol Metabolism	218
	3.3. Regulation of the Flux of Formaldehyde over Dissimilatory and Assimilatory Pathways.............	220
4.	Genetics and Molecular Biology	222
	4.1. Mutagenesis and Auxotrophic Mutants..............	222
	4.2. Classical Genetics................................	223
	4.3. Methanol Pathway Mutants.......................	224
	4.4. Properties of Methanol-Regulated Genes	225
	4.5. Transformation Systems and Vectors	226
	4.6. Expression of Heterologous Genes..................	227
5.	Applications	228
	5.1. General Economics...............................	228
	5.2. Single-Cell Protein	229
	5.3. Enzymes...	231
	5.4. Bulk and Fine Chemicals..........................	232
	5.5. Hosts for Heterologous Gene Expression............	234
6.	Concluding Remarks.....................................	235
	References...	236

Chapter 8

Biotechnological and Applied Aspects of Methane and Methanol Utilizers 245

David J. Leak

1.	Introduction..	245
2.	Biotechnology and Economics...........................	246
3.	Processes Highly Dependent on Substrate Economics.......	247
	3.1. Single-Cell Protein	248
	3.2. SCP from Methanol	249
4.	Metabolic Products from Methylotrophs	256
	4.1. Amino Acids.....................................	257
	4.2. Vitamins and Coenzymes..........................	261
	4.3. Carboxylic Acids	262

5.	Microbial Polymers from C_1 Substrates	262
	5.1. Polysaccharides	263
	5.2. Poly-β-hydroxybutyrate (PHB)	264
6.	Metabolic Activities and Enzymes from Methylotrophs	265
	6.1. Cofactor Regeneration	265
	6.2. Analysis and Biosensors	266
	6.3. Methylotrophs as Biocatalysts. 1. Synthesis	267
	6.4. Methylotrophs as Biocatalysts. 2. Biodegradation	271
7.	The Future Outlook	272
	References	273

Species Index 281

Subject Index 283

Introduction 1

RICHARD S. HANSON

1. DISTRIBUTION IN NATURE OF REDUCED ONE-CARBON COMPOUNDS AND MICROBES THAT UTILIZE THEM

I perceive the purpose of an introductory chapter as an opportunity to integrate the topics of interest to be found later in a volume without unnecessarily duplicating information. It has taken the form of a brief historical perspective, as well as a means of identifying some gaps in information needed to serve the needs of biotechnologists.

Methylotrophs are a very diverse group of microbes united by a common physiological ability to obtain energy from the oxidation of reduced one-carbon compounds and an ability to assimilate formaldehyde derived from the primary substrate for the synthesis of cell material (Anthony, 1982; Zatman, 1981). All methylotrophs also assimilate carbon dioxide via aneplerotic routes and *Methylococcus capsulatus* is capable of limited autotrophic CO_2 fixation via the Calvin–Benson cycle (Anthony, 1982; Zatman, 1981; see also Stanley and Dalton, 1982). Some methylotrophs, particularly *Methylomonas* species, is stimulated by heterotrophic carbon sources (Whittenbury *et al.*, 1970; Zhao and Hanson, 1984). This observation may reflect the difficulty of isolating such strains. Bacteria are known that assimilate CO_2 rather than formaldehyde as the major carbon source but use methanol and methylamines as energy sources (Anthony, 1982; Zatman, 1981).

The traditional definitions of methanotrophs (methane utilizers) and methylotrophs (C_1-utilizing microbes) are adequate for all but a small minority of microbes that utilize reduced C_1 compounds and should not be

---

RICHARD S. HANSON • Gray Freshwater Biological Institute, University of Minnesota, Navarre, Minnesota 55392.

Methane and Methanol Utilizers, edited by J. Colin Murrell and Howard Dalton. Plenum Press, New York, 1992.

discarded. A knowledge of the exceptional cases may be important for certain applications, and for that reason, some are described in this chapter.

Methane is the most abundant one-carbon compound in nature. It is the most stable of carbon compounds in anoxic environments and is produced as an end product of the anaerobic catabolism of recently alive organic matter, by methanogenic bacteria (Dagley, 1975). Methane is produced in soils, ponds, marshes, and the digestive tracts of ruminants and other animals. The annual, global, biogenic production of methane has been estimated to be $0.5-1 \times 10^{15}$ g per year (Ehalt, 1976).

The atmospheric concentration of methane remained constant at 700 mg/liter from 10,000 years ago until approximately 300 years ago and then increased to about 1.7 mg/liter at present (Graedel and Crutzen, 1989; Crutzen and Graedel, 1986). The atmospheric content of methane is estimated to be nearly 5×10^{15} g and is increasing at the rapid rate of 1% per year (Fischer *et al.*, 1985). Its concentration may double in the next century if global sources increase as expected (Graedel and Crutzen, 1989). The impact of methane on global warming is predicted to approach that of carbon dioxide if these predictions are correct (Graedel and Crutzen, 1989; Houghton and Woodwell, 1989; Khalil and Rassmussen, 1986). These predictions may set the tone for future volumes on this group of bacteria.

Methanol is formed by chemical oxidation of methane in the troposphere and descends in rainfall (Ehalt, 1976), as well as from methoxyl groups of plant lignin and pectins (Anthony, 1982). Formaldehyde does not persist in amounts sufficient to account for significant biomass production in nature and is not exclusively metabolized by microbes. Formic acid is produced as an end product of the fermentation of organic compounds and as a by-product of industrial processes. Methylamines (mono-, di-, and trimethylamine), the other major substrates for methylotrophs, are metabolic products of the decomposition of fish and plants (Anthony, 1982). Trimethylamine *N*-oxide, an osmoregulatory compound abundant in marine fish and invertebrates, is reduced to methylamines (Anthony, 1982).

Carbon monoxide and cyanide are substrates utilizable for the growth of methylotrophs according to most of the definitions of this group of microbes. However, the bacteria that utilize carbon monoxide, the *Carboxydobacteria* and the cyanide-utilizing bacteria (Whittenbury and Krieg, 1984), are physiologically and taxonomically different from methylotrophs.

It is known that bacteria are able to use methylated sulfur species as sources of carbon and energy. *Hyphomicrobium* E.G. grows on dimethylsulfoxide and low concentrations of dimethylsulfide (Suylen and Kuenen,

1986). The bacteria of this genus deserve special attention of applied microbiologists because of their versatility and ability to utilize toxic waste compounds that are not metabolized by other methylotrophs. They may be useful in the removal of volatile and soluble dimethylsulfide and methane thiol found in paper mill effluents (Large and Bamforth, 1988). *Hyphomicrobia* can also be used to remove halomethanes, methyl sulfates, methyl-, and ethylamines and for denitrification of streams from aerobic sewerage treatment processes (Claus and Kutzner, 1985; Galli and Leisinger, 1985; Large and Bamforth, 1988).

Isolation and Characteristics of Methylotrophic Bacteria

The first methylotrophic bacterium was described by Leow in 1892 (Quayle, 1987). Quayle has pointed out that this pink-pigmented bacterium named *Bacillus methylicus* that grew on methanol, methylamine, formic acid, and several multicarbon compounds was identical in many characteristics to *Methylobacterium* species strain AM1 and other pink-pigmented facultative methylotrophs isolated many years later in several laboratories (Hanson et al., 1980, 1989; Green and Bousefield, 1983; Jenkins and Jones, 1987).

Söhngen (1906) isolated the first methane-oxidizing bacterium, which he named *Bacillus methanicus*. He recognized that methane was produced in large amounts and suggested that its presence in low concentrations in the atmosphere was due to its oxidation by microorganisms.

Autotrophic metabolism in bacteria had been known for some time before metabolism of reduced one-carbon compounds was studied. J. W. Foster's group at the University of Texas, Austin (Dworkin and Foster, 1956; Leadbetter and Foster, 1958) began studies of the distribution and characteristics of methanotrophs, and Dworkin and Foster reported the isolation of a bacterium that resembled *B. methanicus* 50 years after Söhngen. This bacterium was renamed *Methanomonas methanica* by Orla Jensen (1909) and *Pseudomonas methanica* by Dworkin and Foster (1956). This bacterium was a gram-negative rod motile with a single polar flagellum, which formed pink mucoid colonies on a mineral salts agar under an atmosphere containing methane. Like most methane utilizers described later, this bacterium was an obligate methanotroph. Only methane and methanol supported growth, although formate and formaldehyde were also oxidized by resting cells.

Whittenbury and his co-workers (1970) compared over 100 methane-utilizing bacteria and divided them into five groups (proposed genera): *Methylococcus*, *Methylobacter*, *Methylomonas*, *Methylocystis*, and *Methylosinus*. The assignment of the bacteria into five groups (proposed genera) was based on morphology, resting stages formed, intracytoplasmic membrane structures, and some physiological characteristics (Whittenbury et al., 1970; Whittenbury and Dalton, 1981).

The methanotrophic bacteria have been divided into three groups (I, II, and X) based on resting stages formed, pathways utilized for formaldehyde assimilation, DNA base composition, intracytoplasmic membrane fine structure, and characteristics described later in this volume. Group I and group X methylotrophs assimilate formaldehyde via the efficient ribulose monophosphate pathway, while group II methylotrophs utilize the less efficient serine pathway for the same purpose (Lawrence and Quayle, 1970). The yield of cell carbon (or products derived from assimilated carbon) per mole of substrate utilized is higher in microbes that utilize the Ribulose monophosphate (RuMP) pathway (Anthony, 1982; Higgins et al., 1981).

The methanotrophs differ from the methylotrophic bacteria that do not utilize methane in that the latter bacteria do not contain complex intracytoplasmic membrane structures, lack methane monooxygenase, and few isolates are known to form restings. The morphological and physiological diversity of methanol- and methylamine-utilizing microbes is much greater than that of methanotrophs (Anthony, 1982).

Some methanol- and methylamine-oxidizing bacteria are obligate methylotrophs unable to utilize substrates containing carbon–carbon bonds as energy sources or restricted facultative methylotrophs. Nearly all obligate methanol-utilizing bacteria resemble each other and *Methylophilus methylotrophus* (Anthony, 1982). They are rod-shaped gram-negative bacteria with a mole G + C ratio of 52–56%, utilize the ribulose monophosphate pathway for carbon assimilation, and contain fatty acids with 16 carbon atoms as the predominant phospholipid fatty acids. Restricted facultative methylotrophs grow slowly on only one or two multicarbon compounds (usually sugars), e.g., organism W3A1 (Anthony, 1982).

Other restricted facultative methylotrphs that grow on multicarbon compounds but a more restricted range of compounds than used by facultative methylotrophs include members of the genera *Bacillus* and *Hyphomicrobium*. Species of *Hyphomicrobium* are capable of utilizing nitrate as a terminal electron acceptor for anaerobic respiration (Harder and Attwood, 1978).

A group of related bacteria collectively known as pink-pigmented facultative methylotrophs (PPFMs) (Jenkins and Jones, 1987) include the well-studied *Pseudomonas* AM1, now known as *Methylobacterium extorquens* AM1 (Green and Bousefield, 1983), and *Methylobacterium organophilum* XX (Patt et al., 1974). The bacteria have in common the formation of red pigments, assimilation of formaldehyde via the serine pathway, and growth on a wide variety of multicarbon compounds and rich media. They are aerobic, gram-negative rods containing DNA with a mole G + C content of approximately 62–66%. Several other nonpigmented facultative methylotrophs that grow on methanol, methylamines, or both have been given the generic names *Pseudomonas, Arthrobacter, Klebrsiella, Acinetobacter, Arthrobacter,* and *Alcaligenes*. Some isolates closely related to PPFMs are capable of rapid growth on dichloromethane as a carbon and energy source and contain large amounts of an inducible dichloromethane dehalogenase (Scholtz *et al.*, 1988; Kohler-Staub and Leisinger, 1985; Galli and Leisinger, 1985). Dichloromethane-utilizing bacteria that employ the serine and RuMP pathways for formaldehyde fixation are also known (Scholtz *et al.*, 1988).

The diversity of bacteria that can utilize C_1 energy sources is further illustrated by considering the nutritional versatility of *Paracoccus denitrificans* and *Rhodopseudomonas acidophila*. *P. denitrificans*, a gram-negative, rod-shaped bacterium, can grow heterotrophically on a wide variety of carbon sources or on methanol under aerobic conditions or anaerobically with nitrate, nitrite, or nitrous oxide as electron acceptors. Hydrogen and thiosulfate can be utilized as energy sources for chemolithotrophic growth. Carbon dioxide is fixed via the Calvin–Benson cycle, not via the pathways for formaldehyde assimilation utilized by true methylotrophs (Anthony, 1982). *R. acidophila*, a facultative phototroph, can grow aerobically in the dark using methanol or formate as carbon and energy sources or anaerobically in the light using methanol or other organic compounds as sources of reducing power. In both cases, CO_2 is fixed via the Calvin cycle as a source of carbon. These bacteria and *Pseudomonas oxalaticus*, which can all obtain energy by oxidation of C_1 compounds and fix CO_2 via the Calvin cycle have been called *pseudomethylotrophs* by Zatman (1981) and autotrophic methylotrophs by Large and Bamforth (1988).

A few (three) mycelial fungi and several yeast isolates belonging to the genera *Hansenula, Candida, Pichia, Torulopsis, Kloeckera,* and *Saccharomyces* are able to utilize methanol as a carbon and energy source (Anthony, 1982). The fungi differ in both the properties of the catabolic enzymes and the pathways used for formaldehyde assimilation.

Descriptions of unusual methylotrophs including methanotrophic yeasts (Wolfe, 1981) and facultative methanotrophs have been published (Hanson, 1980; Reed and Dugan, 1987). The methanotrophic yeasts grow very slowly on methane and are of little ecological or biotechnological interest at present. Facultative methanotrophs are difficult to maintain. Unusual physiological variants, including group I methanotrophs that possess a complete tricarboxylic acid cycle, have also been described (Zhao and Hanson, 1984).

Consortia of methanotrophs and higher organisms have been described by Cavanaugh *et al.* (1987) and Childress *et al.* (1986). Characterizations of the methanotrophs involved in these consortia have not been completed.

The descriptions of methylotrophic organisms provided above serve to briefly illustrate the biochemical diversity of this group of microbes. An appreciation of this diversity is important when new biotechnological applications are considered. Each bacterium presumably survived because it adjusted to an ecological niche. The diversity of characteristics is so great that an organism better suited to a specific process than others can be selected. When we undertook a program for the production from methanol of amino acids for animal feeds, we decided that a thermophilic gram-positive bacterium would be most suitable. The choice was made because gram-positive bacteria have been shown to be good excreters of low-molecular-weight compounds, auxotrophic mutants of gram-negative bacteria were difficult to obtain, and a thermophilic bacterium would reduce the expense of cooling fermenters in which methanol was rapidly metabolized. The choice has proven to be a good one. Thermophilic gram-positive, endospore-forming methylotrophs are easily isolated (Dijkhuizen *et al.*, 1988; Schendel *et al.*, 1990). They utilize the efficient RuMP pathway for carbon assimilation, grow to high cell densities, and auxotrophs are isolated at frequencies similar to those observed with other bacteria. Analog-resistant auxotrophic mutants have been shown to excrete relatively large amounts (20 g·l^{-1}) of the amino acid lysine and grow at high cell densities (Schendel, *et al.*, 1990). These bacteria contain a NAD$^+$-linked methanol dehydrogenase (MDH) and yield more biomass per mole of methanol utilized than gram-negative methylotrophs that contain a cytochrome C_L-linked MDH (Alawadhi *et al.*, 1988; Brooke *et al.*, 1989).

For other biotechnological purposes, microbes with different characteristics are more suitable. The classification of methylotrophs according to an officially accepted set of genera and species is an important and uncompleted task (Whittenbury and Dalton, 1981). It is reasonably clear that the division of methanotrophs into three groups (I, II, and X), based on intracytoplasmic membrane structures, the presence or absence of key enzymes of central metabolic routes, dominant phospholipid fatty acids present, pathways for formaldehyde assimilation, base content of DNAs, and other characteristics, as proposed by Whittenbury and Krieg (1984), is justified and widely accepted. Many strains of methanotrophs have been isolated and characterized since 1970 (Hanson *et al.*, 1990). A decision concerning the separation of bacteria within groups I and II into one, two, or more genera remains to be made. New information on phospholipid composition, biochemical characteristics, DNA/DNA homologies, and a limited but ever-increasing number of 5S and 16S ribosomal RNA sequences is available (Wolfrum and Stolp, 1987; Tsuji *et al.*, 1990). The information required to make informed taxonomic decisions should soon be available. The division of methylotrophs into 12 groups that are physiologically related has provided a useful means of comparing them in the absence of officially defined genera (Anthony, 1982).

The oxidation of methane under anaerobic conditions has been reported by several workers (Hanson, 1980; Lidstrom, 1983; Reeburgh, 1980; Remsen *et al.*, 1989). The terminal electron acceptor(s) have not been identified in natural environments and the organisms responsible remain to be isolated in pure culture. The biochemistry of the anaerobic oxidation of fully reduced carbon should provide new insights into microbial biochemistry. The importance of anaerobic methane oxidation in the carbon cycle also needs to be estimated with greater precision.

2. ECOLOGY OF METHYLOTROPHS

Nearly all samples taken from muds, swamps, rivers, rice paddies, oceans, and ponds, soils from meadows and deciduous woods, streams, sewage sludge, and several other habitats yielded methylotrophs in enrichments (Corpe, 1985; Hanson, 1980; Heyer, 1977; Hutton and Zobell, 1949; Whittenbury *et al.*, 1970; Sieburth *et al.*, 1987; Strand and Lidstrom, 1984). Nearly all samples yield a variety of thermophilic methanotrophs (Whittenbury and Dalton, 1981). However, the dominant methanotrophs in any environment have not been identified (Hanson, 1980; Hanson *et al.*, 1989). The number of methanotrophs isolated as colony forming units from pure cultures has been a small fraction of the viable cell population, and the fraction of viable cells recovered from environmental samples is perceived to be smaller (Hanson, 1980). The type of methanotroph isolated from a given sample may reflect the conditions employed for growth more than the dominant type in the original population (Whittenbury *et al.*, 1970; Whittenbury and Dalton, 1981; Whittenbury and Krieg, 1984).

Methanotrophs in freshwater and marine environments examined appear to exist under microaerophilic conditions and require concentrations of methane well below that required for saturation of aqueous solutions (Abramochinka *et al.*, 1987; Hanson, 1980; Laurinavichus *et al.*, 1978; Lidstrom, 1988; Rudd and Taylor, 1980). Growth of some species requires removal of toxic by-products of metabolism and production of cofactors by consorts unless these cofactors have been identified and supplied (Hanson, 1980). It is therefore, not clear that all naturally occurring strains are identified.

One can estimate the relative population sizes of group I and group II methylotrophs using biochemical assays if the sample is sufficient. Assays for enzymes of the RuMP and serine pathways are suitable for these purposes. One can measure the total population of gram-negative methylotrophs because the MDH gene of all gram-negative methylotrophs is highly conserved (Hanson et al., 1989). Assuming a gene copy number (one per genome in all strains examined to date), an average genome molecular weight, and the number of genomes per cell; the amount of an MDH gene probe that hybridizes to total DNA from an environmental sample can be used to estimate the population size.

The diversity of methylotrophs in a population can be estimated using a variety of methods. MDH gene restriction fragment polymorphisms may be useful in estimating the diversity of methylotrophs, the identification of dominant species in environmental samples, and changes in populations with time (Hanson et al., 1989). Fluorescent antibodies prepared against cell surface antigens may be useful for the same purposes (Saralov et al.,1984).

Methanol dehydrogenases can be divided into six groups differing in isoelectric points and mobilities during electrophoresis in polyacrylamide gels used for isoelectric focusing (Anthony, 1982). After transfer to nitrocellulose filters (Western blotting), the proteins can be specifically detected in complex mixtures using antibodies prepared against each antigenic group of MDH's (Hanson et al., 1989). There appear to be no more than two or perhaps three antigenically different groups of MDH proteins (Anthony, 1982). Antibodies prepared against soluble methane monooxygenases can similarly be employed, although it is not clear if soluble MMO is synthesized in environments of interest or if all methanotrophs synthesize soluble MMO. It is important, however, to know if and how much of the enzyme is synthesized for many applications because of its unique ability to catalyze some biotransformations (Dalton and Higgins, 1987; Hou, 1984; Higgins et al., 1981; Stirling et al., 1979).

The use of ribosomal RNA signature probes offers considerable promise. The sequences of 16S ribosomal RNA molecules from the serine pathways methylotrophs and those possessing the RuMP pathway differ considerably (Tsuji et al., 1989). We have synthesized a signature probe that specifically hybridizes to ribosomal RNA of type II methylotrophs and another that hybridizes to ribosomal RNA of type I methanotrophs. These probes may be useful for the enumeration of single cells of each type in complex communities (DeLong et al., 1989) such as those that are expected to be encountered in natural environments and some bioreactors. When a sufficient number of ribosomal RNA sequences are available, it is probable that population size of subgroups of type I and type

II methylotrophs can be estimated by these approaches. The advantages of ribosomal RNA hybridization for the detection and quantitation of groups of organisms in environmental samples have been reviewed by Giovannoni *et al.* (1988).

3. METHYLOTROPHS AND BIOTECHNOLOGY

Interest in the production of single-cell protein (SCP) by methanotrophs and *n*-alkane utilizing microbes in the 1960s and 1970s and the reports of the isolation of many new methanotrophs by Whittenbury and his co-workers (Davies and Whittenbury, 1970; Whittenbury *et al.*, 1970) greatly stimulated funding for academic research on C_1 metabolism and the characteristics of methylotrophs (Large and Bamforth, 1988; Quayle, 1987). Methane as a feedstock is available in high purity; it is relatively inexpensive and abundant (7×10^{13} cubic meters worldwide). Because it is a gas, separation of cells or protein from the substrate is a relatively simple process compared to mixed alkane substrates or molasses. However, the low solubility of methane and the high oxygen demand of methane-based fermentations are severe drawbacks. Methane: air mixtures are explosive over a rather broad range.

Methanol, which is readily obtained by chemical oxidation of methane, is miscible with water, and gas/liquid phase transfer limitations of methane-based fermentations are circumvented with methanol as a substrate. Methanol also has the advantage that it is available in large quantities in excess of demand (Linton and Neikus, 1987; Large and Bamforth, 1988). On a weight basis, it is less expensive and its price has been more stable over time than that of sugars, it can be stored safely for long periods without deterioration, and it is utilized by a wider range of microbes than are able to use methane (Linton and Neikus, 1987). The oxygen demand during methanol metabolism is less than when methane is used as a fermentation feedstock. Methanol is also easily separated from most fermentation products and from biomass in downstream processing (Large and Bamforth, 1988). Therefore, methanol has been the substrate of choice for the ICI process, to produce a SCP product known as Pruteen, using the type I obligately methylophilic bacterium *Methylophilus methlotrophus* AS1 (Large and Bamforth, 1988). This bacterium has a relatively high optimum growth temperature near 40°C and grows rapidly on methanol with a μ_{max} of 0.55 h^{-1}.

SCP from C_1 compounds has failed to compete with other protein products (soybeans, fish, meal, and cottonseed) because the price of agricultural products has remained comparatively inexpensive over the last decade while fermentation plant construction costs and interest costs have increased rapidly.

The methanotrophic bacteria *Methylococcus capsulatus* (Bath) and *Methylosinus trichosporium* OB3b are known to contain both soluble and particulate methane monooxygenase systems (Akent'eva and Gvozdev, 1988; Dalton and Higgins, 1987). The soluble methane monooxygenases have been purified and characterized from these bacteria (Colby *et al.*, 1977; Fox and Lipscomb, 1988; Fox *et al.*, 1988, 1989; Green and Dalton, 1989; Woodland and Dalton, 1984) and from *Methylobacterium* CRL-26 (Patel and Savas, 1987). Resting cells of methanotrophs and soluble methane monooxygenases were capable of monohydroxylating a variety of hydrocarbons and other compounds and they catalyzed the formation of epoxides from several alkenes (Colby *et al.*, 1977; Stirling *et al.*, 1979). Epoxides are important because they can be polymerized to form epoxyhomopolymers and epoxycopolymers like polypropylenes and polyethylenes. The epoxides are excreted from cells and are not rapidly transformed to other products (Large and Bamforth, 1988). The observations that methane monooxygenase catalyzes many oxidations (Colby *et al.*, 1977; Imai *et al.*, 1986) greatly stimulated research on the biochemistry of methane monooxygenase (MMO) and applications of methanotrophs to biotechnology. Recently, it has been observed that methane-amended soils and groundwater (Wilson and Wilson, 1985; Little *et al.*, 1988), mixed and pure cultures of methanotrophs (Fogel *et al.*, 1986; Oldenhuis *et al.*, 1989; Tsien *et al.*, 1989), and soluble MMO can catalyze the destruction of trichloroethylene, dichloroethylene, vinyl chloride, and most of the low-molecular-weight halogenated hydrocarbons and haloforms found as environmental pollutants. These compounds have received attention of regulators because of their toxicity and carcinogenicity, and their removal challenges current water treatment technologies (Sayre, 1988). Dichloromethane is also a frequently used industrial solvent and pollutant that is metabolized by methylotrophic bacteria containing dichloromethane dehalogenase (Galli and Leisinger, 1985).

Tetrachloroethylene and carbon tetrachloride are two commonly occurring pollutants that are not transformed by methanotrophs (Vogel and McCarty, 1985; Tsien *et al.*, 1989). However, these compounds are partially dehalogenated by anaerobic microbes to products that are readily mineralized by methanotrophs or other bacteria that exist in soils and waters (Vogel and McCarty, 1985). Several other microbes containing oxygenases including toluene-oxidizing bacteria (Nelson *et al.*, 1986; Wackett *et al.*, 1989), *Nitrosomonas europea* (Arciero *et al.*, 1989), and propane-utilizing bacteria (Wackett *et al.*, 1989) have been shown to cause the destruction of the incompletely halogenated alkenes. The rates observed with *Methylosinus trichosporium* 0B3b have been at least an order of magnitude greater than any of the rates of trichloroethylene degradation obtained by other bacteria which contain mono- and dioxygenases that we have examined or that have been published to date (Oldenhuis *et al.*, 1989; Tsien *et al.*, 1989).

The rapid oxidation of trichloroethylene (TCE), vinyl chloride, and other halogenated alkenes required the presence of the soluble MMO (Tsien *et al.*, 1989) which was previously shown to be active when the concentration of Cu^{2+} was maintained below 1 µM (Stanley *et al.*, 1983). We (Tsien *et al.*, 1989) have shown that the five protein components of soluble MMO of *M. trichosporium* 0B3b and *Methylosporovibrio methanica* 81Z are synthesized only when the Cu^{2+} concentrations are maintained at low levels and TCE is oxidized only when they are present.

It is appropriate at this point to compare the biochemistry of ammonia-oxidizing bacteria and methanotrophs because of the similarity of their oxygenases. The former are autotrophic bacteria but are capable of oxidizing methane and methanol as well as ammonia (Hyman and Wood, 1983; Bedard and Knowles, 1989). Like methanotrophs, species of the genus *Nitrosomonas* oxidize CO, methane, methanol, bromomethane, propylene, cyclohexane, benzene, and phenol (Bedard and Knowles, 1989). Moreover, methane is oxidized to CO_2 by ammonia oxidizers (Hyman and Wood, 1983). Trichloroethylene is slowly oxidized by *N. europea* (Arciero *et al.*, 1989). Methanotrophs oxidize ammonia to nitrite and possess hydroxylamine oxidoreductase (Bedard and Knowles, 1989). It is not always clear which group of bacteria is responsible for the oxidation of ammonia in freshwaters and soils (Hanson, 1980; Bedard and Knowles, 1989; Ward, 1987).

Properties of methanotrophs other than those described above have received less attention but have considerable commercial potential. Readers are referred to a book by Large and Bamforth (1988) and subsequent chapters of this volume for detailed information about applications employing methylotrophs and their enzymes.

4. GENETICS OF METHYLOTROPHS

The initial attempts at genetic analysis of C_1-utilizing bacteria were frustrated by difficulties encountered in attempts to isolate mutants required for the selection of recombinants and genomic mapping (Holloway, 1984; Holloway *et al.*, 1987). The inability to isolate specific mutants in gram-negative methylotrophs remains an obstacle to their use for certain processes. For example, auxotrophs unable to produce one or more amino acids in a branched biosynthetic pathway are important for production of another amino acid in the same pathway.

The reason for ineffectiveness of mutagenic treatments and enrichment for obtaining auxotrophic mutants of methylotrophs is not understood. It is clear that some classes of mutants including auxotrophs are obtained at frequencies observed with other gram-negative bacteria when a positive selective technique has been available (Bohanon et al., 1987; Harms et al., 1985; Machlin et al., 1987; Nunn and Lidstrom, 1988). The obligate methanotrophs present special problems for the isolation of mutants with defects in genes required for C_1 metabolism. Some strains of methanotrophs are capable of growth as colonies on solid media containing methanol, and MMO mutants can be identified in these bacteria (Nicolaidis and Sargent, 1987). However, mutants of obligate methanotrophs with defects in assimilatory pathways will be difficult to obtain.

Of the three systems traditionally employed for genetic analysis of bacteria; conjugation, transduction, and transformation, only conjugation has proven to be broadly applicable for genetic studies of methylotrophs (Holloway et al., 1987).

The most successful approaches for identifying genes and mapping them have involved cloning DNA fragments in *Escherichia coli* and identifying genes by complementation of mutations of *Pseudomonas aeruginosa*, *E. coli*, or methylotrophs (Allen and Hanson, 1985; Bohanon et al., 1987; Byrom, 1984; Gautier and Bonewall, 1980; Haber et al., 1983; Holloway et al., 1987; Holloway, 1984; Machlin et al., 1987; Levering et al., 1987; Nunn and Lidstrom, 1988; Windass et al., 1980). Windass et al. (1980) isolated conditionally lethal, temperature-sensitive mutants of *Methylophilus methylotrophus*, an obligate methylotroph. One lacked glutamate synthase at the restrictive temperature. The glutamate dehydrogenase (GDH) of *E. coli* was cloned into a broad host range plasmid and was mobilized into the *M. methylotrophus* mutant lacking glutamate synthase. The use of the recombinant organism that possessed the more efficient GDH route for ammonia assimilation resulted in conversion of more methanol to cell carbon than was observed for the wild-type strain (Windass et al., 1980). Cloned genes from methylotrophs that encoded enzymes of biosynthetic pathways have been identified by complementation of *E. coli* and *Pseudomonas* auxotrophs (Holloway et al., 1987; Bohanon et al., 1987), and cloned genes encoding methanol dehydrogenase from a variety of species have been shown to complement mutants of *Methylobacterium extorquens* AM1 and *Methylobacterium organophilum* strain XX. The MDH structural genes of two species have been sequenced (Harms et al., 1987; Machlin and Hanson, 1988). Cytochrome C_L, an electron acceptor for MDH, and genes encoding small basic peptide associated with purified MDHs (from *M. extorquens*

AM1) have also been sequenced (Nunn and Anthony, 1988). The spatial relationships between cloned genes have then been determined by restriction mapping and sequencing (Bastien et al., 1989; Anderson and Lidstrom, 1988; Machlin et al., 1987; Nunn and Anthony, 1988a,b; Nunn and Lidstrom, 1986b; DeVries, 1986; Stephens et al., 1988).

Chromosome-mobilizing plasmids like R68.45 form prime plasmids in vivo which have incorporated chromosomal DNA into the plasmid genome (Holloway et al., 1987). Tatra and Goodwin (1983) and Moore et al. (1983) determined linkage relationships between antibiotic resistance genes and auxotrophic markers using prime plasmids. This approach has been used to map several (20) auxotrophic markers in M. methlotrophus AS1 (Holloway et al., 1987) The segments inserted into prime plasmids are often large, and because the plasmid primes can be mobilized to a variety of gram-negative bacteria, they can be used to map genes over rather large regions of the genome (Holloway et al., 1987).

When mutants cannot be obtained by conventional means, it is often possible to clone the gene of interest, modify them by one of several techniques, and reinsert them into the genome from which they were isolated (Bohanon et al., 1987). Cloned genes can be recognized in a variety of ways. Mullens and Dalton (1987) purified methane monooxygenase components from M. capsulatus (Bath), prepared oligonucleotides that hybridized to portions of the structural genes carried in E. coli clones. Subsequently, Murrell and colleagues have now cloned and sequenced the MMO gene cluster of M. capsulatus (Bath) and Methylosinus trichosporium OB3b (see chapter 4). Genes cloned in expression vectors can be identified by utilizing antibodies prepared against purified proteins.

Several genes are expressed in high levels in methylotrophs. The promoters controlling expression of these genes may be useful for expression of genes foreign to the host. Genes of methylotrophs that are known to be expressed at high levels (10–20% of the soluble protein of the cells) include the hydroxylase components of the soluble methane monooxygenase, methanol dehydrogenase, methanol oxidase, and dichloromethane dehydrogenase (Cregg et al., 1987; Harms et al., 1987; Machlin and Hanson, 1988; Mullens and Dalton, 1987; LaRoche and Leisinger, 1990). Sequences of these genes and upstream regions are known. The dichloromethane dehalogenase is inducible by its substrate or the gratuitous inducer dichloroethane (LaRoche and Leisinger, 1990). The regulation of the expression of methanol dehydrogenases of the pink-pigmented facultative methylotrophs is less well understood and appears to involve large numbers of genes (Bastien et al., 1989; Machlin et al., 1987; Nunn et al., 1987). The amount of MDH in Paracoccus denitrificans cultures was found to vary over 100-fold. MDH synthesis in this bacterium appeared to be repressed by non-C_1 substrates and may be induced by one or more products of methanol metabolism (Harms *et al.*, 1987). It is also possible that the synthesis of the prosthetic group of methanol dehydrogenase, pyrolloquinoline quinone (PQQ), or its incorporation into the apoenzyme may modulate enzymatic activity (Harms *et al.*, 1987).

A complete understanding of metabolic pathways and their regulation obviously requires integrated biochemical and genetic approaches. The introduction of surrogate genetics, the use of broad-host-range conjugative vectors, and the application of recombinant DNA techniques have made it possible to clone-map and analyze the expression of most genes of interest in methylotrophs. The pathway for the synthesis of gloxylate from acetyl coenzyme A in isocitrate lyase minus (ICl$^-$) type II methylotrophs remains to be described. The role of the various formaldehyde dehydrogenases in catabolism and metabolic regulation of some methylotrophs is not understood. The mechanism by which copper limitation and perhaps other growth conditions depress the synthesis of soluble MMO in *M. capsulatus* (Bath) and *Methylosinus trichosporium* OB3b is unknown. Genetic systems need to be developed for the gram-positive, methanol-utilizing bacteria that possess several advantages for some applications.

5. CONCLUSIONS

Many possible applications of methylotrophs, including the use of the bacteria as hosts for expression of recombinant genetic information, the production of vitamins, amino acids, other metabolites, polymers, enzymes, and products of biotransformations, have been investigated (Large and Bamforth, 1988). The proposed applications to bioremediation of water and soils contaminated with toxic chemicals is a recent development. At this time, methylotrophs have not been put to a commercially profitable use. However, optimism for their use and for biotechnology in general remains high.

It will be important to select products that can be produced more economically by methylotrophs than other organisms. The selection of processes will require a detailed knowledge of the diversity of methylotrophs, their physiology, and the regulation of their metabolism. A means of genetic manipulation will be important in nearly all processes. With this information in hand, the proper microbe can be matched with a process to provide the most economical means of generating a product.

There are many gaps in the knowledge required to make well-informed decisions. Regulation of the assimilation and catabolism of C_1 compounds is not well described. We do not understand the role of soluble and particulate MMOs in the metabolism of methanotrophs. It is not known if all methanotrophs carry genetic information for the proteins required for the soluble MMO activity or if these genes are confined to a few species. This enzyme has been found in few methanotrophs. Its absence in some species may reflect its instability, the inability of investigators to induce the bacteria to produce it, or the absence of the genetic information. Although copper limitation results in the synthesis of soluble MMO proteins in some species, it may be an indirect effector of expression. The tools are now in hand to properly ask the questions posed by this lack of information.

Recent studies of soluble methane monooxygenases have provided new insights into the structure of the catalytic sites, the structure of the proteins, and the processes they catalyze. In time, this knowledge may lead to development of biomimetic catalysts that will replace the need for complex and unstable oxygenases and other catalysts in the production of chemicals or destruction of toxic chemicals at low temperatures and at high productivities.

Most of the profitable biotechnology enterprises resulted from research that was not directed at an applied goal. Much of what is known about methylotrophs now was not foreseen in 1970, when the use of methanotrophs for the production of SCP was the application that received the greatest attention. The study of these microbes has in approximately two decades, added much to our knowledge of biochemistry and microbiology. Methane monooxygenases are different in some respects from other mixed-function oxygenases (Ericson *et al.*, 1988; Fox *et al.*, 1988b), PQQ was first discovered in methylotrophs (Duine *et al.*, 1986), novel pathways for the synthesis of intermediates of central metabolic routes were discovered and exist only in the methylotrophic bacteria and fungi (Quayle, 1987; Zatman, 1981). It is likely that further examinations of these microbes with or without applied incentives will be profitable in unforeseen ways. The potential for application of these bacteria in commerce will continue to depend on a strong basic research effort. Studies of these microbes and their ecology will surely be important to understanding global carbon cycling and the influence of events on the chemistry of our planet. We know little of their ecology other than that they are responsible for recycling of enormous amounts of carbon in the biosphere.

REFERENCES

Abramochinka, F. N., Bezrukova, L. V., Koshelev, A. V., Gal'chenko, V. F., and Ivanov, M. V., 1987, Microbial oxidation of methane in a body of freshwater, *Microbiology* **56**:375–382.

Akent'eva, N. P., and Gvozdev, R. I., 1988, Purification and physicochemical properties of methane monooxygenase from membrane structures of *Methylococcus capsulatus, Biochem. USSR* **53**:79–83.

Alawadhi, N., Egli, T. and Hamer, G., 1988, Growth characteristics of a thermotolerant methylotrophic *Bacillus* sp in batch culture, *Appl. Microbiol. Biotech.* **29**:485–493.

Allen, L. H., and Hanson, R. S., 1985, Construction of broad host-range cosmid cloning vectors: identification of genes necessary for growth of *Methylobacterium organophilum* on methanol, *J. Bacteriol.* **161**:955–962.

Anthony, C., 1982, *The Biochemistry of Methylotrophs*, Academic Press, New York.

Arciero, D., Vannelli, T., Logan, M., and Hooper, A. B., 1989, Degradation of trichloroethylene by the ammonia-oxidizing bacterium *Nitrosomonas europea, Biochem. Res. Commun.* **159**:640–643.

Bastien, C., Machlin, S., Zhang, Y., Donaldson, K., and Hanson, R. S., 1989, Organization of genes required for the oxidation of methanol to formaldehyde in three type II methanotrophs, *Appl. Environ. Microbiol.* **55**:3124–3130.

Bedard, C., and Knowles, R., 1989, Physiology, biochemistry, and specific inhibitors of CH_4, NH_4^+, and CO oxidation by methanotrophs and nitrifiers, *Microbiol. Rev.* **53**:68–84.

Bohanon, M. J., Bastien, C. A., Yoshida, R., and Hanson, R. S., 1987, Isolation of auxotrophic mutants of *Methylophilus methylotrophus* by modified-marker exchange, *Appl. Environ. Microbiol.* **54**:271–273.

Brooke, A. G., Watling, E. M., Attwood, M. M., and Tempest, D. W., 1989, Environmental control of metabolic fluxes in thermotolerant methylotrophic *Bacillus* strains, *Arch. Microbiol.* **151**:268–273.

Byrom, D., 1984, Host-vector systems for *Methylophilus methylotrophus*. Microbial growth on C_1 compounds, in: *Microbial Growth on C_1 Compounds* (R. L. Crawford and R. S. Hanson, eds.), *American Society for Microbiology*, Washington, DC, pp. 221–223.

Cavanaugh, C. M., Levering, P. R., Maki, J. S., Mitchell, R., and Lidstrom, M., 1987, Symbiosis of methylotrophic bacteria and deep-sea mussels, *Nature* **325**:346–348.

Childress, J. J., Fisher, C. R., Brooks, J. M., Kennicutt, M. C., II, Bridigare, R., and Anderson, A. E., 1986, A methanotrophic marine molluscan (bivalvia, mytilidae) symbiosis: methane-oxidizing mussels, *Science* **233**(4770):1306.

Claus, G., and Kutzner, H. J., 1985, Denitrification of nitrate and nitric acid with methanol as carbon source, *Appl. Microbiol. Biotech.* **22**(5):378.

Colby, J., Stirling, D. I., and Dalton, H., 1977, The soluble methane monooxygenase of *Methylococcus capsulatus* (Bath). Its ability to oxygenate n-alkanes, n-alkenes, ethers and alicyclic, aromatic and heterocyclic compounds, *Biochem. J.* **165**:395–402.

Corpe, W. A., 1985, A method for detecting methylotrophic bacteria on solid surfaces, *J. Microbiol. Methods.* **3**(3-4):215–223.

Cregg, J. M., Tschoop, J. F., Stillman, C., Siegel, R., Akong, M., Craig, W. S., Buckholz, R. G., Madden, K. R., Kellaris, P. A., Davis, G. R., Smiley, B. L., Cruze, J., Torregrosse, R., Velicelebi, G., and Thill, G. P., 1987, High-level expression and efficient assembly of hepatitis-B surface antigen in the methylotrophic yeast, *Pichia pastoris, Biotechnology* **5**:479–485.

Crutzen, P. J., and Graedel, T. E., 1986, The role of atmospheric chemistry in environment development interactions, in: *Sustainable Development of the Biosphere* (W. C. Clark and R. E. Munn, eds.), Cambridge University Press, Cambridge, MA.

Dagley, S., 1975, A biochemical approach to some problems of environmental pollution, in: *Essays in Biochemistry* (P. N. Campbell and W. N. Aldrige, eds.), Academic Press, London, pp. 81–138.

Dalton, H., and Higgins, I. J., 1987, Physiology and biochemistry of methylotrophic bacteria, in: *Microbial Growth on C_1 Compounds* (H. W. Van Verseveld and J. A. Duine, eds.), Martinus Nijhoff, Dordrecht, pp. 89–94.

Davies, S. L., and Whittenbury, R., 1970, Fine structure of methane and other hydrocarbon-utilizing bacteria, *J. Gen. Microbiol.* **61:**227–232.

DeLong, E. F., Wickham, G. S., and Pace, N. R., 1989, Phylogenetic stains: ribosomal RNA-based probes for the identification of single cells, *Science* **243:**1360–1363.

DeVries, G. E., 1986, Molecular biology of bacterial methanol oxidation, *FEMS Microbiol. Rev.* **39**(3)**:**235–258.

Dijkhuizen, L. N., Arfman, M. M., Attwood, A. G., Brooke, W., Harder, and Watling, E. M., 1988, Isolation and initial characterization of thermotolerant methylotrophic *Bacillus* strains, *FEMS Microbiol. Lett.* **52:**209–214.

Duine, J. A., Frank, J., and Jongejan, J. A., 1986, PQQ and quinoprotein enzymes in microbial oxidation, *FEMS Microbiol. Rev.* **32:**165–178.

Dworkin, M., and Foster, J. W., 1956, Studies on *Pseudomonas methanica*. (Söhngen) Nov. Comb., *J. Bacteriol.* **91:**646–659.

Ehalt, D. H., 1976, The atmospheric cycle of methane, in: *Symposium on Microbial Production and Utilization of Genes* (H. G. Schlegel, G. Gottschalk, and N. Pfennig, eds.), Acad. der Wissenschaften, Gottingen, pp. 13–22.

Ericson, A., Hedman, B., Hodgson, K. O., Green, J., Dalton, H., Bentsen, J. G., Beer, R. H., and Lippard, S. J., 1988, Structural characterization by EXAFS spectroscopy of the binuclear iron center in protein A of methane monooxygenase from *Methylococcus capsulatus* (Bath), *J. Am. Chem. Soc.* **110:**2330–2331.

Fischer, G., Neftel, A., and Oeschger, H., 1985, Increase of atmospheric methane recorded in antarctic ice core, *Science* **299**(4720)**:**1386–1388.

Fogel, M. M., Taddeo, A. R., and Fogel, S., 1986, Biodegradation of chlorinated ethenes by a methane-utilizing mixed culture, *Appl. Environ. Microbiol.* **51**(4)**:**720–724.

Fox, B. G., and Lipscomb, J. D., 1988, Purification of a high specific activity methane monooxygenase hydroxylase component from a type II methanotroph, *Biochem. Biophys. Res. Comm.* **154:**165–170.

Fox, B. G., Surerus, K. K., Munck, E., and Lipscomb, J. D., 1988, Evidence for a μ-oxo-bridged binuclear iron cluster in the hydroxylase component of methane monooxygenase, Mossbauer and EPR studies, *J. Biol. Chem.* **263:**10553–10556.

Fox, B. G., Froland, W. A., Dege, J., and Lipscomb, J. D., 1989. Methane monooxygenase from *Methylosinus trichosporium* OB3b, *J. Biol. Chem.* **264:**10023–10033.

Galli, R., and Leisinger, T., 1985, Specialized bacterial strains for the removal of dichloromethane from industrial waste, *Conserv. Recycling* **8:**91–100.

Gautier, F., and Bonewald, R., 1980, The use of plasmid R1162 and derivatives for gene cloning in the methanol utilizing *Pseudomonas* AM1, *Mol. Gen. Genet.* **178:**375–380.

Giovannoni, S. J., DeLong, E. F., Olsen, G. J., and Pace, N. R., 1988, Phylogenetic group-specific oligodeoxynucleotide probes for identification of single microbial cells, *J. Bacteriol.* **170:**720–726.

Graedel, T. E., and Crutzen, P. J., 1989, The changing atmosphere, *Sci. Am.* **261:**136–143.

Green, J., and Dalton, H., 1989, A stopped-flow kinetic study of soluble methane monooxygenase from *Methylococcus capsulatus* (Bath), *Eur. J. Biochem.* **153**(1)**:**137–144.

Green, P., and Bousefield, I. J., 1983, Emendation of *Methylobacterium*. Patt, Cole and Hanson, 1976; *Methylobacterium rhodium* (Heuman, 1962). Comb. Nov. (corrig: *Methylobacterium radiotolerans* (Ito and Iizuka, 1971) Comb. Nov. corrig., *Int. J. Syst. Bacteriol.* **33:**875–877.

Haber, C. L., Allen, L. N., and Hanson, R. S., 1983, Methylotrophic bacteria: biochemical diversity and genetics, *Science* **221**:1147–1151.

Hanson, R. S., 1980, Ecology and diversity of methylotrophic organisms, *Adv. Appl. Microbiol.* **26**:3–39.

Hanson, R. S., Tsuji, K., Bastien, C., Tsien, H. C., Bratina, B., Brusseau, G., and Machlin, S., 1989, in: *Coal and Gas Biotechnology* (C. Akin, ed.), Institute for Gas Technology, Chicago.

Hanson, R. S., Netrosev, A. I., and Tsuji, K., 1991, The obligate methylotrophic bacteria, in: *The Prokaryotes*, Belowes, A., Trouper, H., Dworkin, M., and Sleifer, K. (eds.), Springer-Verlag, pp. 2350–2364.

Harder, W., and Attwood, M. M., 1978, Biology, physiology and biochemistry of *Hyphomicrobia*, *Adv. Microbiol. Physiol.* **17**:303–359.

Harms, N., DeVries, G. E., Maurer, K., Veltkamp, E., and Stouthamer, A. H., 1985, Isolation and characterization of *Paracoccus denitrificans* mutants with defects in the metabolism of one-carbon compounds, *J. Bacteriol.* **164**(3):1064–1070.

Harms, N., DeVries, G. E., Maurer, K., Hoogendak, J., Stouthamer, A. H., 1987, Isolation and nucleotide sequence of methanol dehydrogenase structural gene from *Paracoccus denitrificans*, *J. Bacteriol.* **169**:3969–3975.

Heyer, J., 1977, Results of enrichment experiments of methane-assimilating organisms from an ecological point of view, in: G. A. Skryabin, Ivanov, M. B., Kondratjeva, E. N., Zavarzin, G. A., Trotsenko, Yu. A., and Netrosev, A. I. (eds.), Microbial Growth on C_1 Compounds, U.S.S.R. Academy of Sciences, Rusching, pp. 19–21.

Higgins, I. J., Best, D. J., Hammond, R. C., and Scott, D., 1981, Methane oxidizing microorganisms, *Microbiol. Rev.* **45**:556–590.

Holloway, B. W., 1984, Genetics of methylotrophs, in: *Methylotrophs: Microbiology, Biochemistry and Genetics* (C. T. Hou, ed.), CRC Press, Boca Raton, FL, pp. 87–104.

Holloway, B. W., Kearny, P. P. and Lyon, B. R., 1987, The molecular genetics of C_1-utilizing organisms: an overview, in: *Microbial Growth on C_1 Compounds* (H. W. Van Verseveld and J. A. Duine, eds.), Martinies Nyhoff, Dordrecht, pp. 223–229.

Hou, C. T., 1984, Propylene oxide production from propylene by immobilized whole cells of *Methylosinus* sp. CRL 31 in a gas-solid bioreactor, *Appl. Microbiol. Biotech.* **19**:1.

Houghton, R. A., and Woodwell, G. M., 1989, Global climatic change, *Sci. Am.* **260**:36–47.

Hutton, W. E., and Zobell, C. E., 1949, The occurrence and characteristics of methane-oxidizing bacteria in marine sediment, *J. Bacteriol.* **58**:463–473.

Hyman, M. R., and Wood, P. M., 1983, Methane oxidation by *Nitrosomonas europea*, *Biochem. J.* **212**:31–37.

Imai, T., Takigawa, H., Nakagawa, S., Tohru Kodama, G.-J., and Minoda, Y., 1986, Microbial oxidation of hydrocarbons and related compounds by whole-cell suspensions of the methane-oxidizing bacterium H-2, *Appl. Environ. Microbiol.* **52**(6):1403–1406.

Jenkins, O., and Jones, D., 1987, Taxonomic studies on some gram-negative methylotrophic bacteria, *J. Gen. Microbiol.* **133**:453–473.

Khalil, M. A. K., and Rassmussen, R. A., 1986, Interannual variability of atmospheric methane: possible effects of El-Nino-Southern oscillation, *Science* **232**:56–58.

Kohler-Staub, D., and Leisinger, T., 1985, Dichloromethane dehalogenase of *Hyphomicrobium* sp. strain DMZ, *J. Bacteriol.* **162**:676–681.

Large, P. J., and Bamforth, C. W., 1988, *Methylotrophy and Biotechnology*, Longman, Wiley, New York.

LaRoche, S. D., and Leisinger, T., 1990, Sequence analysis and expression of the bacterial dichloromethane dehalogenase structural gene, a member of the glutathione–transferase supergene family, *J. Bacteriol.* **172**:164–171.

Laurinavichus, K. S., Belyayev, S. S., and Ivanov, M. V., 1978, A study of microbiological oxidation process of methane in the freshwater lakes of the Mari USSR, *Izv. Akad. Nauk SSR Ser. Biol.* **2**:309–311.

Lawrence, A. J., and Quayle, J. R., 1970, Alternative carbon assimilation pathways in methane-utilizing bacteria, *J. Gen. Microbiol.* **63**:371–374.

Leadbetter, E. D., and Foster, J. W., 1958, Studies of some methane utilizing bacteria, *Arch. Microbiol.* **30**:91–118.

Levering, P. R., Tiesma, L., Woldendrop, J. P., Steensma, M., and Dijkhuizen, L., 1987, Isolation and characterization of mutants of the facultative methylotroph *Arthrobacter* P1 blocked in one-carbon metabolism, *Arch. Microbiol.* **146**(4):346–352.

Lidstrom, M. E., 1983, Methane consumption in Framvaren, an anoxic marine fjord, *Limnol. Oceanogr.* **28**:1247–1251.

Lidstrom, M. E., 1988, Isolation and characterization of marine methanotrophs, *Ant. Van Leeuw.* **54**:189–199.

Linton, J. D., and Neikus, H. G. D., 1987, The potential of one-carbon compounds as fermentation feedstocks, in: *Microbial Growth on C_1 Compounds* (H. W. van Verseveld and J. A. Duine, eds.), Martinus Hijhoff, Dordrecht.

Little, C. D., Palumbo, A. V., Herbes, S. E., Lidstrom, M. E., Tyndall, R. L., and Gilmer, P. J., 1988, Trichloroethylene biodegradation by a methane oxidizing bacterium, *Appl. Env. Microbiol.* **54**:951–956.

Machlin, S. M., and Hanson, R. S., 1988, Nucleotide sequence and transcriptional start site of *Methylobacterium organophilum* XX methanol dehydrogenase structural gene, *J. Bacteriol.* **170**:474–477.

Machlin, S. M., Tam, P. E., Bastien, C. A., and Hanson, R. S., 1987, Genetic and physical analysis of *Methylobacterium organophilum* XX genes encoding methanol oxidation, *J. Bacteriol.* **170**:141–148.

Moore, A. T., Nayudu, M., and Holloway, B. W., 1983, Genetic mapping in *Methylophilus methylotrophus*. ASI, *J. Gen. Microbiol.* **129**:785–799.

Mullens, I. A., and Dalton, H., 1987, Cloning of the gamma-subunit methane monoogygenase from *Methylococcus capsulatus*, *Biotechnology* **5**:490–493.

Nelson, M. J. K., Montgomery, S. O., O'Neil, E. J., and Pritchard, P. H., 1986, Aerobic metabolism of trichloroethylene by a bacterial isolate, *Appl. Environ. Microbiol.* **52**:383–384.

Nicolaidis, A. A., and Sargent, A. W., 1987, Isolation of methane monooxygenase-deficient mutants from *Methylosinus trichosporium* OB3b using dichloromethane, *FEMS Microbiol. Lett.* **41**:47–52.

Nunn, D. N., Anthony, C., 1988, The nucleotide sequence and deduced amino acid sequence of the genes for cytochrome C_L and a hypothetical second subunit of the methanol dehydrogenase of *Methylobacterium* AM1, *Nucl. Acids. Res.* **16**:7722–7723.

Nunn, D. N., and Anthony, C., 1988b, Isolation and complementation analysis of 10 methanol oxidation mutant classes and identification of the methanol dehydrogenase structural gene of *Methylobacterium* sp. strain AM1, *J. Bacteriol.* **166**(2):581–590.

Oldenhuis, R., Vink, R. L., Janssen, D. B., and Witholt, B., 1989, Degradation of chlorinated aliphatic hydrocarbons by *Methylosinus trichosporium* OB3b and toxicity of trichloroethylene, *Appl. Environ. Microbiol.* **53**:2819–2826.

Orla-Jensen, S., 1909, Der hauptlinien des naturlichen bacterian systems, *Zentrbl. Bakteriol. Parasintenkd. Hug. Abt. II* **22**:97–98.

Patel, R. N., and Savas, J. C., 1987, Purification and properties of the hydroxylase component of methane monooxygenase, *J. Bacteriol.* **169**:2313–2317.

Patt, T. E., Cole, G. C., Bland, J., and Hanson, R. S., 1974, Isolation of bacteria that grow on

methane and organic compounds as sole sources of carbon and energy, *J. Bacteriol,* **120**:955–964.
Quayle, J. R., 1987, An eightieth anniversary of the study of C_1 metabolism, in: *Microbial Growth on C_1 Compounds* (H. W. Van Verseveld and J. A. Duine, eds.), Martinus Nijhoff, Dordrecht, pp. 1–5.
Reeburgh, W. S., 1980, Anaerobic methane oxidation: rate depth distributions in Skan Bay sediments, *Earth Planet. Sci. Lett.* **46**:345–352.
Reed, W. N., and Dugan, P. R., 1987, Isolation and characterization of the facultative methylotroph *Mycobacterium* ID-Y, *J. Gen. Microbiol.* **113**:1389–1394.
Remsen, C. C., Minnich, E. C., Stephens, R. S., Bucholz, L. A., and Lidstrom, M. E., 1989, Anaerobic methane oxidation in marine sediments, *J. Great Lakes Res.* **15**:141–146.
Rudd, J. W., and Taylor, C. D., 1980, Methane cycling in aquatic environments, *Adv. Aquat. Microbiol.* **2**:77–150.
Saralov, A. I., Krylova, I. N., Saralova, E. E., and Kusnetsov, S. I., 1984, Distribution and species composition of methane-oxidizing bacteria in lake water, *Microbiology* **53**(5):695–701.
Sayre, I. M., 1988, International standards for drinking water, *J. Am. Water Works Assoc.* **80**:53–60.
Schendel, F. J., Bremmon, C. E., Flickinger, M. C., Guettler, M., and Hanson, R. S., 1990, L-lysine formation at 50°C by mutants of a newly isolated and characterized methylotrophic *Bacillus*, *Appl. Environ. Microbiol.* **56**:963–970.
Scholtz, R., Wackett, L. P., Egli, C., Cook, A. M., and Leisinger, T., 1988, Dichloromethane dehalogenase with improved catalytic activity isolated from a fast-growing dichloromethane-utilizing bacterium, *J. Bacteriol.* **170**:5699–5704.
Sieburth, J. Mc., Johnson, P. W., Eberhardt, M. A., Sieracki, M. E., and Lidstrom, M., 1987, The first methane-oxidizing bacterium from the upper mixing layer of the deep ocean: *Methylomonas pelagica* sp. nov., *Curr. Microbiol.* **14**:285–293.
Sohngen, N. L., 1906, Über bakterien, welche methau ab kohlenstoffnahrung and energiequelle gebrauchen, *Parasitenk Infectionsk. Abt. 2,* **15**:513–517.
Stanley, S. H., and Dalton, H., 1982, Role of ribulose-1,5-bisphosphate carboxylase/oxygenase in *Methylococcus capsulatus* (Bath), *J. Gen. Microbiol.* **128**:2927–2935.
Stanley, S. H., Prior, S. D., Leak, D., and Dalton, H., 1983, Copper stress underlies the fundamental change in intracellular location of methane monooxygenase in methane-oxidizing organisms: studies of batch and continuous cultures, *Biotechnol. Lett.* **5**:487–490.
Stephens, R. L., Haygood, M. G., and Lidstrom, M. E., 1988, Identification of putative methanol dehydrogenase (moxF) structural genes in methylotrophs and cloning of mox F genes from *Methylococcus capsulatus* (Bath) and *Methylomonas albus* BG8, *J. Bacteriol.* **170**:2063–2069.
Stirling, D. L., Colby, J., and Dalton, H., 1979, A comparison of the substrate and electron-donor specifities of the methane monooxygenases from three strains of methane-oxidizing bacteria, *Biochem. J.* **177**:361–364.
Strand, S. E., and Lidstrom, M. E., 1984, Characterization of a new marine methylotroph, *FEMS Microbiol. Lett.* **21**:247–251.
Suylen, G. M. H., and Kuenen, J. G., 1986, Chemostat enrichment and isolation of *Hyphomicrobium* E.G. a dimethyl-sulphide-oxidizing methylotroph and reevaluation of *Thiobacillus* MS1 Antonie van Leeuwenhoek, *J. Microbiol.* **52**:281–293.
Tatra, P. K., and Goodwin, P. M., 1983, R-plasmid mediated chromosome mobilization in the facultative methylotroph *Pseudomonas* AM1, *J. Gen. Microbiol.* **129**:2629–2634.
Tsien, H. C., Brusseau, G. A., Brusseau, R. S., Hanson, R. S., and Wackett, L., 1989,

Biodegradation of trichloroethylene by *Methylosinus trichosporium* OB3b, *Appl. Environ. Microbiol.* **55:**2960–2964.

Tsuji, K., Tsien, H. C., Hanson, R. S., DePalma, S. R., Scholtz, R., and LaRoche, S., 1990, The 16S-like ribosomal RNA sequence analysis for determining phylogenetic relationships among methylotrophs, *J. Gen. Microbiol.* **136:**1–10.

Vogel, T. M., and McCarty, P. L., 1985, Biotransformation of tetrachloroethylene to trichloroethylene, dichloroethylene, vinyl chloride and carbon dioxide under methanogenic conditions, *Appl. Environ. Microbiol.* **49:**1080–1083.

Wackett, L. P., Brusseau, G. A., Householder, S. R., and Hanson, R. S., 1989, Survey of microbial oxygenases: trichloroethylene degradation by propane-oxidizing bacteria, *Appl. Environ. Microbiol.* **55:**2960–2964.

Ward, B. B., 1987, Kinetic studies on ammonia and methane oxidation by *Nitrosococcus oceanus.*, *Arch. Microbiol.* **147:**126–133.

Whittenbury, R., and Dalton, 1981, The methylotrophic bacteria in: *The Procaryotes* (M. P. Starr, H. Stolph, H. G. Truper, A. Balowes, and H. G. Schlegel, eds.), Springer-Verlag, Berlin, pp. 894–902.

Whittenbury, R., Phillips, K. C., and Wilkinson, J. F., 1970, Enrichment, isolation and some properties of methane utilizing bacteria, *J. Gen. Microbiol.* **61:**205–218.

Whittenbury, R., and Krieg, N. R., 1984, *Methylococcaceae* fam. nov., in: *Bergey's Manual of Determinative Bacteriology*, Vol. 1, Williams & Wilkins, Baltimore, pp. 256–262.

Wilson, J. T., and Wilson, B. H., 1985, Biotransformation of trichloroethylene in soil, *Appl. Environ. Microbiol.* **49:**242–243.

Windass, J. D., Worsey, M. J., Pioli, E. M., Pioli, D., Barth, P. T., Atherton, K. T., Dart, E. C., Byrom, D., Powell, K., and Senior, P. J., 1980, Improved conversion of methanol to single cell protein by *Methylophilus methylotrophus, Nature (Lond.)* **287:**396–401.

Wolfe, H. I., 1981, Biochemical characterization of methane oxidizing yeasts, in: *Microbial Growth on C_1 Compounds* (H. Dalton, ed.), Heydon, London, pp. 202–210.

Wolfrum, T., and Stolp, H., 1987, Comparative studies on 5S RNA sequences of RuMP-type methylotrophic bacteria, *Syst. Appl. Microbiol.* **9:**273–276.

Woodland, M. P., and Dalton, H., 1984, Purification of component A of the soluble methane monooxygenase of *Methylococcus capsulatus* (Bath) by high-pressure gel permeation chromatography, *Anal. Biochem.* **139**(2)**:**459.

Zatman, L., 1981, A search for patterns in methylotrophic pathways, in: *Microbial Growth on C_1 Compounds* (H. Dalton, ed.), Heydon, London, pp. 42–54.

Zhao, S.-J., and Hanson, R. S., 1984, Variants of the obligate methanotroph isolate 761M capable of growth on glucose in the absence of methane, *Appl. Environ. Microbiol.* **48:**807–812.

Taxonomy of Methylotrophic Bacteria

PETER N. GREEN

Before delving into the taxonomy of the methylotrophic bacteria, it is worth taking a light-hearted look at the subject of taxonomy itself. What is taxonomy and why is it necessary?

There is much truth to be found in S. T. Cowan's definition of a taxonomist in *A Dictionary of Microbial Taxonomic Usage*. He defines a taxonomist as follows:

> A person who practises or studies taxonomy. As much of taxonomy is highly subjective, taxonomists are individualistic to a degree, and as a group they can be described by the aphorism—as all out of step but our Jock. Taxonomy is as much an art as a science and its interpreters are artists. Most are argumentative; many are aggressive, some are good and are able to make a useful contribution to science. Taxonomists are jealous of their preserve and make things difficult for the newcomer by the use of a complicated (and often unnecessary) terminology. Few will agree on a definition of taxonomy itself, and each is a merry sharp-shooter of his colleagues.

Migula thought that the task of the taxonomist was "to bring order from chaos." Certainly, as far as the methylotrophic bacteria are concerned, that is a fairly apt description of the task ahead.

1. INTRODUCTION

The last 15–20 years have seen an explosion of interest in, and hence isolation of, methylotrophic bacteria. Perhaps not surprisingly, as the main research has centered around the biochemical novelty and commercial viability of such bacteria, the taxonomy of the bewildering variety of organisms that can utilize one-carbon (C_1) compounds has been left flounder-

---

PETER N. GREEN • NCIMB Ltd., Aberdeen AB2 1RY, Scotland.

Methane and Methanol Utilizers, edited by J. Colin Murrell and Howard Dalton. Plenum Press, New York, 1992.

ing. The result, certainly as far as the classification of some groups of methylotrophs is concerned, is a state of semichaotic flux (Green *et al.*, 1984).

The reasons for this are numerous and varied. As far as several facultative methylotrophs are concerned, an overemphasis of methylotrophy *per se,* as a novel taxonomic feature, has contributed to the situation. Equally, in many cases, not enough attention has been paid to sufficiently detailed phenotypic studies or the taxonomic implications resulting from the many new names proposed in the literature.

Having said that, it is only fair to point out that, for the obligate methylotrophs in particular, the lack of physiological and biochemical taxonomic criteria available, resulting from the very restricted nutritional abilities of this group of organisms, has made it difficult to apply "conventional" or "traditional" approaches to their classification. Hence, in many cases, undue taxonomic weight may have been placed on characteristics that are arguably trivial or unstable, such as colony pigmentation, capsule formation, cell size and shape, or the ability to produce enhanced growth on multicarbon compounds. Also, the inability of taxonomists to agree among themselves, as cited above, has not helped. For example, it must be extremely confusing to look down the microscope at a *Methylococcus* to find a rod-shaped organism. *Methylococcus ucrainicus,* for instance, has properties that place it more logically in the genus *Methylomonas* (rod-shaped cells with a mol. % G+C of 52.1) rather than in the genus *Methylococcus* (cocci with a mol. % G+C of 62.5; Whittenbury and Krieg, 1984).

Nevertheless, despite this rather pessimistic introduction, the taxonomic understanding of most groups of methylotrophic bacteria is steadily improving as workers are slowly beginning to recognize, segregate, and study the various groups or organisms in increasing detail. The purpose of this chapter is to discuss and update the taxonomic status of the various groups of methylotrophic bacteria, many of which are, by necessity, arbitrarily grouped.

The following diagram illustrates the major groupings that most workers use to subdivide methylotrophic bacteria.

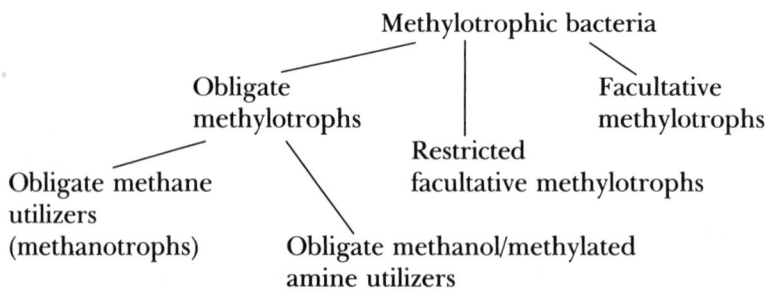

Obligate methylotrophs are defined as organisms that have the ability to utilize, as sole carbon and energy source, compounds that are more reduced than carbon dioxide and contain no carbon–carbon bonds (Colby *et al.*, 1979). Facultative methylotrophs, on the other hand, are organisms that can utilize such C_1 compounds, but are also capable of growth on multicarbon compounds.

A third, presently much smaller group of intermediate organisms, the restricted methylotrophs, is now also recognized. By definition, these bacteria can only utilize a relatively narrow range of multicarbon compounds in addition to C_1 compounds.

Each of these major groups of methylotrophic bacteria will now be discussed in some detail.

2. OBLIGATE METHYLOTROPHS

2.1. Methane-Utilizing Bacteria

This is a diverse group of organisms comprising several genera and possibly more than one family, although evidence for the latter is speculative.

The group encompasses a variety of morphological types, ranging from cocci through straight, curved, and pear-shaped rods, all of which form resting bodies in the form of cysts or exospores (see Table I). To date no authenticated Gram-positive methanotroph has been described in the literature, although it would be surprising were none to exist.

All methanotrophs share the ability to utilize methane, via methane monooxygenase, as sole carbon and energy source under aerobic or microaerophilic conditions. All strains also possess a complex arrangement of intracytoplasmic membranes when grown on methane, have a strictly respiratory type of metabolism with oxygen as the terminal electron acceptor, and are oxidase and catalase positive. All strains can grow on methanol, but some are inhibited by more than trace amounts. Methanotrophs occur in aerobic environments where methane is available and form a vital part of the cycle of methane in nature. In certain aquatic environments, they are found at certain times of the year within a very restricted range of depths (Hanson, 1980), where methane discharged from anaerobic detritus and dissolved oxygen are at levels optimal for their growth.

Methane-utilizing bacteria have been broadly split into two groups (type I and type II), largely on the basis of the major carbon assimilation pathway they use to metabolize methane and on the arrangement of their intracytoplasmic membranes. More recently, one of these groups has been further subdivided (type X), to accommodate strains of *Methylococcus capsu-*

latus, which are the only group of methanotrophs capable of autotrophic CO_2 fixation.

Before 1970, only a handful of methanotrophs had been isolated and described in the literature. The first of these, *Bacillus methanicus*, was isolated by Söhngen in 1906. A similar organism was isolated in 1956 by Dworkin and Foster and named *Pseudomonas methanica*. *Pseudomonas methanitrificans* (Davis et al., 1964), *Methanomonas methano-oxidans* (Brown et al., 1964), and *Methylococcus capsulatus* (Foster and Davis, 1966) were the only other organisms to be described before 1970.

TABLE I. Tentative Scheme for the Primary Categorization of Methanotrophic Bacteria[a]

Characteristic	Group I	Group X	Group II
Morphology	Straight rod	Coccus	Straight, curved or pear-shaped rod
Membrane arrangement			
Bundles of vesicular disks	+	+	−
Paired peripheral membranes	−	−	+
Motility	±	−	±
Resting stage	Azotobacter-type cyst	Azotobacter-type cyst	Lipid cyst or terminal exospore
Rosette formation	−	−	+ (most strains)
Major carbon assimilation pathway	Rump	Rump	Serine
Autotrophic CO_2 fixation	−	+	−
Complete TCA cycle	−	−	+
Nitrogenase	−	+	+
Isocitrate dehydrogenase[a]			
NAD^+ and $NAD(P)^+$ specific	+	−	−
NAD^+ specific	−	+	−
$NAD(P)^+$ specific	−	−	+
Glucose-6-dehydrogenase	+($NADP^+$-specific)	+($NADP^+$-specific)	−[c]
6-Phosphogluconate dehydrogenase	+($NADP^+$-specific)	+($NADP^+$-specific)	−
Predominant fatty acid carbon-chain length	16	16	18
Growth at 45°C	V[b]	+	−
Mol% G+C of DNA	50-54	62.5	61.7 - 63.1

[a]Not all strains classifiable into groups I and II have been shown to possess all the biochemical characteristics outlined in this scheme.
[b]V = variable.
[c]During growth on methane.

1970 was a particularly significant year because of the outstanding contribution made by Whittenbury and his colleagues at Edinburgh. Not only did these workers isolate more than 100 strains of methanotrophs, they also appreciated the taxonomic complexity of the bacteria they had isolated and refrained from proposing a catalogue of new taxa. Instead they merely assigned their strains to informal groups, based mainly on cell morphology, resting stage, and membrane type. Five such groups, or "genera" were suggested: *Methylomonas, Methylobacter, Methylococcus, Methylosinus,* and *Methylocystis.* These same groupings still remain today as the cornerstone of methanotroph taxonomy.

Since the work of Whittenbury and his colleagues, many other methanotrophs have been isolated, mainly by workers in the U.S.S.R. (Malashenko *et al.,* 1972, 1975; Galchenko *et al.,* 1975; Galchenko, 1977). Unfortunately, most of these organisms have not been made available to workers in the West for comparative purposes.

Nevertheless, recent publications by Romanovskaya *et al.* (1980), Galchenko and Andreev (1984), and Urakami and Komagata (1986b, 1987) have greatly contributed to the chemotaxonomic data available on a range of methanotrophic bacteria. Before the individual methanotrophic taxa are discussed in detail, it is necessary to examine some of the criteria that have helped the taxonomy of this group of organisms to evolve.

2.1.1. Cellular Morphology

Strains of obligate methanotrophs have widely varying cellular shapes. These include slender and stout rods of various dimensions, "vibroid" or curved rods, cocci, coccobacilli, pear-shaped cells, and cells that form terminal spores. There is also evidence that some methanotrophs display polar growth or budding (P. N. Green, unpublished observation; Hirsch, 1974).

Care should be taken when examining some of the more recently isolated *Methylococcus* strains where, in many cases, the organism is rod-shaped and does not have a coccal or predominantly coccal morphology.

2.1.2. Resting Stages

Most methanotrophs form an identifiable "resting stage." Whittenbury and his colleagues (1970a,b) observed three distinct types: exospores and two forms of cyst (see Figure 1).

Methylosinus strains are the only group of obligate methylotrophs shown so far to form exospores. Spore formation occurs as the cultures enter the stationary phase of growth, when some of the organisms elongate

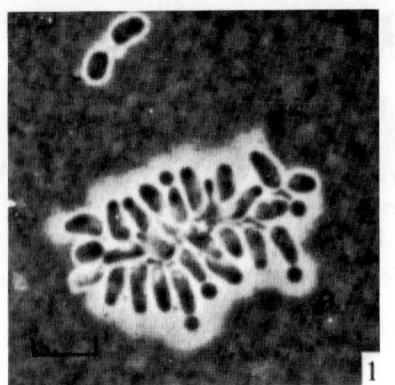

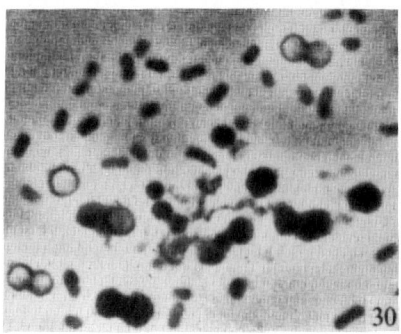

Figure 1. Phase constrast micrographs. A) Indian ink preparation of 'Methylosinus sporium' showing a pair of vegetative organisms (top left) and sporulating organisms budding off spores (center). Bar equals 5 μm; magnification × 1900. B) Preparation of 'Methylobacter vinelandii' showing immature cysts (large dark cells) and mature cysts (refractile cells) and vegetative rods. Magnification × 1900. Both reproduced by kind permission of R. Whittenbury.

and taper to form a pear shape (*M. trichosporium*) or comma shape (*M. sporium*) at the opposite end from the flagella. The round spore is finally budded at this tapered end (Whittenbury et al., 1970b). These exospores are heat-resistant, surviving 85°C for 15 min, and those of *M. trichosporium*, but not *M. sporium*, are capsulated.

Cysts formed by methanotrophs are often referred to by Whittenbury et al. (1970b) as being either lipid cysts or *Azotobacter*-type cysts. Multiple double- and single-bodied cysts morphologically similar to those formed by *Azotobacter* species were shown to exist in strains designated as *Methylobacter* by Whittenbury et al. (1970a). An immature form of this *Azotobacter*-type cyst was also observed in all Whittenbury's *Methylomonas* and *Methylococcus* strains. In many cases, profuse encystment was accompanied by the production of a pigment ranging in color from yellow, to rust, brown, and dark brown, whereas nonencysted cultures remained creamy white (Whittenbury et al., 1970b). In all cases, encysting cells were larger and more refractile in contrast to surrounding vegetative organisms. The immature *Azotobacter*-type cyst was smaller and less refractile than that found in the *Methylobacter* strains.

Lipid cysts contain large lipid inclusions and are formed by only one strain, *Methylocystis parvus*. These lipid cysts, or organisms containing them, are not heat resistant, but are somewhat resistant to desiccation (80% surviving after 7 days; Whittenbury *et al.*, 1970b).

2.1.3. Internal Membranes

All methanotrophs possess complex membranous organelles similar in complexity to those found in nitrifying and photosynthetic bacteria (Whittenbury *et al.*, 1970b). Two basic membrane arrangements have been observed (see Figure 2). Type I methanotrophs (*Methylococcus sensu* Whittenbury, *Methylomonas* and *Methylobacter*) have bundles of disk-shaped vesicles distributed throughout the cell. Type II organisms (*Methylosinus* and *Methylocystis*) have a system of paired peripheral membranes (Whittenbury *et al.*, 1970a,b).

It has been postulated that these internal membranes perform some role uniquely associated with methane assimilation, as methylotrophs able to grow on methanol, but not methane, do not possess such internal membranes (Anthony, 1982).

2.1.4. Major Carbon Assimilation Pathways

Two major carbon assimilation pathways occur in methanotrophs: the ribulose monophosphate (RMP or Rump) pathway, present in type I bacteria and indicated by high levels of hexulosephosphate synthase, and the serine pathway, present in type II bacteria, which have high levels of hydroxypyruvate reductase (Anthony, 1982). However, the mere presence of such indicator enzymes is insufficient evidence for the presence of one pathway or the other, because, for example, type I methanotrophs can possess low levels of hydroxypyruvate reductase in addition to hexulosephosphate synthase.

Methylococcus capsulatus is a unique organism among methanotrophs; hence it merits the creation of a third subgroup—type X. Part of its novelty is attributable to its unique ability to fix CO_2 autotrophically, but not as sole carbon source, via ribulose bisphosphate carboxylase and phosphoribulokinase activity. Most of its carbon, however, is fixed via the RMP pathway and hexulosephosphate synthase (Taylor *et al.*, 1980; Stanley and Dalton, 1982).

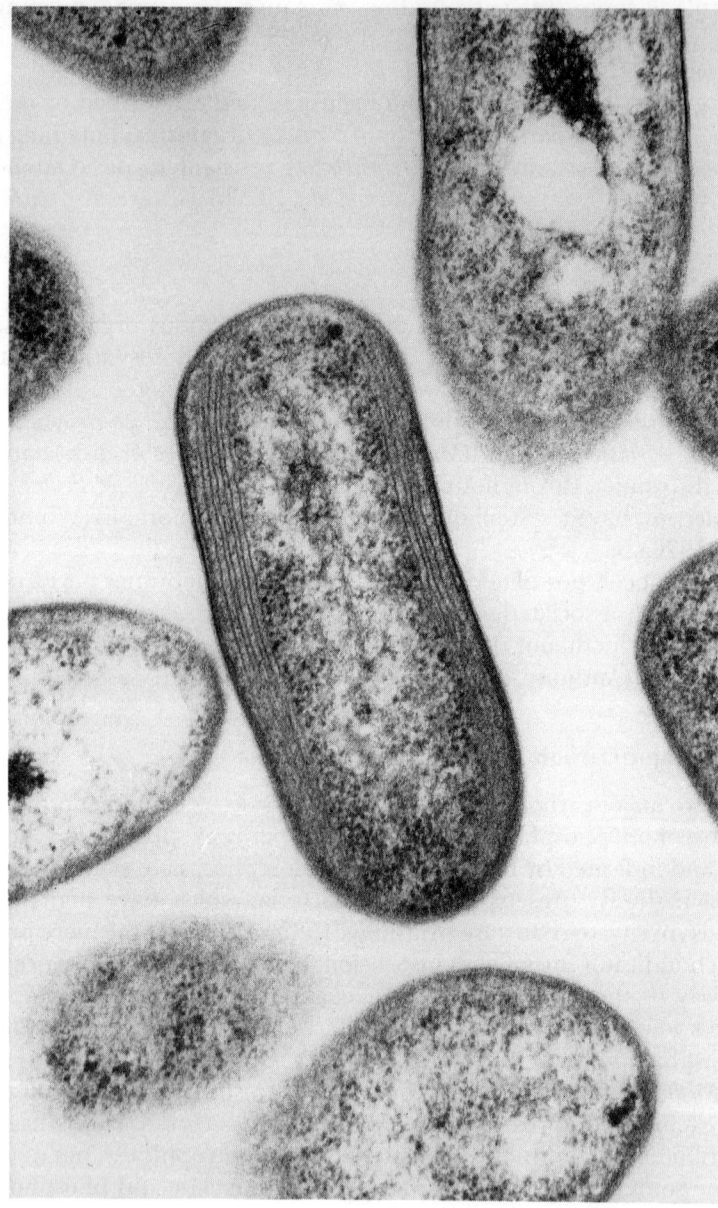

A

Figure 2. Electron micrographs. a) Sections of (A) *'Methylocystis parvus'* (NCIMB 11129) and (B) *'Methylosinus sporium'* (NCIMB 11126), showing a type II membrane system. (C & D) Sections of *Methylomonas methanica* (NCIMB 11130) showing a type I membrane system. Magnification, all × 55,040. Reproduced by kind permission of J. McN. Sieburth.

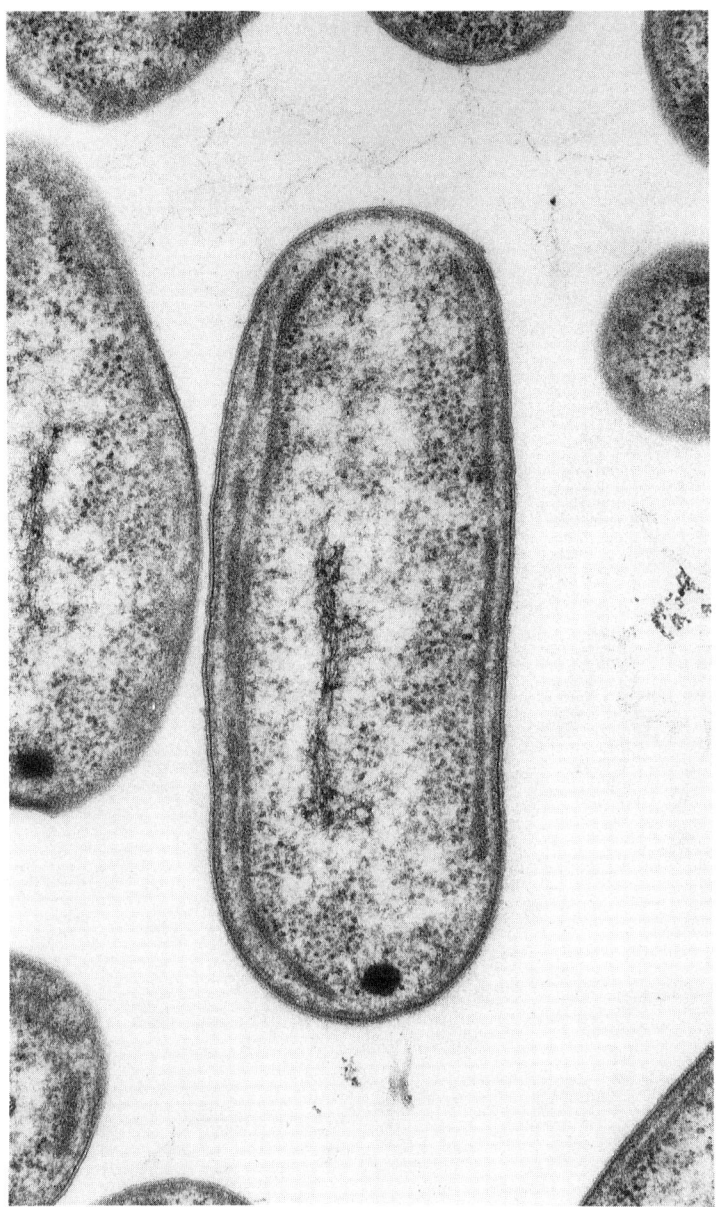

Figure 2. *(continued)*

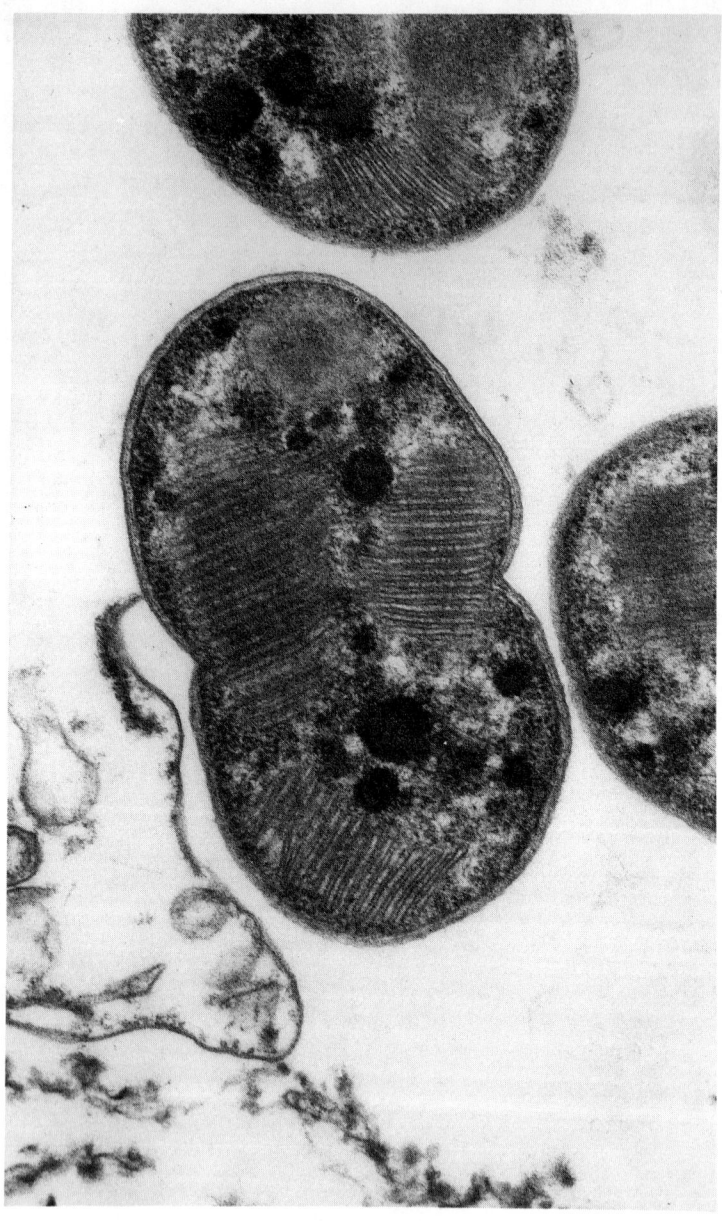

Figure 2. *(continued)*

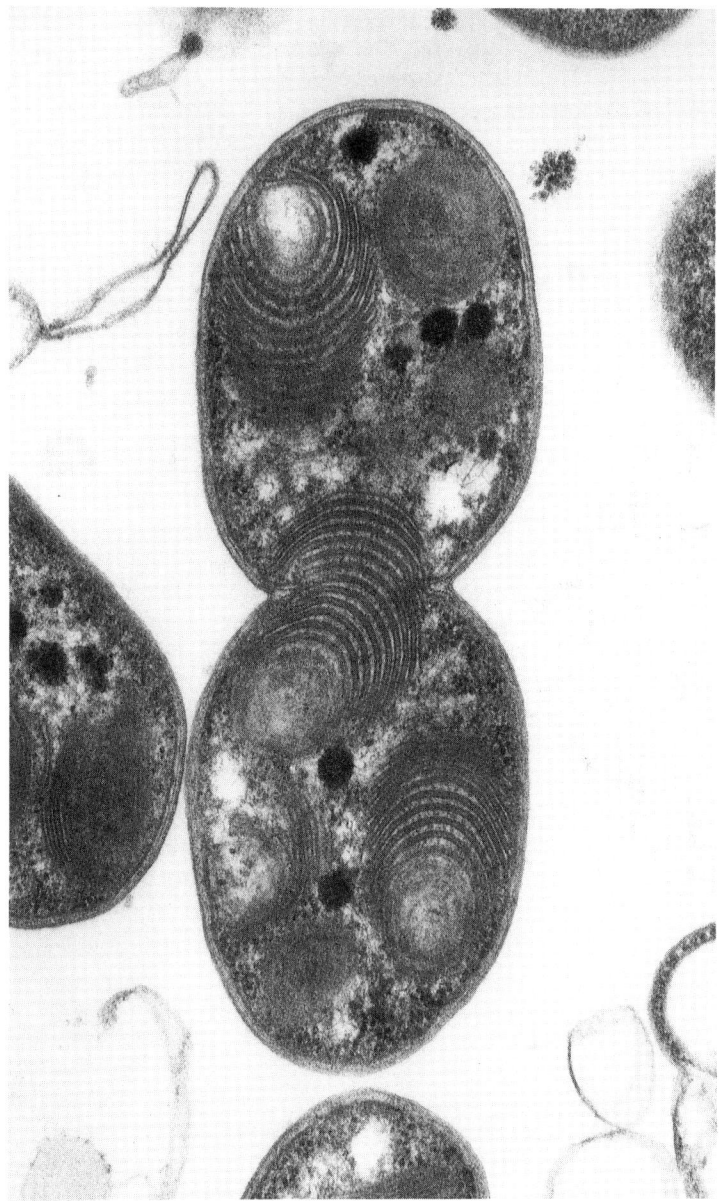

D

Figure 2. *(continued)*

2.1.5. Additional Morphological and Physiological Criteria

2.1.5a. Motility and Flagellar Arrangement. Many, but not all, methanotrophs are motile at some stage during their growth cycle. The majority of the rod-shaped type I organisms (*Methylomonas* and *Methylobacter* spp.) are motile, usually by a single polar flagellum. Most of the coccal methanotrophs, on the other hand, are nonmotile. Type II organisms belonging to the *Methylosinus* group are motile with a polar tuft of flagella, whereas all *Methylocystis* strains described so far are nonmotile.

2.1.5b. Colony Morphology and Pigmentation. Methanotrophs produce a wide range of colony pigments, both soluble and insoluble. Insoluble pigments range from white, off-white, and cream (e.g., *Methylomonas albus*) through yellow (e.g., *Methylococcus luteus*) to light brown (e.g., *Methylobacter vinelandii*) and cinnamon brown (e.g., *Methylococcus gracilis* and *Methylococcus thermophilus*).

In many instances, colony color changes with the age or physiological state of the culture. For example, encysting or sporing organisms often render the colony color darker, usually by the formation of yellowish to brown pigmentation. Not all the colonies may be affected in this way.

Water-soluble pigment production in methanotrophs is rarer, although *Methylomonas methanica* and *Methylosinus sporium* have been reported as producing green to sapphire or brown to black pigments, respectively, on iron-deficient media (Whittenbury *et al.*, 1970a).

Colony size and shape are other morphological features that can be useful in identifying methanotrophs and thus should not be ignored. For example, if methanotrophs are selectively isolated at 45°C, of the two brown-pigmented species that predominate, one will be cinnamon brown with small (1–1.5 mm diameter), smooth, shiny, convex, and entire colonies (*M. thermophilus*), whereas the other will have much larger (≥3 mm diameter) light brown, wrinkled colonies with an erose margin (*M. vinelandii*).

2.1.5c. Capsule Formation. Although capsule formation, due to the production of exopolysaccharide, can sometimes be difficult to check because of the variability of staining techniques between laboratories, it nevertheless has been used as a means of distinguishing between several methanotrophs (see Table II).

2.1.5d. Enhancement of Growth on Various Carbon Sources. Enhanced growth of methane-grown cells in the presence of supplementary carbon sources such as yeast extract, malate, acetate, or succinate was used by Whittenbury and his colleagues (1970a) as a means of differentiating between several of their species groups. However, this has proved a difficult test to interpret (P. N. Green, unpublished observation) and has been excluded from the list of useful differentiating criteria for methanotrophs shown in Table II.

2.1.5e. Growth Temperature. Ability to grow at 37°C has proved a useful feature for separating several of the type I methanotrophs belonging to the *Methylomonas/Methylobacter* groups. However, there appear to be differing views on whether type II organisms grow at this temperature.

Only three groups of methanotrophs grow at 45°C or above, the brown-pigmented *Methylomonas vinelandii* and *Methylococcus thermophilus* and the nonpigmented *Methylococcus capsulatus*. *M. thermophilus* grows at temperatures in excess of 55°C.

At the lower temperature ranges, some of the pink-pigmented taxa, including *Methylomonas methanica*, can grow at 15°C and possibly lower (P. N. Green, unpublished data).

2.1.5f. Growth on Methanol. The ability of methane-utilizing bacteria to grow on methanol was again one of the distinguishing features used initially by Whittenbury and his colleagues to differentiate among the various groups of these organisms. In their study (Whittenbury *et al.*, 1970a), only the *Methylomonas* species groups *albus, agile, rubrum,* and *methanica* were recorded as able to utilize 0.1% (w/v) methanol for growth. While we would corroborate these findings for *M. albus* and *M. agile,* there is some doubt about the ability of all the pink isolates to do so. We were unable to demonstrate methanol utilization in *M. methanica* strain S1 (NCIMB 11130), but methanol present only in the *vapor phase was utilized by a few of our own isolates. Another organism that we found could utilize methanol only in the vapor phase,* presumably because of cellular toxicity at higher levels, was *M. capsulatus* strain Bath.

2.1.5g. Growth on Glucose. Apparent slow growth (3 weeks at 30°C) of several strains of pink-pigmented obligate methane utilizers (*M. methanica, M. rubra,* and *M. rosaceous*) on glucose as sole carbon and energy source has been observed in this laboratory (P. N. Green, unpublished data). This substantiates the claim of Zhao and Hanson (1984), who ob-

---

*Vapor phase growth means growth obtained when bacteria were inoculated onto NMS salts agar containing no added carbon source and the plates incubated were inverted in an enclosed container after one drop of methanol had been placed in the lid of the plate.

TABLE II.
Some Useful Taxonomic Features of the Various Methanotrophic Bacteria

Organism	Growth at °C			Growth on methanol (%)			Motility and flagellation[a]		Capsule formation	Rosette formation
	37	45	55	0.01	0.1	3.0				
Type I bacteria										
Methylomonas methanica	−	−	−	±	±		+	P	+	−
Methylomonas rubra	+	−	−	±	±		+	P	−	
Methylomonas rosaceus	−	−	−	+	−		+	P		
Methylomonas agile	+	−	−		+		+	P	−	
Methylomonas streptobacterium	−	−	−		−		−		+	
Methylomonas albus	+	−	−		+		+	P	−	−
Methylomonas gracilis	+	±			±		+	P	±	
Methylobacter chroococcum	−	−	−	+	−		−		+	−
Methylobacter bovis	+	−	−	+	−		−		−	−
Methylobacter capsulatus	−	−	−	+	−		+		+	−
Methylobacter vinelandii	+	+		+	−		+	P	±	−
Methylococcus minimus	−	−	−		−		−		+	
Methylococcus thermophilus	+	+	+		±		+	P	+	
Methylococcus luteus	+	−	−		−		−		−	
Methylococcus ucrainicus	+	−	−		−		+	P	±	
Type X bacteria										
Methylococcus capsulatus	+	+	−		+		−		+	−
Type II bacteria										
Methylosinus sporium	+	−	−		−		+		[†]+	[†]+
Methylosinus trichosporium	+	−	−		±		+	PT	[†]+	[†]+
Methylocystis parvus	+	−	−		−		−		+	+
Methylocystis minimus	−	−	−		−		−		+	+
Methylocystis methanolicus	±	−	−		+	+	−		+	+
Methylocystis echinoides	±	−	−		+		−		−	+
Methylocystis pyreformis	−	−	−		+		−		+	±

[a] P = polar; PT = polar tuft.
[b] S = straight; C = curved; P = pear shaped.
[c] PP = pink; R = red; W = white; Y = yellow; Gr = green; B = brown; W-B = whitish brown; PB = pale brown; W-P = pinkish white; W-Gr = whitish green.
† See section 2.1.2 for further distinguishing criteria.

TAXONOMY OF METHYLOTROPHIC BACTERIA

Cellular morphology		Colony color	Water-soluble pigments	Principal phospholipids			mol% G+C
Shape[b]	Size (μm)			Methylated derivatives of phosphatidyl-ethanolamine	Cardiolipin	Phosphatidyl-choline	
Rod (S)	0.5–0.5×0.8–2.0	P	$\overset{-}{Gr}$ or	–	–	–	52.4
Rod (S)	0.6–0.8×1.2–1.5	R	–				53.0
Rod (S)	0.6–0.8×1.2	PP	–				
Rod (S)	0.6–0.8×1.2	W	–				
Rod (S)		W	–				
Rod (S)	0.5–0.7×0.8–2.0	W	–	–	–	–	52.4
Rod (S)	0.4–0.5×1.0–1.5	B	$\overset{-}{B}$ or				59.0
Large coccobacillus	4.5×3.2	PP	–	–	+	–	52.1
Rod (S)	0.7–1.0×1.2–2.5	W-B	$\overset{-}{Y}$ or	–	+	–	52.4
Rod (S)	1.0–1.2×1.5–2.5	W-B	–	–	+	–	52.5
Rod (S)	1.0–1.5×1.5–3.0	PB	–	–	+	–	51.4
Small coccobacillus	0.7×0.7–1.0	W	–				
Rod (S)	0.7×0.7×1.2	B	$\overset{-}{B}$ or				63.3
Rod (S)	0.8–1.0×1.5–2.0	Y	–				53.0
Rod (S)	0.7–1.0×1.0–1.8	W	–				52.7
Coccus	0.7–0.9×0.7–0.9	W	–	–	+	+	62.7
Rod (S.C.P.)	0.7–1.0×1.0–2.5	W-B	$\overset{-}{B}$ or	+	+	w+	62.5
Rod (S.C.P.)	0.7–1.0×1.0–4.0	W-B	–	+	+	w+	62.5
Rod (S.C.P.)	0.5–0.7×0.8–2.0	W	$\overset{-}{B}$ or	+	+	+	62.4
Rod (S.C.P.)	0.4–0.5×0.6–0.8	W-Gr		+	+	+	61.7
Rod (S.C.P.)	0.6–0.8×0.9–2.0	W-B	Y	+	+	+	62.5
Rod (S.C.P.)	0.5–0.7×0.8–1.2	W-P	–	+	+	+	63.1
Rod (S.C.P.)	0.7–0.8×0.9–2.5	W	–	+	+	+	62.5

served that their strain *Methylomonas* 761H grew on glucose plus casein hydrolysate. Work is still continuing to investigate this feature.

Other carbohydrates and multicarbon compounds tested (fructose, pyruvate, oxoglutarate, acetate, malate, succinate, citrate, fumarate, and aspartate) did not support growth, nor did any other group of methanotroph appear to grow on glucose.

Growth on glucose is of particular interest, not only metabolically, but because it also occurs in the other groups of obligate methylotrophs, the methanol and methylamine utilizers, where some organisms previously thought to be obligate C_1 utilizers have been shown to utilize glucose (and possibly fructose).

2.1.5h. Rosette Formation. The majority of type II, but not type I or type X, bacteria display rosette formation, where bacteria appear to become attached at their nonflagellated poles. However, this feature is not always seen when microscopically viewing type II organisms, but very much depends on the age and nutritional state of the culture.

2.1.5i. Nitrogen Sources. Whittenbury *et al.*, (1970a) found that all the isolates they examined utilized ammonia as a nitrogen source, and that nitrite and nitrate were used by the majority, and urea, casamino acids, and yeast extract by some.

Of the nitrogen sources tested in our laboratory [ammonia, nitrate, asparagine, cysteine, peptone (Difco), and tryptone (Difco)] against our own isolates and several previously described strains; only *M. thermophilus* failed to use the majority of these compounds, including nitrate. Thus, the use of nitrate rather than ammonia as sole nitrogen source in any isolation procedure could result in failure to isolate this particular group of organisms.

Differences exist among methanotrophs based on their mode of nitrogen metabolism (Murrell and Dalton, 1983; Shishkina and Trotsenko, 1979). One can distinguish between the type I and type II methanotrophs on their ability to fix atmospheric nitrogen (type II only fixes) and their pathways of ammonia assimilation. In all type II methanotrophs examined by Murrell and Dalton (1983), ammonia was assimilated exclusively via the GS/GOGAT* pathway. In type I organisms, the GS/GOGAT pathway operated during growth on nitrate, but enzymes in this pathway were repressed during growth on ammonia, where the nitrogen source was assimilated either by the alanine or glutamate dehydrogenase pathway.

*GS = glutamine synthetase; GOGAT = glutamate synthase.

2.1.5j. Antibiotic Sensitivities. Seventy-four strains, representing eight different groups of methanotrophs isolated in this laboratory, were examined for their antibiotic sensitivities. The following NCIMB cultures were also examined; *Methylomonas albus* NCIMB 11123, *Methylomonas agile* NCIMB 11124, *Methylosinus sporium* NCIMB 11126, *Methylocystis parvus* NCIMB 11129, *Methylomonas methanica* NCIMB 11130, *Methylosinus trichosporium* NCIMB 11131, *Methylococcus capsulatus* (Bath) NCIMB 11132, *Methylomonas rubra* NCIMB 11913, *Methylococcus luteus* NCIMB 11914, *Methylomonas gracilis* NCIMB 11912, *Methylococcus ucrainicus* NCIMB 11915, and *Methylococcus thermophilus* NCIMB 11916.

All strains examined were sensitive to antibiotic disks containing streptomycin (25 µg) and gentamycin (10 µg), and all were resistant to bacitracin (8 units) and novobiocin (5 µg). Although too few strains were examined from several of the representative taxa to produce a reliable taxonomic tool, many of the antibiotic sensitivities obtained were interesting in that, on the limited evidence available, they did appear to form at least a diagnostic aid in several instances.

For example, sensitivity patterns to nalidixic acid and nitrofurantoin appear useful in separating the three pink methanotrophic taxa, whereas *M. thermophilus* was the only strain tested that was sensitive to erythromycin. Also of interest was the sensitivity of mainly the thermophilic *M. vinelandii*, *M. thermophilus*, and *Methylococcus capsulatus* to penicillin G, and the possible use of carbenicillin, cotrimoxazole, framycetin, neomycin, or tetracycline to distinguish between the *Methylosinus* strains (*M. sporium* and *M. trichosporium*) and *Methylocystis parvus*.

This last point highlights the possible use of antibiotics for the selective isolation of specific groups of methanotrophs. For more detailed information see Table III.

2.1.5k. Tricarboxylic Acid (TCA) Cycle and Other Metabolic Enzymes. The levels of most TCA-cycle enzymes are generally lower in type I (and type X) than in type II methanotrophs. In particular, type I bacteria (*Methylomonas* and *Methylobacter*) and *Methylococcus capsulatus* (type X) contain no 2-oxoglutarate dehydrogenase, whereas the levels of this enzyme in type II bacteria (*Methylosinus* and *Methylocystis*) are similar to those measured in nonmethanotrophic bacteria growing on methanol. Thus, it appears that those bacteria in which the major assimilation pathway is the RMP (type I and type X) have an incomplete TCA cycle, whereas those with the serine pathway (type II) have a complete TCA cycle (Anthony, 1982).

Davey *et al.*, (1972) demonstrated that the glucose-6-phosphate and

TABLE III. Antibiotic Sensitivities, Enzyme Activities, and Nitrogen Sources of Some Methanotrophic Bacteria[a]

Feature	Methylomonas methanica	Methylomonas rubra	Methylomonas rosaceus[b]	Methylomonas agile	Methylomonas albus	Methylomonas gracilis
Antibiotic sensitivities						
Erythromycin (5 µg)	−	−	−	−	−	−
Polymyxin B (250 units)	V[c]	+	+	+	w+	
Penicillin G (4 units)	V	−	−	−	−	
Carbenicillin (25 µg)	+		+	+	+	
Colistin sulphate (100 µg)	V		V	+	+	
Chloramphenicol (50 µg)	−	−	−	−	−	
Tetracycline (50 µg)	+	+	+	+	+	
Cephalexin (30 µg)	+		+	+	+	
Cephaloridine (25 µg)	+		+	+	+	
Framycetin (50 µg)	+		+	+	+	
Nalidixic acid (30 µg)	+	+	−	+	+	−
Neomycin (30 µg)	+	+	+	+	+	
Cycloserine (75 µg)	+		−	−	+	
Cotrimoxazole (25 µg)	+		+	+	+	
Nitrofurantoin (200 µg)	−	+	−	−	−	
Sulphafurazole (200 µg)	V		+	−	+	+
Sulphadimethoxine (200 µg)	−	−	−	w+	−	
API Zym enzyme activities						
Lipase	V		V		−	−
Acid phosphatase	V	−	−	−	−	+
Phosphoamidase	w+	−	w+	w+	w+	−
Valine arylamidase	w+	w+	w+	w+	w+	w+
Cystine arylamidase	w+	−	V	−	−	−
β-glucosidase	+	+	−	−	−	−
β-glucuronidase	−	−	−	−	−	−
Trypsin	−	−	−	−	−	−
Nitrogen sources						
Asparagine	+		+	+	+	+
Cysteine	+		−	+	+	−
Peptone	V		−	−	+	+
Tryptone	+	+	+	+	+	+

[a] Taken from Green and Woodford, unpublished data. In many cases results were obtained from a single strain and thus cannot be assumed to be definitive for *all* other strains of that taxon.
[b] Isolated by NCIB staff.
[c] V = variable result; + = sensitive; − = resistant (to antibiotics); w = weakly.

TAXONOMY OF METHYLOTROPHIC BACTERIA   41

			Organism(s)					
Methylo-bacter capsulatus[b]	Methylo-bacter vinelandii[b]	Methyl-ocuccus thermo-philus	Methyl-ococcus luteus	Methyl-ococcus ucrainicus	Methyl-ococcus capsulatus	Methyl-osinus sporium	Methyl-osinus tricho-sporium	Methyl-ocystis parvus
−	−	+	−	−	−	−	−	−
+	+	+	+	+	−	−	−	−
−	+	+	−	−	w+	−	−	−
V	+	V			+	−	−	+
+	−	+			−	−	−	+
−	+	V	−	−	w+	−	+	−
+	+	+	+	+	+	−	−	+
+	+	+	+	+	−	+	+	+
+			+			−	−	−
+	+	+			+	+	+	−
+	−	−	−	−	−	−	−	−
+	+	+	+	+	+	+	+	−
V	−				+	−	−	+
+		+			−	+	+	−
−	+	+	V		+	+	+	+
+	−	+	+	+	−	−	−	
−	−	+	+		−	−	−	−
−	−	−	−	−	w+	w+	w+	w+
−	+	w+	+	+		w+	+	+
w+	+	w+				+	w+	w+
w+	w+	−	w+			w+	w+	+
−		−	−	−	w+	−	w+	w+
−	−	−	w+		−	−	−	−
−	−	−	−	−	−	−	−	−
−	−	−	−	−	−	w+	w+	w+
+	+	−	+	+	+	+	+	+
V	−	−	+	−	−	+	+	+
+	+	V	+	+	+	+	+	+
+	+	V	+	+	+	+	+	+

6-phosphogluconate dehydrogenases that occur in type I organisms are $NADP^+$-specific. These enzymes are absent in type II bacteria.

It has also been shown that type II bacteria have an $NADP^+$-specific isocitrate dehydrogenase, whereas the same enzyme from type I bacteria can reduce either NAD^+ or $NADP^+$. *Methylococcus capsulatus* has an NAD^+-specific isocitrate dehydrogenase.

Pyruvate dehydrogenase, hexokinase, 6-phosphogluconate dehydrogenase, and glucose 6-phosphate dehydrogenase activities were examined by Trotsenko (1976, 1983), who reported that these enzymes were present in all the type I methanotrophs, but absent in all type II methanotrophs studied.

API Zym strips were used to examine several of our laboratory isolates together with some of the documented strains described in the text. Most of the strains examined did not contain any α-glucosidase, α-fucosidase, α-mannosidase, *N*-acetyl-β-glucosamidase, α- or β-galactosidase, or chymotrypsin activity, whereas most strains did possess leucine arylamidase, alkaline phosphatase, and esterase activities. Lipase activity was absent in some of our pink isolates, in *M. thermophilus*, *M. albus*, *Methylobacter capsulatus*, *M. luteus*, *M. vinelandii*, *M. ucrainicus*, and *M. gracilis*, but present (weakly) in *Methylococcus capsulatus* and all type II strains. Acid phosphatase was absent from *M. albus*, *M. agile*, *M. rubra*, and our *Methylobacter capsulatus* strains, but present in most of the other organisms tested (see Table III). A large number of methanotrophs possessed phosphoamidase, with only *M. rubra* and *M. gracilis* definitely showing no activity. Similarly, valine arylamidase was present (often weakly) in most strains except *M. thermophilus*. Only some of the pink isolates, including *M. methanica*, in addition to *Methylococcus capsulatus*, *M. parvus*, and *M. trichosporium* (but not *M. sporium*) demonstrated weak cystine arylamidase activity. Only *M. methanica*, *M. rubra*, and *M. luteus* were weakly β-glucosidase positive, and only our isolate of *M. luteus* was β-glucuronidase positive. Weak trypsin activity was found only in the three type II organisms tested (see Table III).

On agar media, only *M. trichosporium* demonstrated weak hydrolytic activity when gelatin and casein, but not starch or DNA, were added to the basal nitrate mineral salts (NMS) medium. (Plates were incubated in a methane atmosphere.)

2.1.5l. Enhanced Growth of Thermotolerant/Thermophilic Strains. The enhanced growth and pigmentation of *M. vinelandii*, *M. gracilis*, and, in particular, *M. thermophilus* was apparent when these strains were grown in a methane atmosphere on a salts medium containing added copper ions (P. N. Green, unpublished data). This vitamin-copper-rich (VCR) medium was formulated by Shen *et al.* (1982), who considered 0.5–1.5 mg/liter of cupric sulfate ($CuSO_4 \cdot 5H_2$) to be optimal for growth. They also demonstrated that choline and vitamin B_{12} stimulated growth when added to the culture medium.

2.1.5m. DNA Base Ratios (%G+C) and DNA:DNA Homologies. The *Methylobacter* and *Methylomonas* strains isolated by Whittenbury *et al.*, (1970a), i.e., type I methanotrophs, all have a DNA base ratio in the range 50–54% mol G+C, whereas *Methylococcus capsulatus* (type X) and the *Methylosinus* strains (type II) all have base ratios around 62.5% mol G+C.

However, the more recent isolation of several new taxa by Soviet workers has blurred these clear-cut boundaries, in certain cases. New type II *Methylocystis* isolates (*M. minimus, M. methanolicus, M. pyreformis*, and *M. echinoides*) have base ratios in the range 61.7–63.1% mol G+C.

Similarly, the Soviet isolates *Methylococcus luteus* (53% mol G+C), *Methylococcus ucrainicus* (52.7% mol G+C), and *Methylococcus vinelandii* (53% mol G+C), which are all short rods rather than cocci and therefore are most likely to belong to Whittenbury's *Methylobacter* or *Methylomonas* groups, are all within the 50–54% mol G+C originally designated for type I methanotrophs (Whittenbury *et al.*, 1970a).

However, the base compositions of two recent isolates are not within the aforementioned ranges. *Methylococcus thermophilus* and *Methylomonas gracilis* have mol % G+C contents of their DNA of 63.3 and 59, respectively (Romanovskaya *et al.*, 1978). As neither of these organisms is a true coccus like *Methylococcus capsulatus*, and both have a type I membrane system and RMP assimilatory enzymes; it is possible they represent new taxa.

Galchenko and Andreev (1984) observed DNA–DNA homologies of 15–20% between bacteria classified within the groups *Methylococcus, Methylomonas, Methylobacter, Methylosinus*, and *Methylocystis*. The same authors also observed homologies of 50–70% among "species" within these groups and 80–90% among strains of these species.

2.1.5n. Fatty Acids. Methanotrophic bacteria differ in the type of predominant cellular fatty acids they contain. Type I and type X bacteria have saturated and/or monounsaturated fatty acids with a chain length of 16 carbon atoms ($C_{16:0}$ and $C_{16:1}$) as their predominant fatty acids, whereas type II bacteria have a monounsaturated fatty acid with 18 carbon atoms ($C_{18:1}$) predominating (Colby *et al.*, 1979). Other fatty acids found (≤15% of total fatty acids) include tetradecanoic ($C_{14:0}$), which is found largely in the following pigmented taxa: *M. methanica, M. rubra*, and *M. luteus* (Romanovskaya *et al.*, 1980). Smaller amounts of straight-chain $C_{15:0}$, $C_{15:1}$, $C_{17:0}$, and $C_{19:0}$ and hydroxy (3-OH) acids of $C_{10:0}$, $C_{14:0}$, and $C_{16:0}$, as well as

$C_{17:0}$ and $C_{19:0}$ cyclopropane acids, are also found in many methanotrophs (Urakami and Komagata, 1987). Indeed, Andreev and his co-workers have increased the resolution of fatty acid analyses in their laboratories to enable differentiation between methanotrophs at the subgenus level. For example, as many as eight structural isomers (mainly associated with different double-bond positions) can be found in certain strains. In type I methanotrophs, the isomeric composition of hexadecenoic ($C_{16:1}$) acids as well as the contents of $C_{16:0}$ and $C_{14:0}$ straight-chain acids vary markedly (Galchenko and Andreev, 1984).

Another interesting difference in fatty acid content was observed between the thermophilic/thermotolerant *Methylococcus capsulatus* and *M. thermophilus* and mesophilic strains. The thermophilic organisms had much higher content (ca. 40–50%) of $C_{16:0}$ than the mesophiles (ca. 10–30%).

2.1.5o. Phospholipid Composition. Variation in the phospholipid composition allows differentiation between the three major types of methanotrophs, possibly at the group or generic level. All type II bacteria examined contain the methylated derivatives of phosphatidylethanolamine (mono- and dimethylphosphatidylethanolamine) in addition to phosphatidylcholine. These phospholipids are absent from the lipid pool of type I methanotrophs. *Methylococcus capsulatus* (type X) contains phosphatidylcholine like the type II methanotrophs, but no methylated phosphatidylethanolamines.

Bacteria grouped by Whittenbury and his colleagues (1970a) as *Methylomonas* differ from those grouped as *Methylobacter* by the lack of cardiolipin, which is present in the latter in substantial amounts (Galchenko and Andreev, 1984). Type II methanotrophs *Methylocystis* and *Methylosinus* can be differentiated by a two-dimensional chromatography of their phospholipids. Strains belonging to *Methylosinus* contain only trace amounts of phosphatidylcholine, one of the main components of *Methylocystis*.

2.1.5p. Isoprenoid Compounds. In a study (Urakami and Komagata, 1986b) on the occurrence of isoprenoid compounds in a variety of methylotrophic bacteria, the respiratory quinone, ubiquinone, with eight isoprene units (Q-8, coenzyme Q), was found to be the major isoprenoid quinone present in methanotrophic bacteria.

Subsequent studies (Collins and Green, 1985; Collins *et al.*, 1986) have shown that in addition to Q-8, two other novel quinones exist in the methanotrophs.

The first, 18-methylene-ubiquinone-8 (MQ-8), the principal quinone component in *Methylococcus capsulatus* NCIMB 11132, *M. thermophilus* strain 109, *M. gracilis* NCIMB 11912, *M. methanica* NCIMB 11130, and *Methylomonas* 761H, may not be of taxonomic significance, but the second compound, 11-methylene-18 dimethyl-ubiquinone-6 (MDMQ-6), was found only in *Methylomonas rubra,* and not the phenotypically very similar *M. methanica,* which contained MQ-8.

Sterols (4α-methylsteroids) and the sterol precursor squalene have been found in large amounts in *Methylococcus capsulatus* (Bird *et al.,* 1971), but other methanotrophs have yet to be examined in detail for these compounds.

2.1.5q. Electrophoretic Comparison of Total Soluble Proteins. Soviet workers (Galchenko and Andreev, 1984) have shown that protein patterns obtained from polyacrylamide gel electrophoresis (PAGE) of total soluble cellular proteins from methanotrophs are a useful taxonomic tool. Such systems have been successfully used before to probe the taxonomic structure of other groups of methylotrophic bacteria (Hood *et al.,* 1988), where large numbers of strains sharing the same gel can have their protein fingerprints compared by, for example, scanning densitometry. The result can be expressed diagrammatically in the form of a dendogram showing percentage similarities between individual strains and groups of organisms. Such a dendogram is illustrated in Figure 3.

2.2. Taxonomic Structure within the Methanotrophs

The lack of molecular biological studies performed on the methanotrophs, coupled with the plethora of nonvalidated species and genera, and the scant or inadequate descriptions of individual organisms or groups of organisms, make it very difficult, if not impossible, to propose a sound taxonomic structure for this group of organisms with any degree of confidence.

Therefore, until evidence is found to the contrary, perhaps it is prudent to make the assumption, as did Whittenbury and Krieg (1984), that all obligate methane-utilizing bacteria belong to a single family, the *Methylococcaceae*. A description of this family (taken, in part, from Whittenbury and Krieg, 1984) together with limited descriptions of the five loosely named genera and the species groups within each now follows.

2.2.1. The *Methylococcaceae*

A diverse group of rod- and coccal-shaped organisms having in common the ability to utilize methane (via methane monooxygenase) *as a sole carbon and energy source under aerobic or microaerophilic conditions.* Facultative methane

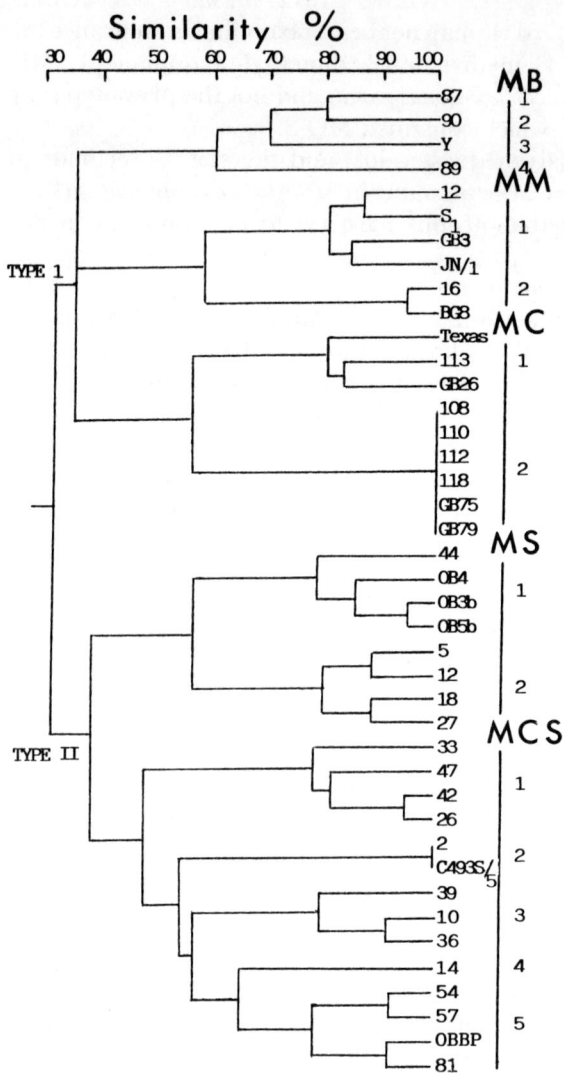

Figure 3. The taxonomic dendrogram based on protein gel electrophoretograms of methanotrophs. MB, *Methylobacter*; MM, *Methylomonas*; MC, *Methylococcus*; MS, *Methylosinus*; MCS, *Methylocystis*. MB: 1. *Methylobacter vinelandii*; 2. *Methylobacter chroococcum*; 3. *Methylobacter capsulatus*; 4. *Methylobacter hovis*. MM: 1. *Methylomonas methanica*; 2. *Methylomonas albus*. MC: 1. *Methyloccus capsulatus*; 2. *Methylococcus thermophilus*. MS: 1. *Methylosinus trichosporuim*; 2. *Methylosinus sporuim*. MCS: 1. *Methylocystis minumus*; 2. *Methylocystis echinoides*; 3. *Methylocystis methanolicus*; 4. *Methylocystis pyreformis*; 5. *Methylocystis parvus*. (Taken from Galchenko and Andreev, 1984.)

oxidizers (i.e., organisms able to use a wide range of complex organic compounds as alternatives to methane for carbon and energy) are excluded from the family. *Gram negative.* All strains possess a *complex arrangement of intracytoplasmic membranes when grown on methane.* These membrane arrangements may consist of bundles of vesicular disks or paired membranes aligned to the cell periphery. All strains have a *strictly respiratory type of metabolism,* with oxygen as the terminal electron acceptor. Those strains tested are *catalase and oxidase positive.* Occur in aerobic environments where methane is available, i.e., in soil, mud, or water adjacent to or overlaying anaerobic environments.

Type genus: *Methylococcus* (Foster and Davis, 1966), 1929 AL (approved lists).

2.2.2. Further Descriptive Information

Some rod and coccal forms are motile, the rods possessing a single polar flagellum or a tuft of polar flagella. Strains of methanotrophs may be pigmented (yellow, ocher, pink, red, or brown). Encystment of many vegetative cultures that produce white or off-white colonies can result in varying degrees of browning of colonies, especially in older cultures. Some strains produce water-soluble pigments (blue, yellow, or brown) under certain conditions, e.g., iron limitation. Increasing amounts of copper in the growth medium may also result in a deepening of the brown pigmentation formed by some strains. Resting stages may be formed by some strains. These can be: (1) *cysts,* indistinguishable from those formed by *Azotobacter* species; (2) *lipid cysts,* characterized by large lipid inclusions (mainly composed of poly-β-hydroxybutyrate), absence of intracellular membranes, and formation of a thick layer on the outer surface of the cell; or (3) *exospores,* formed by budding at one cell pole. All three kinds are resistant to desiccation, and the exospores are also resistant to heat (85°C for 15 min).

Growth temperature ranges vary among strains. Some strains are thermophilic or thermotolerant, failing to grow at 30°C, but growing at 45–55°C. Many strains are sensitive to the normal oxygen tension of an air atmosphere and require decreased oxygen levels (microaerophilic conditions) for growth, and several groups of organisms form extracellular polysaccharides (capsules and/or slimes).

Two major types of membrane arrangements are associated with methane-grown cells as described above. Cells containing vesicular disks have been designated the type I arrangement, and cells with paired membranes extending throughout the cell or running parallel to the cytoplasmic membrane as type II.

The pathway for methane oxidation is as follows: (1) oxidation to methanol, via methane monooxygenase; (2) methanol to formaldehyde, via methanol dehydrogenase; (3) formaldehyde to formate, via formaldehyde dehydrogenase; and (4) formate to carbon dioxide, via formate dehydrogenase. Formaldehyde formed in step (2) is assimilated by either the serine pathway (in type II bacteria) or the ribulose monophosphate pathway (in type I bacteria). At least one species, *Methylococcus capsulatus* (type X), has the potential ability to fix CO_2 by the Benson–Calvin cycle.

Differences in the tricarboxylic acid (TCA) cycle occur among methane oxidizers. Some strains have a complete cycle, but others lack the enzyme 2-oxoglutarate dehydrogenase. Other differences among strains occur with respect to isocitrate dehydrogenase activity and the pyridine nucleotide requirements of this enzyme (see Section 2.1.5k).

Most strains can utilize methyl formate, dimethyl carbonate, and methanol as sole source of carbon and energy, but some find methanol a toxic substrate if supplied in other than trace concentrations. Carbon dioxide, acetate, formate, some amino acids, and the following cooxidizable substrates can serve as supplementary carbon sources for obligate methanotrophs, provided formaldehyde is available via methane or methanol oxidation or is added separately: primary alcohols, alkanes, dimethyl- and diethylether, alicyclic, aromatic, and heterocyclic compounds, and carbon monoxide.

Most strains utilize ammonia and nitrate as nitrogen source, and some can also utilize nitrite, urea, amino acids, and yeast extract. In addition, some strains fix nitrogen under microaerophilic conditions.

Fatty acid and phospholipid content varies among methanotrophs. Type I bacteria have a predominance of fatty acids of the $C_{16:0}$ and $C_{16:1}$ type, but do not contain methylated derivatives of phosphatidylethanolamine or phosphatidylcholine. Type II bacteria, on the other hand, possess these phospholipids and have $C_{18:1}$ as their principal fatty acid component.

Most methanotrophs contain isoprenoid quinones of the ubiquinone type with eight isoprene units (Q-8), although groups containing novel quinones have recently been discovered (see Section 2.1.5p). In addition, some methanotrophs contain sterols and the sterol precursor squalene.

DNA base ratios vary among methane oxidizers from 50 to 63.1% mol G+C, with type II organisms and type X (*Methylococcus capsulatus*) within the range 61.7–63.1 and the majority of type I organisms within the range 50–54% mol G+C.

2.2.3. Taxonomic Comments

Many of the named methanotrophs mentioned in the literature, while having had their descriptions validly published, have nevertheless never been formally validated by publication in the *International Journal of Systematic Bacteriology* and hence have no standing in nomenclature. The only nomenclaturally valid names of obligate methane-utilizing bacteria are the following: the family *Methylococcaceae*, the genus *Methylococcus* (including the species *M. luteus, M. mobilis* Hazeu *et al.*, 1980, *M. bovis, M. chroococcus, M. whittenburyi, M. vinelandii*), and the genus *Methylomonas* (including the species *M. methanica*).

However, many organisms that have either been placed in, or transferred to, the genus *Methylococcus*, in particular, may well be improperly classified. On morphological grounds alone, it is doubtful whether several organisms should belong to that genus. This concern is obviously shared by Romanovskaya (1984), who proposed the new genus *Methylovarius* in an attempt to resolve the heterogeneity within *Methylococcus*.

Consequently, the following genus, group, and species descriptions reflect the present author's personal taxonomic bias, as is bound to be the case when discussing a group of organisms whose taxonomic structure is still in its infancy. Only when taxonomists begin to compare the genomic relatedness of this varied group of organisms will a sound taxonomic structure begin to emerge. (See phylogenetic discussion at the end of this chapter.)

2.2.4. The Genus *Methylococcus*

Cells *spherical*, often occurring in pairs. *Nonmotile*. Resting stage is a *cyst*. Gram negative. Aerobic, having a strictly respiratory type of metabolism with oxygen as the terminal electron acceptor. *Methane, methanol, and formaldehyde are the principal compounds serving as sole carbon and energy sources.* The mol% G+C of the DNA is 62.5.

Type species: *Methylococcus capsulatus* (Foster and Davis, 1966), 1929 AL.

Cocci, 1.0 µm in diameter. Capsules are formed. Poly-β-hydroxybutyrate is not formed. Nitrate is the preferred nitrogen source. No growth factors are required. Growth occurs between 30 and 50°C, but not at 55°C. Other characteristics are as indicated in Table II for group X organisms. (See also Tables II and III).

The type strain is *M. capsulatus* (strain Texas, Foster and Davis, 1966), NCIMB 11853, ATCC 19069, although *M. capsulatus* (Bath) NCIMB 11132 is the experimental workhorse.

2.2.5. The Genus *Methylomonas*

Coccobacilli or straight rods, 0.5–0.7 × 0.7–2.0 µm. *Motile or nonmotile.* Motile cells have a single polar flagellum. Gram negative. Aerobic, having a strictly respiratory type of metabolism with oxygen as the terminal electron acceptor. Oxidase and catalase positive. Methane, methanol, and formaldehyde are the principal sources of carbon and energy. Organic growth factors are not normally required. Growth occurs at ≤ 20–37°C, but not at 45°C. The mol% G+C of the DNA is 50–54. *Methylated derivatives of phosphatidylethanolamine, phosphatidylcholine, and cardiolipin are usually absent from cellular lipids.*

Type species: *Methylomonas methanica;* type strain not extant. Suggested neotype; S1 (NCIMB 11130). Other characteristics are as indicated in Table II for group I organisms. See Figure 2 (C & D).

The following species groups belong to *Methylomonas* (see Tables II and III for differentiating features):

1. *Methylomonas methanica* (Whittenbury *et al.*, 1970a).
2. *Methylomonas rubra (Methylomonas rubrum)* (Whittenbury *et al.*, 1970a).
    Suggested reference strain: 15m (NCIMB 11913).
3. *Methylomonas rosaceus* (Whittenbury *et al.*, 1970a).
    No extant reference strain.
4. *Methylomonas agile* (Whittenbury *et al.*, 1970a).
    Suggested reference strain: A30 (NCIMB 11124, ATCC 35068).
5. *Methylomonas streptobacterium* (Whittenbury *et al.*, 1970a).
    No extant reference strain.
6. *Methylomonas alba (Methylomonas albus)* (Whittenbury *et al.*, 1970a).
    Suggested reference strain: BG8 (NCINB 11123).

2.2.6. *Methylobacter*

Description as for *Methylomonas,* except for the presence of *cardiolipin.* This phospholipid is absent from *Methylomonas* strains (see Tables II and III for differentiating features).

1. *Methylobacter bovis (Methylococcus bovis)* (Whittenbury *et al.*, 1970a).
    Suggested reference strain: CM.
2. *Methylobacter capsulatus (Methylococcus whittenburyi)* (Whittenbury *et al.*, 1970a).
    Suggested reference strain: Y (NCIMB 11128).
3. *Methylobacter vinelandii (Methylococcus vinelandii)* (Whittenbury *et al.*, 1970a). See Figure 1(B). Grows at 45°C.
    Suggested type strain: VKM-53B.

2.2.7. *Methylosinus*

Straight, curved, or pear-shaped rods, which *sometimes show budding,* 0.7–1.0 × 1.0–4.0 μm. *Motile, usually due to a polar tuft of flagella.* Form *exo-*

spores as a resting stage, which are resistant to drying and heat. Encysting cells can result in pale-brown pigmentation of the normally nonpigmented colonies. *Cells contain methylated derivatives of phosphatidylethanolamine, cardiolipin,* and trace amounts of phosphatidylcholine in their phospholipids. Other characteristics as indicated in Table II for group II organisms.

The following species belong to *Methylosinus:*

1. *Methylosinus sporium* (Whittenbury *et al.*, 1970a). See Figures 1(A) and 2(A).
   Suggested reference strain: 5 (NCIMB 11126, ATCC 35069).
2. *Methylosinus trichosporium* (Whittenbury *et al.*, 1970a).
   Suggested reference strain: OB3B (NCIMB 11131, ATCC 35070).

2.2.8. Methylocystis

Description as for *Methylosinus,* apart from the *lack of motility* and the sensitivity of strains to various antibiotics (see Table II). Although Whittenbury and his colleagues originally described *Methylosinus* as having rod- or pear-shaped cells and *Methylocystis* as having rod or vibroid (curved) cells, in fact all three types of cellular shape appear to occur in both genera (P. N. Green, personal observation).

The following species groups belong to *Methylocystis:*

1. *Methylocystis parvus* (Whittenbury *et al.*, 1970a). See Figure 2(A).
   Suggested reference strain: oBBP (NCIMB 11129, ATCC 35066).
   2. *Methylocystis minimus*.
   Suggested reference strain: 42 (Galchenko, 1977).
3. *Methylocystis methanolicus*.
   Suggested reference strain: 10 (Galchenko, 1977).
4. *Methylocystis pyreformis*.
   Suggested reference strain: 14 (Galchenko, 1977).
5. *Methylocystis echinoides* (= *M. fistulosa*).
   This species is unusual in that it has rigid tubular structures located radially on the cell surface, observable in thin section under the electron microscope (Galchenko, 1977).

2.2.9. Species Incertae Sedis

Species or groups of methanotrophic bacteria whose taxonomic affiliations are uncertain are listed below, together with comments where appropriate.

1. *Methylobacter chroococcum (Methylococcus chroococcus)* (Whittenbury *et al.*, 1970a). This organism is characterized by its large coccoba-

cillary cell shape (3.2 × 4.5 µm), which is atypically larger than any of the other methanotrophs described above. It forms pale-pink colonies and contains cardiolipin like other *Methylobacter* species. No easily available reference strain.

2. *Methylomonas gracilis* (Romanovskaya et al., 1978). This organism, whose upper growth limit is around 45°C, while morphologically and physiologically appearing similar to *Methylomonas*, has a mol% of 59.0, which excludes it from that genus as presently defined. Suggested reference strain: 14L (NCIMB 11912).

3. *Methylococcus minimus* (Whittenbury et al., 1970a). Small coccal or coccobacillary cells not as spherical as those of *M. capsulatus*, suggesting that this organism may not belong to the genus *Methylococcus*. No easily available reference strain exists.

4. *Methylococcus thermophilus* (Malashenko et al., 1975). Apart from its high growth temperatures of 50–55°C, this organism is unusual in that its rod-shaped morphology excludes it from the genus *Methylococcus*, yet its mol% G+C value of 63.3 excludes it from the rod-containing genera: *Methylomonas* and *Methylobacter*. Suggested reference strains: VKM-2Yu and 111.

5. *Methylococcus luteus* (Romanovskaya et al., 1978). The only yellow-pigmented methanotroph. This rod-shaped organism with a mol% G+C value of 53.0 probably belongs to *Methylomonas* or *Methylobacter*. Suggested reference strain: VKM-53B.

6. *Methylococcus ucrainicus* (Malashenko et al., 1972). Like *M. luteus*, this rod-shaped organism with a mol % G+C value of 52.7 probably belongs to *Methylomonas* or *Methylobacter*. Suggested reference strain: VKM-160 (NCIMB 11915).

7. *Methylococcus fulvus* (Malashenko et al., 1972).

8. *Methylococcus mobilis* (Hazeu et al., 1980).

9. *Methylomonas flagellata* (Morinaga et al., 1976).

10. *Methylomonas margaritae* (Takeda et al., 1974).

11. *Methylomonas methanitrificans* (Davis et al., 1964). Synonym: *Pseudomonas methanitrificans*.

12. *Methylomonas methanooxidans* (Brown and Strawinski, 1958). Synonym: *Methanomonas methanooxidans*.

13. *Methylovibrio soehngenii* (Hazeu and Steenis, 1970). Different strains probably belong to either *M. sporium* and/or *M. parvus* (Romanovskaya et al., 1978).

14. *Soehngenia thermomethanica* (Shen et al., 1982). Name tentatively proposed for a thermophilic strain (H-2). Likely to be a strain of *M. thermophilus* (P. N. Green, unpublished observation).

15. *Methylovarius* (Romanovskaya, 1984). A genus created to attempt

to address the problem of heterogeneity within the genus *Methylococcus*. Romanovskaya suggests *Methylovarius* should consist of *M. ucrainicus, M. chroococcum, M. vinelandii, M. whittenburyi, M. luteus,* and *M. minimus,* leaving *M. capsulatus, M. gracilis, M. thermophilus,* and *M. mobilis* in the genus *Methylococcus*.

2.2.10. Marine Methane-Utilizing Bacteria

Reports of obligate methane-utilizing bacteria from marine environments were described in the literature as early as 1949 by Hutton and Zobell. They found these bacteria to be quite common in the topmost layers of marine sediments, especially where methane and free oxygen were present. Of 150 organisms they isolated from plates seeded with enrichment cultures, 14 isolates appeared to be pure methane utilizers. The isolates were all Gram-negative, nonsporeforming rods (0.6–1.0 × 2.0–3.3 μm). In older cultures some strains produced chocolate-colored growth. The authors found that methane and either ethane and/or propane supported growth of the bacteria in the presence of oxygen and CO_2. Ammonium salts, nitrate, glutamic acid, or peptone served as nitrogen sources.

More recently, obligate methane oxidizers have been reported in the thermocline of offshore seawaters (Sargasso Sea), where small amounts of methane are present (Sieburth, personal communication). A single group of organisms was enriched from several samples (type I obligate methanotrophs), which consisted of nonpigmented, motile, Gram-negative rods that required seawater (Na^+ ions) for growth and grew on either methane or methanol, but not on other C_1 or multicarbon compounds. They could also use either ammonia or nitrate as sole nitrogen source and grew at 20 or 30°C, but not at 10°C. They were inhibited by natural sunlight. The name *Methylomonas pelagica* was proposed for these organisms (Sieburth *et al.*, 1987), which had a mol% G+C content of 49.1. The type strain was deposited in the National Collection of Marine Bacteria as NCIMB 2265.

2.3. Obligate Methanol and Methylated Amine Utilizers

This group of organisms consists largely of a variety of Gram-negative, aerobic, rod-shaped bacteria that superficially resemble pseudomonads (Jenkins *et al.*, 1987).

While many isolates utilize only C_1 compounds, such as methanol and/or methylated amines, as sole carbon and energy source, others have been isolated and described which, in addition to these C_1 compounds, can also utilize a restricted range of multicarbon compounds such as glucose or fructose.

In 1975, Colby and Zatman isolated three types of methanol/methylated amine utilizers. A group that could utilize only methanol and/or methylated amines; a second group that could, in addition, utilize glucose; and a third group that could utilize a restricted range of multicarbon compounds and could also grow on nutrient agar. Colby and Zatman proposed these last two groups should be termed type M (more restricted) and type L (less restricted) facultative methylotrophs.

As many of the so called obligate methanol/methylated amine utilizers discussed in the literature have been shown to be able to assimilate glucose and/or fructose as sole carbon source, it perhaps makes more sense to discuss these and Zatman's similar type M restricted strains (W3A1 and W6A) together. The type L restricted facultative methylotrophs, which are in effect heterotrophs, will be discussed separately.

Pseudomonas C (Chalfan and Mateles, 1972) and *Pseudomonas* (organism) W1 (Dahl *et al.*, 1972) were among the first obligate methanol utilizers to be described in the literature. However, there must remain some doubt about *Pseudomonas* C representing a pure culture of a *bona fide* obligate methylotroph. Its initial description lists, in addition to glucose, several other multicarbon compounds as capable of supporting growth as sole carbon source. Also, in our hands, *Pseudomonas* C was capable of growth on nutrient agar and may, at that time, have been a mixed culture.

These, and similar obligate methanol and methylated amine utilizers are listed in Table IV.

In 1977, *Methylobacillus* became the first validly published genus proposed to accommodate such organisms (Yordy and Weaver, 1977). A type species *M. glycogenes* was also proposed and was defined as containing nonmotile rods, which formed colorless to pale-yellow colonies on methanol salts media and which could utilize only methanol and methylamine to support growth.

Unfortunately, like other species that have been proposed on the strength of a single strain, obvious problems arose. As similar organisms were examined in parallel, it quickly became apparent that both the species and the genus descriptions were too restrictive. This led Urakami and Komagata (1986a) to propose the genus description of *Methylobacillus* should be emended to accommodate the more diverse organisms being isolated and the increasing range of phenotypic features being examined.

As with the obligate methane utilizers, the lack of easily observable phenotypic features hampered initial taxonomic studies of the obligate methanol/methylated amine utilizers. Similarly, it was not until chemotaxonomic and genotaxonomic techniques were used to probe this group of organisms (Jenkins, 1984; Jenkins *et al.*, 1984; Jenkins and Jones, 1987) that a taxonomic structure began to emerge. Indeed, using such tech-

TABLE IV. Obligate Methanol/Methylated Amine Utilizers[a]

Organism(s)	Reference
Pseudomonas C	Chalfan and Mateles (1972)
Pseudomonas (organism) W1	Dahl *et al.*, (1972)
strain 4B6	Colby and Zatman (1973)
strain C2A1 (*Aminomonas aminovorus*)	Colby and Zatman (1973)
Pseudomonas RJ 3	Mehta (1973)
Methylomonas methylovora	Kouno and Ozaki (1975)
Pseudomonas W6	Babel and Miethe (1974)
Methylophilus methylotrophus (*Pseudomonas methylotropha*)	Byrom and Ousby (1975)
Methylomonas M15	Sahm and Wagner (1975)
strains W3A1 and W6A	Colby and Zatman (1975)
Methylomonas aminofaciens 77a	Ogata *et al.*, (1977)
Methylomonas P11	Drabikowska (1977)
strain BC3	Chen *et al.*, (1977)
Methylomonas methanolophila	Suzuki *et al.*, (1977)
Methylobacillus glycogenes	Yordy and Weaver (1977)
Methylomonas (*Pseudomonas*) *methanolica* (*methanovorans*)	Amano *et al.*, (1975)
Pseudomonas J	Matsumoto and Tobari (1978)
strain L3	Hirt *et al.*, (1978)
Methylomonas clara	Drozd and Linton (1981)

[a]The following organisms have also appeared in the literature: *Achromobacter methanolophila, Methanomonas methylovora, M. methylovora* ss. *thiaminophia, Protaminobacter candidus, Protaminobacter thiaminophagus, Pseudomonas inaudita, Pseudomonas insueta, Pseudomonas methanolica, Methylomonas espexii, Methylomonas methanocatalesslica, Methylomonas methanofructolica, Pseudomonas methanomigas, Pseudomonas methylonica* (See Urakami and Komagata, 1986a) and *Aeromonas methanocola, Aeromonas methanophilum, Corynebacterium yamanasiensis, Methylomonas probus, Pseudomonas flavomethanolophila, Pseudomonas utilis* and *Pseudomonas viscogena* (see Urakami and Komagata, 1986b).

niques in addition to a detailed phenotypic study, these authors demonstrated a fair degree of heterogeneity within this complex of organisms.

Thus, based on the results of a numerical taxonomic study, polar lipid analyses, DNA base composition (Jenkins and Jones, 1987), polyacrylamide gel electrophoresis (PAGE) of whole cell protein (Byrom, 1981), electrophoretic patterns of various enzymes and DNA–DNA homology studies (Urakami and Komagata, 1981; Urakami *et al.*, 1985), Jenkins *et al.* (1987) proposed a second genus of obligate methanol/methylated amine utilizers: *Methylophilus*. They considered that the obligate methylotrophs exemplified by *Methylophilus methylotrophus* (NCIMB 10515) and *Methylomonas clara* (NCIMB 11809) are sufficiently distinct from the type strain of *Methylobacillus glycogenes* (NCIMB 11375, ATCC 29475) to merit separate genus status.

Methylophilus methylotrophus and *Methylomonas clara* differ from *Methylobacillus glycogenes* in utilizing glucose as the sole carbon and energy source and in polar lipid composition. *M. glycogenes* (NCIMB 11375) contains diphosphatidylglycerol (DPG), phosphatidylglycerol (PG), phosphatidylethanolamine (PE), and one unidentified glycolipid, while *M. methylotrophus* (NCIMB 10515) contains PG (but *no* DPG), PE, and two unidentified glycolipids, one of which may be identified with that from *M. glycogenes* (Jenkins and Jones, 1987). The same authors also showed that there were low (ca. 22%) levels of DNA–DNA homology between *M. glycogenes* (NCIMB 11375) and both *M. methylotrophus* (NCIMB 10515) and *M. clara* (NCIMB 11809).

Urakami *et al.* (1985), who performed additional DNA–DNA homology studies on a number of strains, produced data that suggest there may be several species groups within *Methylobacillus* and *Methylophilus*. However, no reliable phenotypic or chemotaxonomic tests are available yet to differentiate at the species level.

Descriptions of the two major genera now follow. The genus description of *Methylophilus* is taken largely from Jenkins *et al.* (1987). The description of *Methylobacillus* is taken from the published works of Yordy and Weaver (1977) and Urakami and Komagata (1986a), modified in the light of the removal of some strains now assigned to *Methylophilus*, the polar lipid analyses of Jenkins and Jones (1987), and some unpublished data. Key differentiating criteria are highlighted.

2.3.1. *Methylobacillus*

Short Gram-negative rods. Motile with a single polar flagellum or nonmotile. Obligate methylotrophs that can utilize methanol, but not methane as sole carbon and energy source. Some strains can also utilize methylated amines, formate, and fructose. *Glucose is not utilized* as a carbon source. Strict aerobes with a respiratory metabolism. DNA base composition ranges from 50.0 to 56.0 mol% G+C. The major cellular fatty acids are straight-chain saturated $C_{16:0}$ and monounsaturated $C_{16:1}$. The major quinone is Q-8, with Q-7 and Q-9 as minor components. Cells contain PG, DPG, and PE (see Section 2.3) as principal polar lipids. The type species is *Methylobacillus glycogenes*.

2.3.2. Description of *Methylobacillus glycogenes*

Cells are rods with rounded ends measuring 0.3–0.5 × 0.8–2.0 µm. No endospores are present. Cells occur singly, rarely in pairs, and are motile by means of a single polar flagellum or are nonmotile. Capsules are not formed and granules of poly-β-hydroxybutyrate do not accumulate. Does not grow in nutrient or peptone broth. Colonies on methanol-salts agar are shiny, smooth, raised, entire, white to light yellow, and 1–3 mm in diameter after 3 days' incubation at 30°C. Water-soluble pigments are not produced. Nitrate is reduced to nitrite. The methyl red and Voges-Proskauer tests are negative. Indole, hydrogen sulfide, and ammonia are not produced. Hydrolysis of gelatin and starch and denitrification are not observed. Acid is not produced from D-glucose or D-fructose. L-arabinose, D-xylose, D-glucose, D-mannose, galactose, maltose, sucrose, lactose, trehalose, D-sorbitol, D-mannitol, inositol, glycerol, starch, succinic acid, citric acid, acetic acid, ethanol, and hydrogen are not utilized as sole carbon source. Utilization of D-fructose and methylated amines differs among strains. Thiamine is required as a growth factor for some strains. Ammonia, urea, and nitrate are used as sole nitrogen sources. Urease and oxidase are produced. Catalase is produced by most strains. Good growth occurs between pH 6.0 and 8.0. All strains grow at 30°C and most strains grow at 37°C. Some strains can grow at 42°C, but most strains do not grow in media containing 3% (w/v) NaCl.

DNA base composition, cellular fatty acid composition, and quinones present are as in the genus description. Methanol is utilized via the ribulose monophosphate (RMP) pathway.

The type strain T-11 (= NCIMB 11375 = ATCC 29475 = JCM 2850) is nonmotile and was isolated from soil by Yordy and Weaver in 1977. The DNA base composition of the type strain is 53.2 mol% G+C (Urakami and Komagata, 1986a).

If glucose utilization is adopted as a key differentiating criterion between the genera *Methylobacillus* and *Methylophilus*, then the following strains are likely to belong to the genus *Methylobacillus* in addition to the type strain: *Pseudomonas insueta* (strains ATCC 21276 = JCM 2854 and ATCC 21453 = JCM 2855); *Pseudomonas methanolica* (strain ATCC 21704 = JCM 2857); *Methanomonas methylovora* subsp. *thiaminophila* (ATCC 21370 = JCM 2849); *Methylomonas methylovora* (strains ATCC 21369 = JCM 2844, ATCC 21963 = JCM 2848, and ATCC 21852 = JCM 2840); *Protaminobacter thiaminophagus* (ATCC 21371 = JCM 2853); *Aminomonas aminovorus* C2A1 (NCIMB 11268 = JCM 2866); *Protaminobacter candidus* (ATCC 21372 = JCM 2852), and *Achromobacter methanolophila* (ATCC 21275 = JCM 2841) (Jenkins, 1984; P. N. Green, unpublished data).

2.3.3. *Methylophilus*

Straight or slightly curved, Gram-negative rods, (0.3–0.6 × 0.8–1.5 μm) when grown on methanol mineral salts agar. Motile by polar flagella or nonmotile. Endospores absent. No cellular inclusions. No sheath or prosthecae detected. No capsules formed, but slime may be produced by some strains. Colonies on methanol–mineral salts agar plates incubated for 2 days at 30 or 37°C are circular, 1–2 mm in diameter, with entire edge, convex, and translucent to opaque. No, or extremely poor, growth obtained on nutrient agar. Optimum growth temperature 30–37°C; no growth occurs at 4 or 45°C. Optimum pH for growth, 6.5–7.2. Aerobic. Metabolism respiratory; very little or no acid is produced from glucose. Methanol and *glucose*, but not methane, are utilized as sole carbon and energy sources by all strains. In addition, methylated amines, formate, and fructose may be utilized as sole carbon and energy source. Nitrate and ammonium salts are utilized as nitrogen sources. Oxidase and catalase positive.

The predominant fatty acid composition is of the nonhydroxylated straight-chain saturated and monounsaturated types, with $C_{16:0}$ and $C_{16:1}$ predominating (Ikemoto *et al.*, 1978; Urakami and Komagata, 1979). Methanol is assimilated via the RMP pathway.

The major isoprenoid quinone components are ubiquinones, with eight isoprene units (Q-8) (Urakami and Komagata, 1979, 1986b; Jenkins and Jones, 1987). Strains belonging to *Methylophilus* contain the following major polar lipids: phosphatidylglycerol, phosphatidylethanolamine, and two unidentified glycolipids. They do not contain diphosphatidylglycerol (Jenkins *et al.*, 1987). DNA base composition ranges from 50 to 53 mol % G+C.

Bacteria of this kind have been isolated from activated sludge, mud, river, and pond water. The type species is *Methylophilus methylotrophus*.

2.3.4. Description of *Methylophilus methylotrophus*

This description is based on studies of the type strain NCIMB 10515 and on NCIMB 11809 (*Methylomonas clara* = ATCC 31226). Morphology and general characteristics are as given for the genus. Cells are motile by a single flagellum. In addition to methanol, *glucose* is utilized as sole carbon and energy source. Methylated amines are also utilized by some strains. Variable results are obtained with fructose as growth substrate (Jenkins and Jones, 1987). Acid is not produced from glucose. Acetoin may or may not be produced. Tween 20, 40, and 60 are hydrolyzed, but Tween 80 is not. Urease is produced, as well as leucine arylamidase. Phosphatase production is weak or variable. Sulphatase and H_2S are not produced. Gelatin is not liquefied. Extracellular deoxyribonuclease and ribonuclease are not produced. No growth occurs in the presence of 0.01% (w/v) potassium tellurite or with 5% (w/v) NaCl. Resistant to penicillin, oleandomycin, sensitive to naldixic acid, streptomycin, and a number of other antibiotics.

The mol % G+C of the DNA is 49.8–50.3. DNA–DNA homology values of 83–88% have been reported between *M. methylotrophus* NCIMB 10515 and *M. clara* NCIMB 11809 (Byrom, 1981; Jenkins, 1984). The type strain, which was isolated from activated sludge, is AS1 (= NCIMB 10515).

If glucose utilization is adopted as a key differentiating criterion between the genera *Methylobacillus* and *Methylophilus*, then the following strains are likely to belong to the genus *Methylophilus*, in addition to NCIMB 10515 and *M. clara:* strain W3A1 (NCIMB 11348), *Methanomonas methylovora* (NCIMB 11376 = ATCC 21852 = JCM 2840), *Ps. methylotropha* (JCM 286), *Pseudomonas* C (JCM 2865) (Jenkins, 1984; P. N. Green, unpublished data).

2.3.5. Additional Information

Features that *may* be useful for strain differentiation at the intrageneric level are growth on monomethylamine, trimethylamine, and fructose as sole carbon and energy sources; motility; catalase; growth at 42 and 47°C; thiamine requirement; and tolerance of 3% NaCl.

The genera *Methylobacillus* and *Methylophilus* are almost certainly heterogeneous, each containing more than one species, but at present, they form the taxonomic cornerstone of this group of organisms. Previous studies on electrophoretic patterns of enzymes (Urakami and Komagata, 1979) and phenotypic characteristics (Jenkins *et al.* 1984) have indicated that four subgroups exist. More recently, Urakami *et al.*, (1985) described six DNA homology groups.

5S RNA sequencing performed on a limited number of organisms (Bulygina *et al.*, 1989) has more recently elucidated three separate taxonomic groups, possibly at the genus level. Govorukhina and Trotsenko (1989, 1991) have added weight to these findings by publishing some descriptive information on a possible new group of obligate methanol-utilizing bacteria that had very low (5–10%) levels of DNA–DNA homology with the type strains of *Methylobacillus* and *Methylophilus*. They have formally proposed the new genus *Methylovorus* for these organisms, which utilize glucose and contain cardiolipin and diphosphatidylglycerol.

2.3.6. Marine Methanol-Utilizing Bacteria

Yamamota *et al.* (1978, 1980) isolated 65 marine bacteria that could utilize methanol as sole carbon and energy source. All their isolates were Gram-negative, strictly aerobic motile rods that required vitamin B_{12} and Na^+ ions for growth. Their cultures were split into two major groups based on their ability to utilize fructose as a growth substrate. Strains that could utilize fructose were tentatively assigned to a new species, *Alteromonas thalassomethanolica,* and strains that could not were assigned to a second species, *Methylomonas thalassica.* (Neither of these species has since been validated.) Strains within these two species groups were further subdivided depending on their ability to utilize mono-, di-, and trimethylamine. The mol% G+C of Yamamoto's isolates ranged from 43.8 to 47.6.

In 1984, Strand and Lidstrom described a marine obligate methylotroph (strain FMD) that utilized methanol via the RMP pathway. Although they did not formally propose a name for their strain, they suggested that it might belong to the genus *Methylophilus.*

Similar organisms were isolated and described the following year by Janvier *et al.* (1985). Their isolates, like those of Yamamoto, were strictly aerobic, Gram-negative, motile rods that required vitamin B_{12} for growth. None of their isolates grew on methane or complex nutrient media, and all but two grew on methanol, methylamine, and fructose. Seventeen strains also grew on monomethylamine (MMA) and 10 on trimethylamine (TMA). Fructose was the only non-C_1 substrate tested that supported growth of these organisms. Representative strains examined used the RMP pathway for C_1 assimilation and MMA was oxidized through either methylamine dehydrogenase or methylglutamate dehydrogenase. The mean G+C content of 33 strains examined was 43 mol%, and based on DNA–DNA hybridization data, two related groups were identified among 11 representative strains examined.

Based on their findings, and in particular their homology data, which demonstrated there was no significant ($\leq$ 8%) hybridization between their isolates and the type strains of *Methylobacillus glycogenes* and *Methylophilus methylotrophus,* they proposed the new genus *Methylophaga* be created to accommodate their new marine isolates. Two new species, *M. marina* and *M. thalassica,* were proposed based on the homology data, although at present they cannot be distinguished using "routine" phenotypic tests. A description of this new genus follows: *Methylophaga* (taken from Janvier *et al.,* 1985). Cells are Gram-negative rods that are motile by means of single polar flagellum. Very thick (20–30 nm) periplasmic space. Cells can be broken by osmotic shock after washing with 0.5M NaCl. Strictly aerobic, moderately halophilic, and auxotrophic for vitamin B_{12}. Strains do not grow on peptone yeast extract medium containing (or not containing) NaCl. Except fructose, the only growth substrates that are used are C_1 compounds, such as methanol and methylamine, which are dissimilated by the RMP pathway. Do not grow on methane. The range of G+C contents of the DNA is 38–46 mol%. Isolated from marine environments. The type species is *Methylophaga marina.*

2.3.7. Description of *Methylophaga marina*

Cells are short, straight rods 0.2×1.0 μm. Colonies on an artificial seawater–methanol agar are pale pink. Catalase and oxidase positive; reduction of nitrate negative. Optimum growth temperature, 30–37°C. Grows at 10 and 40°C. Na^+ and Mg^{2+} ions are required for growth. Grows on fructose and monomethylamine (MMA). Some strains grow on dimethylamine, but none grow on trimethylamine. MMA is dissimilated by an N-methylamine dehydrogenase or through a methylglutamate dehydrogenase. In polyacrylamide gels, the electrophoretic migration distances relative to bromophenol blue (Rf values) are 0.33 for glucose-6-phosphate dehydrogenase and 0.30 or 0.19 for methanol dehydrogenase. The G+C content of the DNA is 43 mol%. The type strain is 222 (= NCIMB 2244 = ATCC 35842).

2.3.8. Description of *Methylophaga thalassica*

Description as for *M. marina*, except for the following: All strains grow on MMA and most grow on fructose and dimethylamine. Some strains grow on trimethylamine. Most dissimilate MMA through N-methylglutamate dehydrogenase. In polyacrylamide gels, the electrophoretic migration distance relative to bromophenol blue (Rf values) for glucose-6-phosphate dehydrogenase is 0.35 and for methanol dehydrogenase is 0.22 or 0.27. The G+C content of the DNA is 44 mol%. The level of hybridization with DNA of the type species *M. marina* is ca. 30–35% (S1 nuclease method). The type strain is YK-4015 (= NCIMB 2163 = ATCC 33146).

2.3.9. Restricted Facultative Methylotrophs

As discussed earlier in this chapter, "restricted facultative methylotroph" was the name coined by Colby and Zatman (1975) for organisms that were nutritionally intermediate between obligate and facultative C_1utilizers.

The more restricted type M strains, such as W3A1 and the pink methanotrophs, which can utilize only glucose in addition to C_1 substrates, have already been discussed and are probably synonymous with what were previously considered to be the classical "obligate" methanol/methylated amine utilizers: *Methylobacillus/Methylophilus/Methylororus* or obligate methane utilizers, e.g., *Methylomonas/Methylobacter*.

The taxonomic position of the less restricted facultative methylotrophs (type L), which can grow on a limited number of multicarbon compounds and on nutrient agar, is less clear, although in the present author's opinion, as a group, they are essentially facultative methylotrophs.

The two such organisms described by Colby and Zatman in 1975 (strains S2A1 and PM6) were *Bacillus* spp. This is interesting, because recently Dijkhuizen *et al.* (1988) isolated what appear to be a novel group of thermotolerant methylotrophic bacilli that have a restricted nutritional range of growth substrates. These organisms are Gram-positive sporeformers, and while they undoubtedly belong to the genus *Bacillus* as presently described, based on limited studies (P. N. Green, unpublished data), they do not appear to belong to any of the thermophilic or mesophilic *Bacillus* species currently described in the literature (see Chapter 4).

These organisms can grow at 35°C, with an upper limit of 55–60°C. They fix methanol via the RMP pathway. In addition to methanol, all strains grow well in mineral medium with maltose, mannitol, glucose, and pyruvate, as well as nutrient agar, although all strains require growth factors.

The thermotolerant bacilli appear to be taxonomically distinct from the Colby and Zatman strains, which are nonauxotrophic mesophiles.

3. FACULTATIVE METHYLOTROPHS

Facultatively methylotrophic bacteria are ubiquitous and are a very cosmopolitan group of organisms comprising a wide variety of genera, linked only by their ability to assimilate, as sole carbon source, a C_1 substrate. As a group, they constitute an unnatural classification, based on a single overweighted feature. Even individual organisms all too often are looked upon as being unique, simply because of their methylotrophy. For example, a *Pseudomonas fluorescens* that utilizes methanol has no immediate justification in being regarded as taxonomically distinct from a *Ps. fluorescens* that does not utilize methanol, or does utilize phenol, or is a nonfluorescent variant, and so forth.

Tables V to IX give examples of various groups of facultatively methylotrophic bacteria. These tables are not intended to represent an exhaustive list of all such organisms ever isolated, but merely cite a range of bacteria with appropriate references.

Various groups of facultative methylotrophs, many of which are groups of convenience having little taxonomic relatedness, will now be discussed in some detail:

3.1. The Genus *Methylobacterium*

Pink-pigmented facultative methylotrophs (PPFM) are widely distributed in nature (Green and Bousfield, 1981). The first unequivocal published evidence demonstrating a pure culture of such an organism was that of Bassalik (1913), who isolated *Bacillus extorquens*. As can be seen from Table V and in a number of publications (Green and Bousfield, 1982, 1988), these Gram-negative to variable, motile, strictly aerobic, rod-shaped bacteria, which exhibit polar growth, have previously been classified in a number of genera. However, recent numerical taxonomic (Green and Bousfield, 1982), electrophoretic (Hood *et al.*, 1988), and genotypic (Hood *et al.*, 1987) studies led to a proposal (Green and Bousfield, 1983) to emend the description of the genus *Methylobacterium* (Patt *et al.*, 1976) to accommodate these organisms. Subsequently, three new species of *Methylobacterium* (Green and Bousfield, 1988) have been proposed.

The genus *Methylobacterium* was originally proposed to accommodate a single strain (XX) isolated from lake water (Patt *et al.*, 1974). This strain was reported to be able to utilize methane, as well as methanol and other complex organic molecules, as sole carbon source, and the name *Methylobacterium organophilum* was proposed for it (Patt *et al.*, 1976). Unfortunately, this key generic feature of facultative methanotrophy, which was thought to be plasmid borne (Hanson, 1980), has since been lost.

The result of this, an excellent example of monothetic classification, was that dozens of other PPFM that were phenotypically very similar to *M. organophilum* (Green and Bousfield, 1982) were excluded from the genus because of a single feature; i.e., they could not utilize methane.

A subsequent emendation of the genus description (Green and Bousfield, 1983), so that methane assimilation is no longer an essential feature, has resolved the problem. The emended genus description is as follows:

Methylobacterium (from Green and Bousfield, 1983). Rods, 0.8–1.0 × 1.0–8.0 µm, occurring singly or occasionally in rosettes; occasionally branched and pleomorphic (see Figure 4). Motile by single polar, subpolar, or lateral flagella, although some strains are not vigorously motile. Cells often contain large sudanophilic inclusions and sometimes volutin granules. Gram negative, although many strains are Gram variable; representative strains have the multilayered cell wall structure and the type of citrate synthase (Green and Bousfield, 1982) characteristic of Gram-negative bacteria. Most strains grow slowly (and some do not grow at all) on nutrient agar. Colonies on glycerol–peptone agar are ≤1–3 mm in diameter and pale pink to bright orange–red; colonies on methanol-salts agar

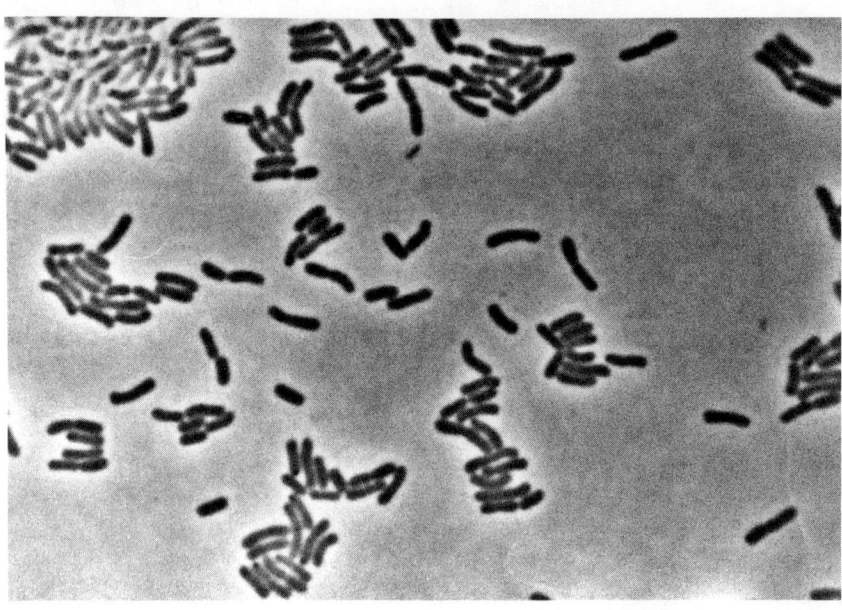

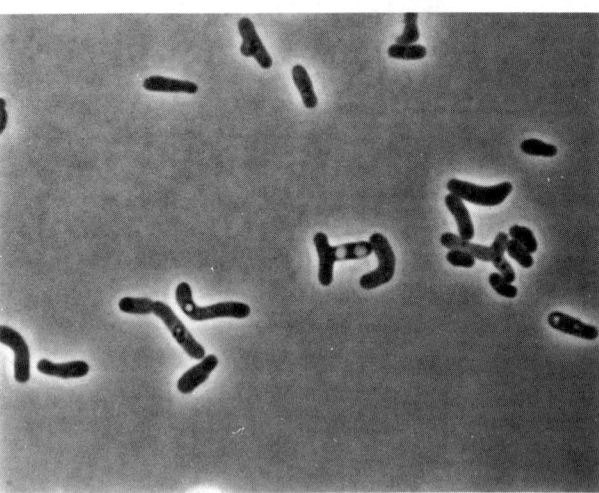

Figure 4. A) Phase constrast micrograph of *Methylobacterium extorquens* (NCIMB 9399) exhibiting branching. 10mm = 5 μm. B) Micrograph of a Gram stained preparation of *Pseudomonas aminovorans*. 10mm = 3 μm. C) *Hyphomicrobium* sp. (reproduced by kind permission of C. S. Dow).

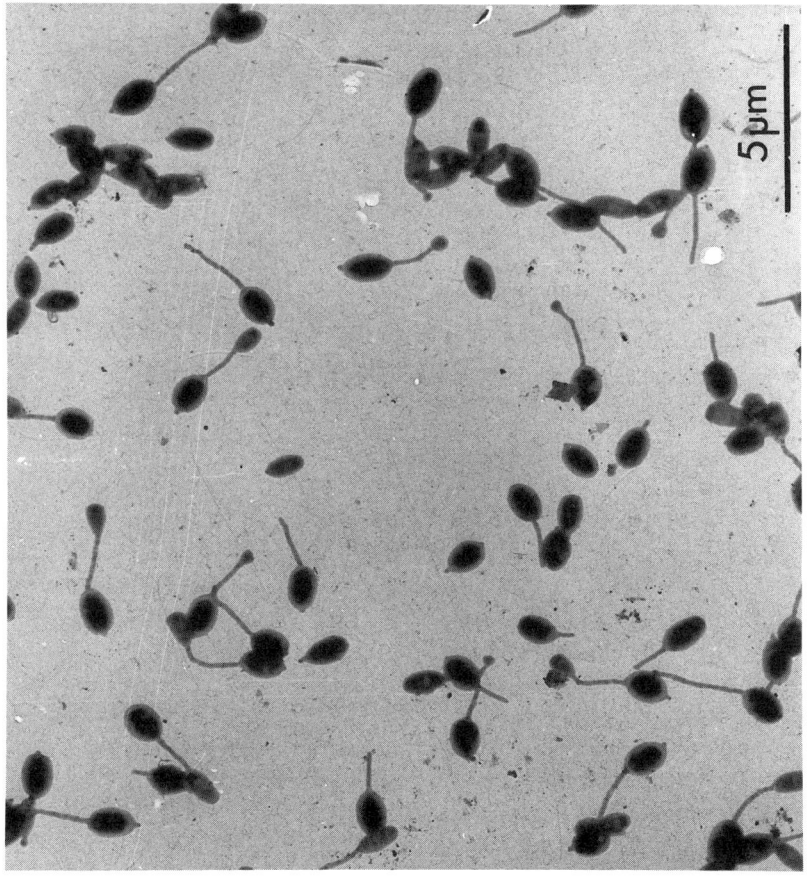

Figure 4. *(continued)*

TABLE V. Pink-Pigmented Facultative Methylotrophs Assigned to the Genus *Methylobacterium*[a]

Organism	Strain or collection no.	Reference
M. organophilum	XX (NCIMB 11278^T)	Patt et al., (1976)
M. extorquens	NCIMB 9399^T. See figure 4(A)	Janota-Bassalik and Pedyk (1961)
(Bacillus extorquens		
Vibrio extorquens	Pseudomonas AM1 (NCIMB 9133)	Peel and Quayle (1961)
Pseudomonas extorquens		
Flavobacterium extorquens)[b]	Pseudomonas M27 (NCIMB 9686)	Anthony and Zatman (1964)
	Protaminobacter ruber (NCIMB 2879)	den Dooren de Jong (1927)
	Pseudomonas methylica strain 2	Kirikova (1970)
M. rhodinum	NCIMB 9421^T	Heumann (1962)
M. radiotolerans	0-1 (NCIMB 10815^T)	Ito and Iizuka (1971)
	Pseudomonas spp. (NCIMB 9142 and 9143)	Hayward (1960)
M. mesophilicum	A47 (NCIMB 11561^T)	Austin and Goodfellow (1979)
M. rhodensianum	Pseudomonas 1 (NCIMB 12249^T)	Rock et al., (1976)
	Corynebacterium rubrum	Graf and Bauer (1973)
	Protaminobacter ruber	Sato (1978)
M. zatmanii	Pseudomonas 135 (NCIMB 12243^T)	Rock et al., (1976)
M. fujisawaense	0-31 (NCIMB 12417^T)	Kouno and Ozaki (1975)
	Unassigned *Methylobacterium* strains	
Pseudomonas sp.	PRL-W4	Kaneda and Roxburgh (1959)
Pseudomonas sp.	AM2	Blackmore and Quayle (1970)
Pseudomonas sp.	PP	Ladner and Zatman (1969)
Pseudomonas sp.	3A2	Colby and Zatman (1972)
Pseudomonas sp.	RJ1	Mehta (1973)
Pseudomonas sp.	TP1	Sperl et al., (1974)
Organism	FM02T	Toraya et al., (1975)
Pseudomonas sp.	YR, JB1 and PCTN	Bellion and Spain (1976)
Pseudomonas sp.	2941	Yamanaka and Matsumoto (1977)
Pseudomonas sp.	AT2	Boulton et al., (1980)
Pseudomonas sp.	80	Kortstee (1980)

[a] In addition to the above, Austin *et al.* (1978)(pink chromogens), Urakami and Komagata (1979) (*Ps. rosea* strains), and Kouno and Ozaki (1975) (*59* pink pseudomonads) have isolated and described *Methylobacterium* strains, some of which have since been assigned to species groups (see Green and Bousfield, 1988).
T = type strain.
[b] Species names in parentheses are all synonyms of *M. extorquens*.

are a more uniform pale pink. The pigment is insoluble and probably carotenoid (Downs and Harrison, 1974; Ito and Iizuka, 1971). In static liquid media strains grow as a pink surface ring or pellicle. Strictly aerobic; catalase and oxidase (often weakly) positive. Chemoorganotrophs, facultative methylotrophs, and occasionally facultative methanotrophs. The ability of some strains to utilize methane as a sole source of carbon and energy is easily lost if strains are not maintained on an inorganic medium in a methane atmosphere (R. S. Hanson, personal communication). Representative strains have been shown to assimilate C_1 compounds via the serine pathway (Quayle, 1972) and to have a complete tricarboxylic acid cycle when they are grown on complex organic substrates (Colby and Zatman, 1975).

Members of the genus have been isolated from soil, dust, fresh water, lake sediments, leaf surfaces and nodules, rice grains, air, and hospital environments. The optimum growth temperature is in the range 25–30°C, and the DNA base composition ranges between 68.4 and 72.4 mol% (Hood *et al.*, 1987). The type species is *M. organophilum (Patt et al.*, 1976) NCIMB 11278 (= ATCC 27886 = JCM 2833).

Features useful for differentiating among the eight species of *Methylobacterium* and other strains that represent probable centers of variation within the genus are given in Table VI.

Two additional species of methane-utilizing organisms previously assigned to *Methylobacterium*, *M. ethanolicum* and *M. hypolimneticum* (Lynch *et al.*, 1980), but which do not fit the present genus description, have since been shown to have either lost their ability to utilize methane or to be mixed cultures (M. E. O'Connor, personal communication; Lidstrom-O'Connor *et al.*, 1983).

3.2. *Pseudomonas aminovorans* and Related Strains

In 1927, den Dooren de Jong published the description of several facultative methylotrophs that could utilize methylated amine(s), but not methanol, as sole carbon and energy source. One of these organisms, *Pseudomonas aminovorans* (NCIMB 9039) has been shown (P. N. Green, unpublished data) to have a high phenotypic similarity to *Pseudomonas* MA (Shaw *et al.*, 1966), *Pseudomonas* MS (Kung and Wagner, 1970), strain 2A3 (Colby and Zatman, 1973), and organism 5H2 isolated by Hampton and Zatman in 1973. L. B. Perry (unpublished observation) has since shown that representative organisms, in particular *Ps. aminovorans*, exhibit budding or polar growth. As Perry (personal communication) had also observed that various members of the ribosomal ribonucleic acid (rRNA) superfamily IV sensu De Ley (1978) reproduced by budding division, it

TABLE VI. Features that Differentiate among Species and Some Unassigned *Methylobacterium*[a] Strains

Species or strain	Substrates utilized as sole C source[b]						
	D-Glucose	D-Fucose	D-Xylose	L-Arabinose	Fructose	L-Aspartate/ L-Glutamate	Citrate
M. zatmanii	−	−	−	−	+	−	−
M. extorquens	−	−	−	−	−	V	−
M. rhodesianum	−	−	−	−	+	V	−
NCIMB 9141 and 9145[d]	−	−	−	−	+	−	−
N-6[d]	−	−	−	−	+	−	+
M. rhodinum	W+	−	−	−	+	+	−
M. organophilum	+	−	−	−	+	−	+
602 and 317[d]	−	−	−	−	V	+	+
790[d]	−	−	−	−	−	+	+
M. radiotolerans	+	+	+	+	−	+	+
M. fujisawaense	+	+	+	+	V	+	+
M. mesophilicum	+	+	+	+	−	+	+
N-2[d]	+	+	+	+	+	+	+
R14[d]	−	−	−	−	+	+	+
D12[e]	−	−	−	−	+	+	+
B46[d]	−	−	−	−	+	+	+
7 and 35[f]	+	−	−	+	+	V	−

[a] Owing to the slow growth of some strains on certain substrates, carbon utilization tests were read after 14 days of incubation at 30°C (Green and Bousfield, 1982). Doubtful results were checked by twice subculturing in liquid medium.
[b] V = variable result; W = weak growth.
[c] Most strains that utilize sebacate can also utilize pimelate, suberate, azelate, and adipate.
[d] For source of these strains see Green and Bousfield (1982).
[e] Isolated by P. N. Green from roadside dust.
[f] Isolated by H. Stolp, University of Bayreuth, Bayreuth, Germany.

was not surprising that De Ley *et al.* (1987) placed *Ps. aminovorans* NCIMB 9039 in the *Rhizobium–Agrobacterium* complex of rRNA superfamily IV [see Figure 4(B)].

Another group of budding bacteria that contain facultative methylotrophs is the genus *Blastobacter*. Three species within this genus—*B. viscosus* (methanol), *B. denitrificans* (methanol), and *B. aminooxidans* (methylamines)—can all grow on C_1 compounds (see Table IX). However, all three species differ phenotypically from the *Ps. aminovorans* group of organisms. In addition, recent work (Green and Gillis, 1989) suggests that *Ps. aminovorans* and the related strains discussed represent a new taxon, distinct

	Substrates utilized as sole C source[b]							
Sebecate[c]	Acetate	Betaine	Tartrate	Ethanol	Methyl-amine	Trimethyl-amine	Methane	Growth on peptone-rich nutrient agar (Oxoid CM55)
−	+	−	V	+	+	V	−	+
−	+	+	V	+	+	−	−	V
−	+	+	−	+	+	−	−	+
−	+	+	+	+	+	−	−	+
−	+	+	−	+	+	−	−	+
−	+	+	−	+	+	−	−	+
−	+	−	−	+	+	+	V	+
+	+	−	−	+	−	−	−	+
+	−	−	−	−	−	−	−	−
+	+	+	−	V	−	−	−	+
+	+	−	V	V	−	−	−	+
V	−	−	−	+	−	−	−	−
+	−	−	−	−	−	−	−	+
+	+	−	−	+	−	−	−	+
−	+	−	−	+w	−	−	−	−
−	−	−	−	+	+	−	−	−
−	+	+	−	+	+	−	−	+

from other members of the *Rhizobium–Agrobacterium* complex and from the genus *Blastobacter*.

3.3. Other Non-Pink-Pigmented Motile Rods

Apart from *Ps. aminovorans* and related organisms discussed above, several other groups of non-pink-pigmented motile rods have been isolated and described that can utilize C_1 compounds facultatively (see Table VII). These have included Colby and Zatman's strains 1B1, 2A3, 7B1, and 8B1, which were isolated from soil, water, and fish, and all of which could utilize methylated amines. Some could, in addition, utilize methanol.

Pseudomonas oleovorans, isolated by Longinova and Trotsenko (1977b), is unusual among facultative methylotrophs and Gram-negative organisms in particular, in that it is one of the few such organisms that utilize C_1 compounds via the RMP pathway.

A strain of *Pseudomonas fluorescens* isolated by Kirikova (1970) on a formate salts medium has not been examined in sufficient detail to determine whether it represents a new biotype or biovar of this species.

More recently, *Rhizobium meliloti* NCIMB 12075^T was shown to be capable of growth on methanol during a screening of several rhizobia for methylotrophy (P. N. Green, unpublished data).

3.4. Nonmotile Rods and Coccobacilli

Organism 5B1 was isolated by Colby and Zatman in 1973 and described by them as a nonmotile rod. However, in our hands (Green and Bousfield, 1982), this organism was polarly flagellated and most likely a pseudomonad. Other nonmotile facultative methylotrophs have included organisms as diverse as *Klebsiella* and *Acinetobacter* strains (see Table VII).

The diplococcus strain designated PAR by Leadbetter and Gottlieb

TABLE VII. Facultatively methylotrophic Gram-Negative Non-Pink-Pigmented Rods or Coccobacilli

Organism	Reference
Motile rods	
Pseudomonas aminovorans (NCIMB 9039)	den Dooren de Jong (1927)
Pseudomonas MA (NCIMB 11590)	Shaw et al., (1966)
Pseudomonas MS (NCIMB 11591)	Kung and Wagner (1970)
strain 5H2 (NCIMB 11347)	Hampton and Zatman (1973)
Pseudomonas 1A3	Colby and Zatman (1973)
Pseudomonas strains 1B1, 2A3, 2B1, 8B1	Colby and Zatman (1973)
Pseudomonas oleovorans	Longinova and Trotsenko (1977b)
Pseudomonas fluorescens	Kirikova (1970)
Acetobacter/Gluconobacter MB58	Steudel et al., (1980)
Rhizobium meliloti NCIMB 12075^T	
strain 5B1 (NCIMB 11346)	Colby and Zatman (1973)
Nonmotile rods/coccobacilli	
Klebsiella 101	Nishio et al., (1975)
Acinetobacter	Kouno and Ozaki (1975)
diplococcus 'PAR'	Leadbetter and Gottlieb (1967)
Acidomonas methanolica	Urakami et al., (1989a)
Paracoccus alcaliphilus	Urakami et al., (1989b)
Paracoccus aminophilus	
Paracoccus aminovorans	
Paracoccus kocurrii	
Appendaged bacteria	
Hyphomicrobium app.	(see Anthony, 1982)

(1967) may be similar to the autotrophic methanol utilizer *Paracoccus denitrificans*, but this is speculative, as direct comparisons have not been made.

More recently, a new genus, *Acidomonas*, was created (Urakami *et al.*, 1989a) to accommodate nonmotile, rod-shaped, facultative methanol-utilizing acidophiles. These organisms, which grow between pH 2.0 and 5.5, have a DNA base composition of 63–65 mol% G+C. Based on phenotypic, chemotaxonomic, and DNA–DNA homology studies, the organisms examined were considered sufficiently different from other acidophilic taxa, such as *Acetobacter*, *Gluconobacter*, *Acidiphilium*, and *Thiobacillus*, to merit separate taxonomic status. *Acidomonas methanolicus* (formerly *Acetobacter methanolicus*) was proposed as the type strain.

At the opposite end of the pH spectrum, another new species has recently been proposed for alkaliphilic facultative methanol utilizers. These coccobacillary organisms grew at pH 7.0–9.5, but not below pH 6.5 or above 10.0. The 15 strains examined had a DNA base composition of 64–66 mol% G+C, but were considered not to belong to the most closely related species, *Paracoccus denitrificans*. *P. denitrificans* does not grow in alkaline conditions and has a low (<44%) DNA–DNA homology with the alkaliphilic methanol utilizers. For these organisms the name *Paracoccus alcaliphilus* was proposed (Urakami *et al.*, 1989b).*

3.5. The Hyphomicrobia

The hyphomicrobia are appendaged bacteria [see Figure 4(C)] that reproduce by budding and have a complex life cycle involving nonmotile prosthecate mother cells and motile swarmer cells. The biology, physiology, and biochemistry of these organisms, many of which share the feature of facultative methylotrophy, have been reviewed by Harder and Attwood (1978). Strains can utilize methanol, methylated amines, and formate. In addition, some strains can also utilize dichloromethane, dimethylsulphoxide, and dimethyl sulphide as sole source of carbon and energy. Many are effectively restricted facultative methylotrophs, growing only on C_1 compounds and on those substrates metabolized exclusively by way of acetyl-CoA, such as acetate, ethanol, and 3-hydroxybutyrate, but not compounds with three or more carbon atoms. They are also able to grow oligocarbophilically at the expense of carbon compounds in the atmosphere, and this often leads to the erroneous conclusion that these organisms are able to grow, albeit poorly, on a wide range of carbon sources (Anthony, 1982).

---

*Since then, two new species of *Paracoccus* (*P. aminophilus* and *P. aminovorane*) have been described in the literature (Urakami *et al.*, 1990) which can utilize N, N-dimethylformamide, and one (*P. kocurii*) which can utilize tetramethylammonium hydroxide (Ohara *et al.*, 1990).

An important characteristic of the hyphomicrobia is their ability to grow anaerobically on a methanol salts medium using nitrate as the terminal electron acceptor, thus providing a means of selective isolation (Sperl and Hoare, 1971).

Urakami and Komagata (1979) have shown that in the strains of hyphomicrobia they examined, coenzyme Q-9 was present and $C_{18:1}$ was the predominant fatty acid.

3.6. Gram-Positive Facultative Methylotrophs

Gram-positive bacteria growing on methanol and/or methylated amines may well, in the course of time, be shown to be as diverse an assemblage of organisms as the Gram-negative facultative methylotrophs. However, to date, most Gram-positive bacteria that can utilize these C_1 compounds belong to one of three groups: the coryneforms, the actinomycetes, or the genus *Bacillus* (see Table VIII); most of the bacilli have already been discussed as restricted facultative methylotrophs.

Where information is available, the majority of Gram-positive facultative methylotrophs isolated so far incorporate their C_1 substrates almost exclusively via the RMP pathway.

3.7. Facultative Autotrophs and Phototrophs

Table IX lists a variety of bacteria that can assimilate methanol, usually formate, and occasionally methylated amines via their oxidation to CO_2 and the Benson–Calvin cycle.

TABLE VIII. Gram-Positive Facultative Methylotrophs

Organism	Reference
Arthrobacter rufescens	Akiba *et al.*, (1970)
Arthrobacter strains 1A1, 1A2 and 2B2	Colby and Zatman (1973)
Arthrobacter strains	Kouno and Ozaki (1975)
Arthrobacter globiformis strains B-175, B-126 and B-53	Longinova and Trotsenko (1976)
Arthrobacter P1	Levering *et al.*, (1981)
Bacillus cereus M-33-1	Akiba *et al.*, (1970)
Mycobacterium vaccae	Longinova and Trotsenko (1977c)
Brevibacterium fuscum 24	Longinova and Trotsenko (1977c)
Amycolatopsis methanolica (formerly *Nocardia/Streptomyces* 239)	De Boer *et al.*, (1990)
Sporomusa acidovorans	Ollivier *et al.*, (1985)

TABLE IX. Facultative Autotrophs[a] **and Phototrophs**

Organism	Reference
Rhodopseudomonas palustris	Quadri and Hoare (1969)
Rhodopseudomonas acidophila (and some other *Rhodospirillaceae*)	Seifert and Pfennig (1979)
Thiobacillus A2 (formate only)	Kelly *et al.*, (1979)
Thiobacillus novellus	Chandra and Shethna (1977)
Nitrobacter agilis	Ida and Alexander (1965)
Alcaligenes eutrophus H-16 (formate only)	Friedrich *et al.*, (1979)
Paracoccus denitrificans	Cox and Quayle (1975)
Pseudomonas oxalaticus (formate only)	Khambata and Bhat (1953)
Pseudomonas gazotropha	Romanova and Nozhevnikova (1977)
Achromobacter 1L	Longinova and Trotsenko (1979)
Microcyclus aquaticus (NCIMB 1801^T)	Urakami and Komagata (1979)
Microcyclus eburneus (ATCC 21373)	Kouno and Ozaki (1975)
Blastobacter viscosus	Longinova and Trotsenko (1979)
Blastobacter aminooxidans (grows on methylated amines)	Doronina *et al.*, (1983)
Blastobacter denitrificans (autotrophic growth unconfirmed)	Hirsch and Muller (1985)
Xanthobacter autotrophicus (growth on formate not tested)	Wiegel *et al.*, (1978)
Hydrogenomonas eutropha Z-11 (formate only)	Namsaraev *et al.*, (1971)

[a]Unless otherwise stated, all strains are Gram negative and grow on methanol and formate.

The rhodopseudomonads are red, motile (at some stage of their growth) prosthecate phototrophs that reproduce by budding. They grow anaerobically in the light using methanol or formate as a reductant, or aerobically in the dark using methanol or formate as sole carbon and energy source (Seifert and Pfennig, 1979).

In contrast, *Paracoccus denitrificans* (formerly *Micrococcus denitrificans*) is an oxidase-positive, nonmotile coccus or coccobacillus. It is a facultative chemoautotroph that can grow anaerobically on methanol with nitrate as terminal electron acceptor.

Some microcycli, which form curved, doughnut- or horse-shoe-shaped cells, have also been shown to utilize methanol facultatively via CO_2. Green and Bousfield (1982) examined eight strains in their numerical taxonomic study and found all, including *Microcyclus eburneus* (Kouno and Ozaki, 1975) and *M. polymorphum* NCIMB 10516 (ICI Isolate), to be strains of the species *Microcyclus aquaticus* (now called *Ancylobacter aquaticus*). Urakami and Komagata (1979) have shown that strains of *M. aquaticus* have $C_{18:1}$ as their principal fatty acid, a feature common to methylotrophs using the serine pathway. Interestingly, Longinova *et al.* (1978) have demonstrated hydroxypyruvate reductase as well as ribulose biphosphate carboxylase activity in their strain (Z-238) of *M. aquaticus*.

Xanthobacter autotrophicus (formerly *Corynebacterium autotrophicum*) is a nitrogen-fixing, hydrogen-oxidizing bacterium that can also utilize methanol autotrophically as sole carbon and energy source.

Pseudomonas oxalaticus, shown by Green and Bousfield (1982) to be a peritrichously flagellated rod, is more likely to belong to the genus *Alcaligenes*.

3.8. Facultatively Methylotrophic Marine Bacteria

Although a number of obligate methane and methanol utilizers have been isolated from marine sources (see Section 2.3.6.), very few, if any, *bona fide* heterotrophic facultative methylotrophs have been isolated.

The marine hyphomicrobia described by Attwood and Harder (1972) and Hirsch (1974) and the so-called facultative methanol utilizers described by Yamamoto *et al.* (1980) are either restricted or obligate methylotrophs, respectively.

3.9. Anaerobic Methanol-Utilizing Bacteria

Obligately anaerobic methanol-utilizing bacteria are found among the methanogenic and the acidogenic bacteria (Dijkhuizen *et al.*, 1985). Methanogenic bacteria such as *Methanosarcina barkeri* convert methanol to CH_4 and CO_2. Methylotrophic acidogens such as *Butyribacterium methylotrophicum*, which can also grow on methanol, are reviewed by Zeikus (1983a,b).

Sporomusa acidovorans is a recently proposed species comprising Gram-negative, spore-forming homoacetic acid–producing anaerobes that can use methanol or H_2 as an energy source (Ollivier *et al.*, 1985).

3.10. Methylotrophic Halophilic Methanogens

Recently, Paterek and Smith (1988) proposed a new genus and species (*Methanohalophilus mahii*) for such organisms. Their isolates required NaCl concentrations in the moderately halophilic range (1.0–2.5M) for optimal growth and methanogenesis. These coccoid organisms, which were isolated from sediments of the Great Salt Lake (Utah, USA), can utilize methanol and methylated amines as sole source of carbon and energy.

4. PHYLOGENY OF METHYLOTROPHS

This chapter has described the historical background and impact of recent taxonomic studies on the various groups of methylotrophic bacteria discussed. However, because of their restricted mode of metabolism, many of the groups of organisms examined have remained, to some extent, taxonomic enigmas.

The relationships within and between the identifiable groups of obligate methane and methanol utilizers have remained the subject of debate and speculation. Similarly, it has remained far from clear what the relationship of some of the well-described facultatively methylotrophic taxa is, both to the obligate methylotrophs and to other heterotrophic taxa or autotrophic organisms.

Recent work by Tsuji *et al.* (1990) involving 16S ribosomal RNA sequence analyses has already begun to answer many of these questions. In the "phylogenetic tree" shown in Figure 5, a number of very interesting features are shown. First, with regard to the methanotrophs, all three type II methanotrophs examined (*Methylocystis parvus, Methylosinus trichosporium,* and *Methylosporovibrio methanica*) were located on a branch of the *Rhodopseudomonas/Agrobacterium* complex that equates to part of the rRNA superfamily IV (*sensu* De Ley, 1978). Also related to this complex but remote from the type II methanotrophs were the pink facultatively methylotrophic *Methylobacterium* strains, thus suggesting a common ancestry (see also Sato, 1978). This ancestry is further substantiated by the shared morphological feature of polar growth or "budding" common to members of both these groups of organisms (P. N. Green, personal observation). Indeed, it has also been shown that other groups of facultative methylotrophs that exhibit polar growth, such as *Pseudomonas aminovorans, Xanthobacter autotrophicus,* and possibly some strains of *Blastobacter,* also belong to rRNA superfamily IV (Green and Gillis, 1989), and by implication the alpha purple eubacteria.

At the other end of the phylogenetic tree, the only other two obligate methanotrophs examined, viz., *Methylococcus capsulatus* (type X) and *Methylomonas methanica* (type I), were distantly related to the type II organisms and to each other. This may be reassuring in that the three groups of obligate methanotrophs formed on the basis of previous phenotypic and chemotaxonomic studies are confirmed by this genotypic analysis.

In addition, it appears likely that the type II methanotrophs, along with several facultative methylotrophs, all with the common morphological feature of polar growth, evolved together with the photosynthetic organisms and are probably quite distinct in evolutionary terms from the other obligate methane and methanol-utilizing bacteria.

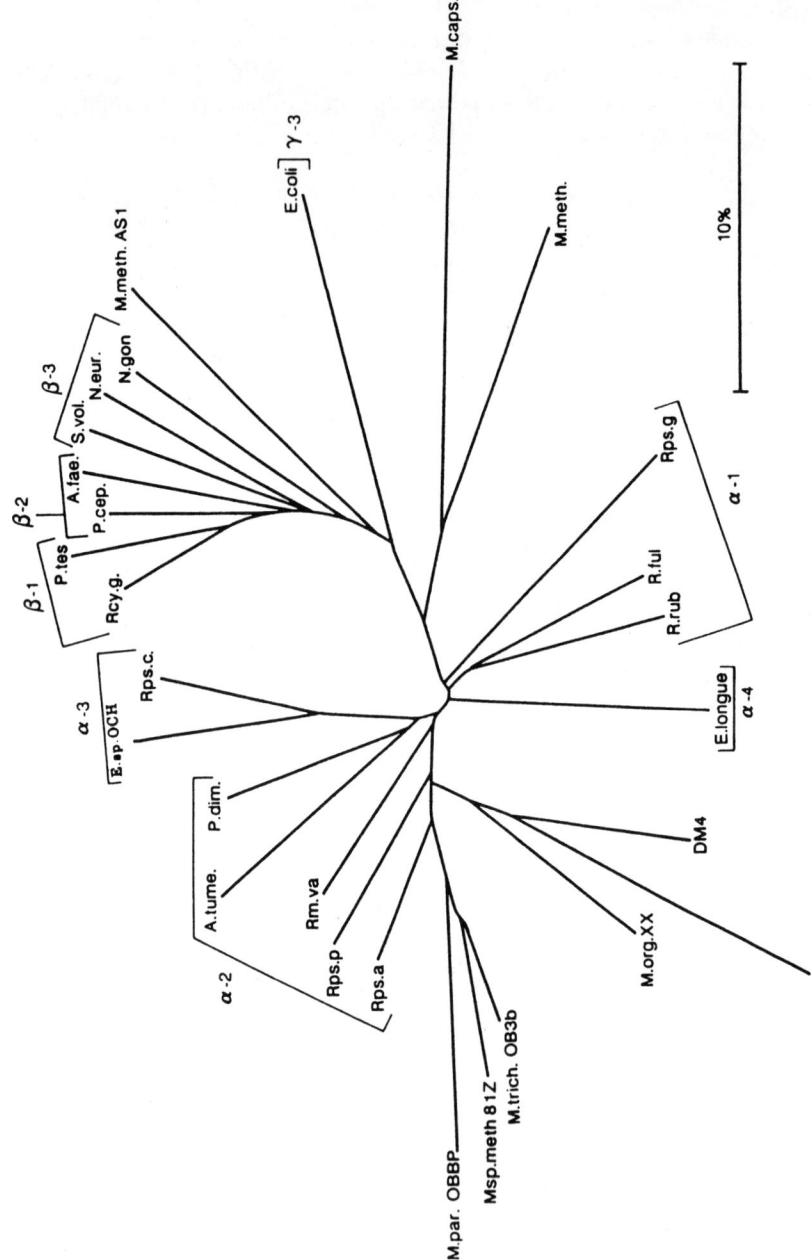

Figure 5. Phylogenetic relationships among purple eubacteria and some methylotrophs based on 16 S ribosomal RNA sequences. Reproduced by kind permission of R. S. Hanson.

The only obligate methanol-utilizing bacterium examined was distantly related to both the methanotrophs and facultative methylotrophs. This study by Tsuji et al., (1990) has demonstrated that it is now possible to distinguish and broadly classify many methylotrophic bacteria using 16S rRNA analysis. In addition, unique sequences within the 16S rRNA molecules of the organisms examined by these authors may be utilized to construct group- or species-specific probes to facilitate taxonomic, commercial, and environmental applications.

ACKNOWLEDGMENT. I wish to thank R. S. Hanson for making this phylogenetic data available before it appeared in the literature.

REFERENCES

Akiba, T., Ueyama, H., Seki, M., and Fukimbara, T., 1970, Identifications of lower alcohol-utilizing bacteria, *J. Fermentation Technol.* **48**:323–328.

Amano, Y., Sawada, H., Takada, N., and Terui, G., 1975, Isolation and characterization of *Methylomonas methanolica* nov. sp., *J. Fermentation Technol.* **53**:315–326.

Anthony, C., 1982, *The Biochemistry of Methylotrophs*, Academic Press, London.

Anthony, C., and Zatman, L. J., 1964, The microbial oxidation of methanol. 1. Isolation and properties of *Pseudomonas* M27, *Biochem. J.* **92**:609–614.

Attwood, M. M., and Harder, W., 1972, A rapid and specific enrichment procedure for *Hyphomicrobium* spp., *Antonie van Leeuwenhoek* **38**:369–378.

Austin, B., and Goodfellow, M., 1979, *Pseudomonas mesophilica*, a new species of pink bacteria isolated from leaf surfaces, *Int. J. Systematic Bacteriol.* **29**:373–378.

Babel, W., and Miethe, D., 1974, Kurze originalmitteilungen, *Ztschr. Alleg. Mikrobiol.* **14**:153–156.

Bassalik, K., 1913, Uber die Vererbeitung der Oxalsaure durch *Bacillus extorquens* n. sp., *Jahr. Wissenschaft. Botan.* **53**:255–302.

Bellion, E., and Spain, J. C., 1976, The distribution of the isocitrate lyase serine pathway amongst one-carbon utilizing organisms, *Can. J. Microbiol.* **22**:404–408.

Bird, C. W., Lynch, J. M., Pirt, F. J., and Reid, W. W., 1971, Steroids and squalene in *Methylococcus capsulatus* grown on methane, *Nature* **230**:473.

Blackmore, M. A., and Quayle, J. R., 1970, Microbial growth on oxalate by a route not involving glyoxylate carboligase, *Biochem. J.* **118**:53–59.

Boulton, C. A., Haywood, G. W., and Large, P. J., 1980, N-methylglutamate dehydrogenase, a flavohaemoprotein purified from a new pink trimethylamine-utilizing bacterium, *J. Gen. Microbiol.* **117**:293–304.

Brown, L. R., and Strawinski, R. J., 1958, Intermediates in the oxidation of methane, *Bacteriol. Proc.* **58**:96–132.

Brown, L. R., Strawinski, R. J., and McCleskey, C. S., 1964, The isolation and characterization of *Methanomonas methanooxidans* Brown and Strawinski, *Can. J. Microbiol.* **10**:791–799.

Byrom, D., 1981, Taxonomy of methylotrophs: a reappraisal, in: *Microbial Growth on C1-Compounds* (H. Dalton, ed.), Heyden, London, pp. 278–284.

Byrom, D., and Ousby, J. C., 1975, Identification of a methanol oxidizing pseudomonad, in:

Microbial Growth on C1-Compounds (edited by the organizing committee), Society of Fermentation Technology, Osaka, Japan, pp. 23-27.
Bulygina, E. S., Galchenko, V. F., Govorukhina, N. I., Netrusov, A. I., Nikitin, D. I., Romanovskaya, V. A., Trotsenko, Yu. A., and Chumakov, K. M., 1990, The taxonomic studies of methylotrophic bacteria by comparison of nucleotide sequences of 5S ribosomal RNA, *J. Gen. Microbiol.* **136**:441-446.
Chalfan, Y., and Mateles, R. I., 1972, New pseudomonad utilizing methanol for growth, *Appl. Microbiol.* **23**:135-140.
Chandra, T. S., and Shethna, Y. I., 1977, Oxalate, formate, formamide and methanol metabolism in *Thiobacillus novellus*, *J. Bacteriol.* **131**:389-398.
Chen, B. J., Hirt, W., Lim, H. C., and Tsao, G. T., 1977, Growth characteristics of a new methylomonad, *Appl. Environ. Microbiol.* **33**:269-274.
Colby, J., and Zatman, L. J., 1972, Hexose phosphate synthase and tricarboxylic acid cycle enzymes in bacterium 4B6, an obligate methylotroph, *Biochem. J.* **128**:1373-1376.
Colby, J., and Zatman, L. J., 1973, Trimethylamine metabolism in obligate and facultative methylotrophs, *Biochem. J.* **132**:101-112.
Colby, J., and Zatman, L. J., 1975, Tricarboxylic acid cycle and related enzymes in restricted facultative methylotrophs, *Biochem. J.* **148**:505-511.
Colby, J., Dalton, H., and Whittenbury, R., 1979, Biological and biochemical aspects of microbial growth on C1 compounds, *Annu. Rev. Microbiol.* **33**:481-517.
Collins, M. D., and Green, P. N., 1985, Isolation and characterization of a novel coenzyme Q from some methane oxidizing bacteria, *Biochem. Biophys. Res. Commun.* **133**:1125-1131.
Collins, M. D., Howarth, O. W., and Green, P. N., 1986, Isolation and structural determination of a novel coenzyme from a methane oxidizing bacterium, *Arch. Microbiol.* **146**:263-266.
Cowan, S. T., 1968, A Dictionary of Microbial Taxonomic Usage, Oliver and Boyd, Edinburgh, p. 105.
Cox, R. B., and Quayle, J. R., 1975, The autotrophic growth of *Micrococcus denitrificans* on methanol, *Biochem. J.* **150**:569-571.
Dahl, J. S., Mehta, R. J., and Hoare, D. S., 1972, New obligate methylotroph, *J. Bacteriol.* **109**:916-921.
Davey, J. F., Whittenbury, R., and Wilkinson, J. F., 1972, The distribution in the methylobacteria of some key enzymes concerned with intermediary metabolism, *Arch. Mikrobiol.* **87**:359-366.
Davis, J. B., Coty, V. F., and Stanley, J. P., 1964, Atmospheric nitrogen fixation by methane oxidizing bacteria, *J. Bacteriol.* **88**:468-472.
De Boer, L., Dijkhuizen, L, Grobben, G., Goodfellow, M., Stackebrandt, E., Parlett, J. H., Whitehead, D., and Witt, D., 1990, *Amycolatopsis methanolica* sp. nov., a facultatively methylotrophic actinomycete, *Int. J. Systematic Bacteriol.* **40**:194-204.
De Ley, J., 1978, Modern molecular methods in bacterial taxonomy: evaluation, application, prospects, in: *Proceedings of the 4th International Conference on Plant Pathogenic Bacteria* (I.N.R.A., Angers, ed.), Gibert-Clarey, Tours, pp. 347-357.
De Ley, J., Mannheim, W., Segers, P., Lievens, A., Denyn, M., Van Houcke, M., and Gillis, M., 1987, Ribosomal ribonucleic acid cistron similarities and taxonomic neighbourhood of *Brucella* and CDC Group Vd, *Int. J. Systematic Bacteriol.* **37**:35-42.
den Dooren de Jong, L. E., 1927, Uber protaminophage bakterien, *Zentralbl. Bakteriol. Parasit. Infektionskrankh. Hygiene* (Abteilung II) **71**:193-232.
Dijkhuizen, L., Hansen, T. A., and Harder, W., 1985, Methanol, a potential feedstock for biotechnological processes, *Trends Biotechnol.* **3**:262-267.

Dijkhuizen, L., Arfman, N., Attwood, M. M., Brook, A. G., Harder, W., and Watling, E. M., 1988, Isolation and initial characterization of thermotolerant methylotrophic *Bacillus* strains, *FEMS Microbiol. Lett.* **52**:209–214.

Doronina, N. V. M., Govurukhina, N. I., and Trotsenko, Yu. A., 1983, *Blastobacter aminooxidans*, a new species of bacteria growing autotrophically on methylated amines, *Mikrobiologiya* **52**:709–715.

Downs, J., and Harrison, D. E. F., 1974, Studies on the production of pink pigment in *Pseudomonas extorquens* NCIMB 9399 growing in continuous culture, *J. Appl. Bacteriol.* **37**:65–74.

Drabikowska, A. K., 1977, The respiratory chain of a newly isolated *Methylomonas* P11, *Biochem. J.* **168**:171–178.

Drozd, J. W., and Linton, J. D., 1981, in: *Continued Culture of Cells* (P. H. Calcott, ed.), CRC Uniscience Publications, Boca Raton, Florida, pp. 113–141.

Dworkin, J. W., and Foster, J. W., 1956, Studies on *Pseudomonas methanica* (Sohngen) nov. comb., *J. Bacteriol.* **72**:646–659.

Foster, J. W., and Davis, R. H., 1966, A methane dependent coccus, with notes on classification and nomenclature of obligate methane utilizing bacteria, *J. Bacteriol.* **91**:1924–1931.

Friedrich, C. G, Bowien, B., and Friedrich, B., 1979, Formate and oxalate metabolism in *Alcaligenes eutrophus, J. Gen. Microbiol.* **115**:185–192.

Galchenko, V. F., 1977, New species of methanotrophic bacteria, in: *Microbial Growth on C1-Compounds* (G. K. Skryabin *et al.*, eds.), Scientific Centre for Biological Research, USSR Academy of Sciences, Pushchino, p. 10.

Galchenko, V. F., and Andreev, L. V., 1984, Taxonomy of obligate methanotrophs, in: *Microbial Growth on C1-Compounds* (R. A. Crawford and R. S. Hanson, eds.), American Society for Microbiology, Washington, DC, pp. 269–275.

Galchenko, V. F., Shishkina, V. N., Tyurin, V. S., and Trotsenko, Yu. A., 1975, Isolation of pure cultures of methanotrophic bacteria and their properties, *Mikrobiologiya* **44**:844–850.

Govorukhina, N. I., and Trotsenko, Yu. A., 1989, *Methylovorus*—a new genus of restricted facultatively methylotrophic bacteria, in: *Abstracts of the 6th International Symposium on Microbial Growth on C1-Compounds*, Gottingen, Germany.

Govorukhina, N. I., and Trotsenko, Yu. A., 1991, *Methylovorus*—a new genus of restricted facultatively methylotrophic bacteria, *J. Systematic Bacteriol.* **41**:158–162.

Graf, W., and Bauer, L., 1973, Red bacterial growth (*Corynebacterium rubrum* n. spec.) in tap water systems., *Zentralbl. Bakteriol. Parasitenk. Infektionskrankh. Hygiene* (Abteilung II) **73**:74–96.

Green, P. N., and Bousfield, I. J., 1981, The taxonomy of pink pigmented facultatively methylotrophic bacteria, in: *Microbial Growth on C1-Compounds* (H. Dalton, ed.), Heyden, London, pp. 285–293.

Green, P. N., and Bousfield, I. J., 1982, A taxonomic study of some Gram negative facultatively methylotrophic bacteria, *J. Gen. Microbiol.* **128**:623–638.

Green, P. N., and Bousfield, I. J., 1983, Emendation of *Methylobacterium* Patt, Cole and Hanson 1976; *Methylobacterium rhodinum* (Heumann 1962) comb. nov. corrig.; *Methylobacterium radiotolerans* (Ito and Iizuka, 1971) comb. nov. corrig.; and *Methylobacterium mesophilicum* (Austin and Goodfellow, 1979) comb. nov., *Int. J. Systematic Bacteriol.* **33**:875–877.

Green, P. N., and Bousfield, I. J., 1988, Three new *Methylobacterium* species: *M. rhodesianum* sp. nov., *M. zatmanii* sp. nov., and *M. fujisawaense* sp. nov., *Int. J. Systematic Bacteriol.* **38**:124–127.

Green, P. N., and Gillis, M., 1989, Classification of *Pseudomonas aminovorans* and some related methylated amine utilizing bacteria, *J. Gen. Microbiol.* **135**:2071–2076.

Green, P. N., Hood, D., and Dow, C. S., 1984, Taxonomic status of some methylotrophic bacteria, in: *Microbial Growth on C1-Compounds* (R. L. Crawford and R. S. Hanson, eds.), American Society for Microbiology, Washington, DC, pp. 251–254.

Hampton, D., and Zatman, L. J., 1973, The metabolism of tetramethylammonium chloride by bacterium 5H2, *Biochem. Soc. Trans.* **1**:667.

Hanson, R. S., 1980, Ecology and diversity of methylotrophic organisms, *Adv. Appl. Microbiol.* **26**:3–39.

Harder, W. and Attwood, M. M., 1978, Biology, physiology and biochemistry of hyphomicrobia, *Adv. Microbial Physiol.* **17**:303–359.

Hayward, A. C., 1960, Relationship between *Protaminobacter ruber* and some red pigmented pseudomonads, *J. Appl. Bacteriol.* **23**:ii.

Hazeu, W., and Steenis, P. J., 1970, Isolation and characterization of two vibrio-shaped methane oxidizing bacteria, *Antonie Van Leeuwenhoek* **36**:67–72.

Hazeu, W., Batenburg-Van Der Vegte, W. H., and De Bruyn, J. C., 1980, Some characteristics of *Methylococcus mobilis* sp. nov., *Arch. Microbiol.* **124**:211–220.

Heumann, W., 1962, Die Methodik der Kreuzung sternbildener Bakterien, *Biol. Zentralbl.* **81**:341–354.

Hirsch, P., 1974, Budding bacteria, *Ann. Rev. Microbiol.* **28**:391–444.

Hirsch, P., and Muller, M., 1985, *Blastobacter aggregatus* sp. nov., *Blastobacter capsulatus* sp. nov., and *Blastobacter denitrificans* sp. nov., new budding bacteria from freshwater habitats, *Systematic Appl. Microbiol.* **6**:281–286.

Hirt, W., Papoutsakis, E., Krug, E., Lim, H. C., and Tsao, G. T., 1978, Formaldehyde incorporation by a new methylotroph (L3), *Appl. Environ. Microbiol.* **36**:56–62.

Hood, D. W., Dow, C. S., and Green, P. N., 1987, DNA:DNA hybridization studies on the pink pigmented facultative methylotrophs, *J. Gen. Microbiol.* **133**:709–720.

Hood, D. W., Dow, C. S., and Green, P. N., 1988, Electrophoretic comparison of total soluble proteins in the pink pigmented facultative methylotrophs, *J. Gen. Microbiol.* **134**:2375–2383.

Hutton, W. E., and Zobell, C. E., 1949, The occurrence and characteristics of methane oxidizing bacteria in marine sediments, *J. Bacteriol.* **58**:463–473.

Ida, S., and Alexander, M., 1965, Permeability of *Nitrobacter agilis* to organic compounds, *J. Bacteriol.* **90**:151–156.

Ikemoto, S., Katoh, K., and Komagata, K., 1978, Cellular fatty acid composition in methanol utilizing bacteria, *J. Gen. Microbiol.* **24**:41–49.

Ito, H. and Iizuka, H., 1971, Part XIII. Taxonomic studies on a radio-resistant *Pseudomonas*, *Agric. Biol. Chem.* **35**:1566–1571.

Janota-Bassalik, L., and Pedyk, D., 1961, Ability of *Flavobacterium extorquens* Bassalik to utilize various sources of carbon with particular reference to glucose, *Acta Microbiol. Polon.* **10**:225–238.

Janvier, M., Frehel, C., Grimont, F., and Gasser, F., 1985, *Methylophaga marina* gen. nov., sp. nov. and *Methylophaga thalassica* sp. nov., marine methylotrophs, *Int. J. Systematic Bacteriol.* **35**:131–139.

Jenkins, O., 1984, Numerical taxonomic and chemical studies on obligate methanol utilizing bacteria, Ph.D. thesis, University of Leicester.

Jenkins, O. and Jones, D., 1987, Taxonomic studies on some Gram negative methylotrophic bacteria, *J. Gen. Microbiol.* **133**:453–473.

Jenkins, O., Byrom, D., and Jones, D., 1984, Taxonomic studies on some obligate methanol utilizing bacteria, in: *Microbial Growth on C1-Compounds* (R. L. Crawford and R. S.

Hanson, eds.), American Society for Microbiology, Washington, DC, pp. 255–261.
Jenkins, O., Byrom, D., and Jones, D., 1987, *Methylophilus:* a new genus of methanol utilizing bacteria, *Int. J. Systematic Bacteriol.* **37:**446–448.
Kaneda, T., and Roxburgh, J. M., 1959, A methanol utilizing bacterium. A. Description and nutritional requirements, *Can. J. Microbiol.* **5:**87–98.
Kelly, D. P., Wood, A. P., Gottschal, J. C., and Kuenen, J. G., 1979, Autotrophic metabolism of formate by *Thiobacillus* strain A2, *J. Gen. Microbiol.* **114:**1–13.
Khambata, S. R., and Bhat, J. V., 1953, Studies on a new oxalate-decomposing bacterium, *Pseudomonas oxalaticus, J. Bacteriol.* **66:**505–507.
Kirikova, N. N., 1970, Properties of two strains of *Pseudomonas* utilizing one carbon compounds, *Mikrobiologiya* **39:**18–23.
Kortstee, G. J. J., 1980, The homoisocitrate–glyoxylate cycle in pink facultative methylotrophs, *FEMS Microbiol. Lett.* **8:**59–65.
Kouno, K., and Ozaki, A., 1975, Distribution and identification of methanol utilizing bacteria, in: *Microbial Growth on C1-Compounds* (edited by the organizing committee), Society of Fermentation Technology, Osaka, Japan, pp. 11–21.
Kung, H-S, and Wagner, C., 1970, Oxidation of C1 compounds by *Pseudomonas* sp. MS., *Biochem. J.* **116:**357–365.
Ladner, A., and Zatman, L. J., 1969, Formaldehyde oxidation by the methanol dehydrogenase of *Pseudomonas* PP, *J. Gen. Microbiol.* **55:**xvi.
Leadbetter, E. R., and Gottlieb, J. A., 1967, On methylamine assimilation in a bacterium, *Arch. Mikrobiol.* **59:**211–217.
Levering, P. R., Van Dijken, J. P., Veenhuis, M., and Harder, W., 1981, *Arthrobacter* P1, a fast growing versatile methylotroph with amine oxidase as a key enzyme in the metabolism of methylated amines, *Arch. Microbiol.* **129:**72–80.
Lidstrom-O'Connor, M. E., Fulton, G. L., and Wopat, A. E., 1983, "*Methylobacterium ethanolicum*": a syntrophic association of two methylotrophic bacteria, *J. Gen. Microbiol.* **129:**3139–3148.
Longinova, N. V., and Trotsenko, Yu. A., 1976, Facultative methylotroph belonging to the genus *Arthrobacter, Microbiology* **44:**892–896.
Longinova, N. V., and Trotsenko, Yu. A., 1977a, *Blastobacter viscosus*—a new species of autotrophic bacteria utilizing methanol, *Mikrobiologiya* **48:**785–792.
Longinova, N. V., and Trotsenko, Yu. A., 1977b, Methanol metabolism in *Pseudomonas oleovorans, Mikrobiologiya* **46:**210–216.
Longinova, N. V., and Trotsenko, Yu. A., 1977c, Abstract of the Second International Conference on Microbial Growth on C_1 Compounds, pp. 37–39. Published by the Scientific Centre for Biological Research, U.S.S.R. Academy of Science, Pushchino.
Longinova, N. V., and Trotsenko, Yu. A., 1979, Autotrophic growth on methanol by bacteria isolated from activated sludge, *FEMS Microbiol. Lett.* **5:**239–243.
Longinova, N. V., Namsaraev, B. B., and Trotsenko, Yu. A., 1978, Autotrophic metabolism of methanol in *Microcyclus aquaticus, Mikrobiologiya* **47:**168–170.
Lynch, M. J., Wopat, A. E., and O'Connor, M. L., 1980, Characterization of two new facultative methanotrophs, *Appl. Environ. Microbiol.* **40:**400–407.
Malashenko, Yu. R., Romanovskaya, V. A. and Kvasnikov, E. I., 1972, Taxonomy of bacteria utilizing gaseous hydrocarbons, *Mikrobiologiya* **41:**871–879.
Malashenko, Yu. R., Romanovskaya, V. A., Bogachenko, V. N., and Shved, A. D., 1975, Thermophilic and thermotolerant methane assimilating bacteria, *Mikrobiologiya* **44:**855–862.
Matsumoto, T., and Tobari, J., 1978, Methylamine dehydrogenase of *Pseudomonas* sp. J, *J. Biochem.* **83:**1591–1597.

Mehta, R. J., 1973, Studies on methanol oxidizing bacteria. 1. Isolation and growth studies, *Antonie Van Leeuwenhoek* **39**:295-302.
Morinaga, Y., Yamanaha, S., Otsuka, S., and Hirose, Y., 1976, Characteristics of a newly isolated methane oxidizing bacterium, *Methylomonas flagellata* nov. sp., *Agric. Biol. Chem.* **40**:1539-1545.
Murrell, J. C., and Dalton, H. 1983, Nitrogen fixation in obligate methanotrophs, *J. Gen. Microbiol.* **129**:3481-3486.
Namsaraev, B. B., Nozhevnikova, A. N., and Zavarzin, G. A., 1971, Utilization of formic acid by hydrogen bacteria, *Mikrobiologiya* **40**:772-776.
Nishio, N., Yano, T., and Kamikubo, T., 1975, Isolation of methanol utilizing bacteria and its vitamin B12 production, *Agric. Biol. Chem.* **39**:21-27.
Ogata, K., Izumi, Y., Kawamori, M., Asano, Y., and Tani, Y., 1977, Amino acid formation by methanol utilizing bacteria, *J. Fermentation Technol.* **55**:444-451.
Ohara, M., Katayama, Y., Tsuzaki, M., Nakamoto, S., and Kuraishi, H., 1990, *Paracoccus kocurrii* sp. nov., A tetramethylammonium-assimilating bacterium, *Int. J. Systematic Bacteriol.* **40**:292-296.
Ollivier, B., Cordruwisch, R., Lombardo, A., and Garcia, J-L., 1985, Isolation and characterization of *Sporomusa acidovorans* sp. nov., a methylotrophic homoacetongenic bacterium, *Arch. Microbiol.* **142**:307-310.
Paterek, J. R., and Smith, P. H., 1988, *Methanohalophilus mahii* gen. nov., sp. nov., a methylotrophic halophilic methanogen, *Int. J. Systematic Bacteriol.* **38**:122-123.
Patt, T. E., Cole, G. C., Bland, J., and Hanson, R. S., 1974, Isolation and characterization of bacteria that grow on methane and organic compounds as sole sources of carbon and energy, *J. Bacteriol.* **120**:955-964.
Patt, T. E., Cole, G. C., and Hanson, R. S., 1976, *Methylobacterium*, a new genus of facultatively methylotrophic bacteria, *Int. J. Systematic Bacteriol.* **26**:226-229.
Peel, D., and Quayle, J. R., 1961, Microbial growth on C1 compounds. 1. Isolation and characterization of *Pseudomonas* AM1, *Biochem. J.* **81**:465-469.
Quadri, S. M. H., and Hoare, D. S., 1969, Formic hydrogenlyase and the photoassimilation of formate by a strain of *Rhodopseudomonas palustris, J. Bacteriol.* **95**:2344-2357.
Quayle, J. R., 1972, The metabolism of one-carbon compounds by microorganisms, *Adv. Microb. Physiol.* **7**:119-203.
Rock, J. S., Goldberg, I., Ben-Bassat, A., and Mateles, R. I, 1976, Isolation and characterization of two methanol utilizing bacteria, *Agric. Biol. Chem.* **40**:2129-2135.
Romanova, A. K., and Nozhevnikova, A. N., 1977, Assimilation of one-carbon compounds by carboxydobacteria, in: *Microbial Growth on C1-Compounds* (G. K. Skryabin *et al.*, eds.), Scientific Centre for Biological Research, U.S.S.R. Academy of Sciences, Pushchino, pp. 109-110.
Romanovskaya, V. A., 1984, *Methylovarius* gen. nov., a new genus. *Mikrobiologiya* **53**:777-784.
Romanovskaya, V. A., Malashenko, Yu. R., and Bogachenko, V. N., 1978, Corrected diagnosis of the genera and species of methane utilizing bacteria, *Mikrobiologiya* **47**:120-130.
Romanovskaya, V. A., Malashenko, Yu. R., and Grishchenko, N. I., 1980, Diagnosis of methane oxidizing bacteria by numerical methods based on cell fatty acid composition, *Mikrobiologiya* **49**:969-975.
Sahm, H., and Wagner, F., 1975, Isolation and characterization of an obligate methanol utilizing bacterium *Methylomnas* M-15, *Eur. J. Appl. Microbiol.* **1**:147-158.
Sato, K., 1978, Bacteriochlorophyll formation of facultative methylotrophs, *Protaminobacter ruber* and *Pseudomonas* AM1, *FEMS Lett.* **85**:207-210.
Seifert, E., and Pfennig, N. F., 1979, Chemoautotrophic growth of *Rhodopseudomonas* species

with hydrogen and chemotrophic utilization of methanol and formate, *Arch. Microbiol.* **122:**177-182.
Shaw, W. V., Tasi, L., and Stadtman. E. R., 1966, The enzymatic synthesis of *N*-methyl-glutamic acid, *J. Biol. Chem.* **241:**935-945.
Shen, G-J., Kodama, T., and Minoda, Y., 1982, Isolation and culture conditions of a thermophilic methane oxidizing bacterium, *Agric. Biol. Chem.* **46:**191-197.
Shishkina, V. N., and Trotsenko, Yu. A., 1979, Pathways of ammonia assimilation in obligate methane utilizers, *FEMS Microbiol. Lett.* **5:**187-191.
Sieburth, J. McN., Johnson, P. W., Eberhardt, M. A., Sieracki, M. L., Lidstrom, M. E., and Laux, D., 1987, The first methane oxidizing bacterium from the upper mixing layer of the deep ocean: *Methylomonas pelagia* sp. nov., *Curr. Microbiol.* **14:**285-293.
Sohngen, N. L., 1906, Ueber bakterien, welche methan als kohlenstoffnahrung und energiequelle gebrauchen, *Zentralbl. Bakteriol. Parasitenk. Infektionskrankh. Hygiene,* (Abteilung II) **15:**513-517.
Sperl, G. T., and Hoare, D. S., 1971, Denitrification with methanol: a selective enrichment for *Hyphomicrobium* species, *J. Bacteriol.* **108:**733-736.
Sperl, G. T., Forrest, H. S., and Gibson, D. T., 1974, Substrate specificity, of the purified primary alcohol dehydrogenases from methanol oxidizing bacteria, *J. Bacteriol.* **118:**541.
Stanley, S. H., and Dalton, H., 1982, Role of ribulose-1.5-biphosphate carboxylase/oxygenase in *Methylococcus capsulatus* (Bath), *J. Gen. Microbiol.* **128:**2927-2935.
Steudel, A., Miethe, D., and Babel, W., 1980, Bacterium MB 58, ein methylotrophes "essigsaurebakterium," *Ztschr. Allgemeine Mikrobiol.* **20:**663-672.
Strand, S. E., and Lidstrom, M. E., 1984, Characterization of a new marine methylotroph, *FEMS Microbiol. Lett.* **21:**247-251.
Suzuki, M., Kuhn, I., Berglund, A., Unden, A., and Heden, C-G, 1977, Identification of a new methanol utilizing bacterium and its characteristic responses to some chemicals, *J. Fermentation Technol.* **55:**459-465.
Takeda, K., Motomatsu, Y., Fukuoka, S., and Takahara, Y., 1974, Characterization and culture conditions for a methane oxidizing bacterium, *J. Fermentation Technol.* **52:**793-798.
Taylor, S., Dalton, H., and Dow, C. S., 1980, Purification and initial characterization of ribulose 1.5-bisphosphate carboxylase from *Methylococcus capsulatus* (Bath), *FEMS Microbiol. Lett.* **8:**157-160.
Toraya, T., Yongsmith, B., Tanaka, A., and Fukui, S., 1975, Vitamin B_{12} production by a methane utilizing bacterium, *Appl. Microbiol.* **30:**477-479.
Trotsenko, Yu. A., 1976, Isolation and characterization of obligate methanotrophic bacteria, pp. 329-336, in: *Microbial Production and Utilization of Gases* (H. G. Schlegel, G. Gottschalk, and N. Pfennig, eds.), E. Goltze, Gottingen.
Trotsenko, Yu. A., 1983, Metabolic features of methane and methanol utilizing bacteria, *Acta Biotechnol.* **3:**269-277.
Tsuji, K., Tsien, H. C., Hanson, R. S., Depalma, S. R., Scholtz, R., and Laroche, S., 1990, 16S ribosomal RNA sequence analysis for phylogenetic relationship among methylotrophs, *J. Gen. Microbiol.* **36:**(In Press).
Urakami, T., Araki, H., Oyanogi, H., Suzuki, K. I., and Komagala, K., 1990, *Paracoccus aminophilus* sp. nov. and *Paracoccus aminovorans* sp. nov., which utilize N,N-dimethylformamide, *Int. J. Systematic Bacteriol.* **40:**287-291.
Urakami, T., and Komagata, K., 1979, Cellular fatty acid composition and coenzyme Q system in Gram negative methanol utilizing bacteria, *J. Gen. Appl. Microbiol.* **25:**343-360.

Urakami, T., and Komagata, K., 1981, Electrophoretic comparison of enzymes in the Gram negative methanol utilizing bacteria, *J. Gen. Appl. Microbiol.* **27**:381–403.

Urakami, T., and Komagata, K., 1986a, Emendation of *Methylobacillus* Yordy and Weaver 1977, a genus for methanol utilizing bacteria, *Int. J. Systematic Bacteriol.* **36**:502–511.

Urakami, T., and Komagata, K., 1986b, Occurrence of isoprenoid compounds in Gram negative methanol, methane and methylamine utilizing bacteria, *J. Gen. Microbiol.* **32**:317–341.

Urakami, T., and Komagata, K., 1987, Cellular fatty acid composition with special reference to the existence of hydroxy fatty acids in Gram negative methanol, methane and methylamine utilizing bacteria, *J. Gen. Microbiol.* **33**:135–165.

Urakami, T., Tamaoka, J., and Komagata, K., 1985, DNA base composition and DNA-DNA homologies of methanol utilizing bacteria, *J. Gen. Appl. Microbiol.* **31**:243–253.

Urakami, T., Tamaoka, J., Suzuki, K-I., and Komagata, K., 1989a, *Acidomonas* gen. nov., incorporating *Acetobacter methanolicus* as *Acidomonas methanolica* comb. nov., *Int. J. Systematic Bacteriol.* **39**:50–55.

Urakami, T., Tamaoka, J., Suzuki, K-I., and Komagata, K., 1989b, *Paracoccus alcaliphilus* sp. nov., an alkaliphilic and facultatively methylotrophic bacterium, *Int. J. Systematic Bacteriol.* **39**:116–121.

Whittenbury, R., and Krieg, N. R., 1984, Family IV. *Methylococcaceae*, in: *Bergey's Manual of Systematic Bacteriology* (N. R. Krieg and J. G. Holt, eds.), Williams & Wilkins, Baltimore, pp. 256–261.

Whittenbury, R., Phillips, K. C., and Wilkinson, J. F., 1970a, Enrichment, isolation and some properties of methane utilizing bacteria, *J. Gen. Microbiol.* **61**:205–218.

Whittenbury, R., Davies, S. L., and Davey, J. F., 1970b, Exospores and cysts formed by methane utilizing bacteria, *J. Gen. Microbiol.* **61**:219–226.

Wiegel, J., Wilke, D., Baumgarten, J., Opitz, R., and Schlegel, H. G., 1978, Transfer of the nitrogen fixing hydrogen bacterium *Corynebacterium autotrophicum* Baumgarten et al., to *Xanthobacter* gen. nov., *Int. J. Systematic Bacteriol.* **28**:573–581.

Yamamoto, M., Seriu, Y., Kouno, K., Okamoto, R., and Insui, T., 1978, Isolation and characterization of marine methanol utilizing bacteria, *J. Fermentation Technol.* **56**:451–458.

Yamamoto, M., Iwaki, H., Kouno, K., and Inui, T, 1980, Identification of marine methanol utilizing bacteria, *J. Fermentation Technol.* **58**:99–106.

Yamanaka, K., and Matsumoto, K., 1977, Purification, crystallization and properties of primary alcohol dehydrogenase from a methanol oxidizing *Pseudomonas* sp. no. 2941, *Agric. Biol. Chem.* **41**:467–475.

Yordy, J. R., and Weaver, T. L., 1977, *Methylobacillus*: a new genus of obligately methylotrophic bacteria, *Int. J. Systematic Bacteriol.* **27**:247–255.

Zeikus, J. G., 1983a, Chemical and fuel production from one-carbon fermentations: a microbiological assessment, in: *Organic Chemicals from Biomass* (D. L. Wise, ed.), Benjamin/Cummings, Menlo Park, CA, pp. 359–383.

Zeikus, J. G., 1983b, Metabolism of one-carbon compounds by chemotrophic anaerobes, *Adv. Microb. Physiol.* **24**:215–299.

Zhao, S-J., and Hanson, R. S., 1984, Variants of the obligate methanotroph isolate 761M capable of growth on glucose in the absence of methane, *Appl. Environ. Microbiol.* **48**:807–812.

Methane Oxidation by Methanotrophs

3

Physiological and Mechanistic Implications

HOWARD DALTON

1. INTRODUCTION

Methane-oxidizing bacteria (methanotrophs) are similar biochemically to their methanol-utilizing relatives (see Chapter 5) in that the carbon growth substrate is oxidized to carbon dioxide via formaldehyde at which level various assimilatory pathways are employed to synthesize biomass. Where the methanotrophs differ, at the biochemical level, resides in the possession of the enzyme methane monooxygenase (MMO), which is responsible for the oxidation of methane to methanol. This enzyme, which can exist in one of two forms depending on the level of copper ions in the environment (Stanley *et al.*, 1983), requires reduced pyridine nucleotide as an electron donor and therefore must place some energetic demand on the cells that possess it. The consequence of this requirement is that although the oxidation of methane to methanol is exergonic on thermodynamic grounds, cells that grow on methane show similar molar growth yields when methanol is used as a substrate so that the methane-to-methanol step becomes energetically neutral *in vivo*. No evidence of energetic coupling (to, say, the formation of ATP) has been adduced to date.

At the morphological level there are considerable difference between methanotrophs and other methylotrophs. The complex series of intra-

---

HOWARD DALTON • Department of Biological Sciences, University of Warwick, Coventry CV4 7AL, England.

Methane and Methanol Utilizers, edited by J. Colin Murrell and Howard Dalton. Plenum Press, New York, 1992.

cytoplasmic membranes so elegantly shown 20 years ago (Davies and Whittenbury, 1970; Smith *et al.*, 1970; Proctor *et al.*, 1969) in the methanotrophs appeared to be present only when the bacteria were grown on methane. However, cells grown on methanol have also been shown to contain membranes (Davies and Whittenbury, 1970; De Boer and Hazeu, 1972; Linton and Vokes, 1978; Best and Higgins, 1981; Prior and Dalton, 1985), so one cannot be certain that methane *per se* is necessary for intracytoplasmic membrane formation. The fact that cells grown on methanol do possess MMO activity [since methanol can be a substrate for the enzyme (Colby *et al.*, 1977)] could still indicate that it is the presence of the enzyme which is responsible for the levels of membranes observed. Certainly, such a complex series of intracytoplasmic membranes is relatively rare among the prokaryotes, with only the photosynthetic bacteria, ammonia and nitrite oxidizers, and cyanobacteria showing anything like these structures. Studies by Prior and Dalton (1985) do appear to indicate that copper ions play an important role in the synthesis of membranes in these bacteria even when grown on methanol. Increasing levels of copper ions to steady-state chemostat cultures of *Methylococcus capsulatus* (Bath) were strongly correlated with the intracytoplasmic membrane content of the cells. A similar finding had been noted earlier by Takeda and Tanaka (1980), who reported that the presence of intracytoplasmic membranes in *Methanomonas margaritae* was dependent on the addition of copper sulfate to the growth medium. Cells grown in copper-deficient media were devoid of these membranes and only vesicles were observed at the periphery of the cell. Similar vesicles were also observed by Best and Higgins (1981), but they discounted copper as having an important role to play in determining the presence of either vesicles or membranes. They were unable to pinpoint the exact environmental or physiological parameters that led to the formation of internal structures but did suggest that the dissolved oxygen concentration may be important.

In the context of this chapter, it is therefore important to explore the biochemical characteristics of the MMO system and its implications on the physiology of these bacteria. Leak (see Chapter 8) has indicated the important biotechnological features of MMO, so these will only be touched on in passing. Probably the most fertile area for methane oxidation research today, however, concerns the elucidation of the structural and mechanistic characteristics of MMO that permit the facile oxidation of methane to methanol.

2. CHEMICAL VERSUS BIOLOGICAL CATALYSIS

It is becoming abundantly clear that the demand for methanol will be increasing dramatically in the coming years if it is to be used as an alternative to gasoline in motor cars. The existing feedstock source of methanol is methane, which is converted to methanol in an indirect three-stage process involving the production synthesis gas (reaction 1). This reaction is thermodynamically endergonic, whereas the formation of methanol from carbon monoxide and dihydrogen (reaction 3) is thermodynamically exergonic. Reaction 2 is necessary to correct for the imbalance in the 1:3 ration of $CO:H_2$ generated in reaction 1 such that a ratio of 1:2 $CO:H_2$ is produced, which is the stoichiometric optimum for methanol synthesis in reaction 3. Overall this indirect process is energy requiring ($\Delta H° = +27.6$ kcal/mole), whereas a direct route as shown by reaction 4 would be exergonic ($\Delta H° = -30.7$ kcal/mole) and therefore does not require a high energy input. It is clearly reaction 4 that more closely resembles the biological catalyst.

$$CH_4 + H_2O \xrightarrow[\substack{700-900°C \\ 1-25 \text{ bar}}]{\substack{15-20\% \text{ Ni catalyst} \\ \text{on } Al_2O_3 \text{ or } SiO_2}} CO + 3H_2 \quad (1)$$

$$CO_2 + H_2 \xrightarrow{\text{Ni catalyst}} CO + H_2O \quad (2)$$

$$CO + 2H_2 \xrightarrow[\substack{250-280°C \\ 70-110 \text{ bar}}]{\text{Cu/Zn catalyst}} CH_3OH \quad (3)$$

$$CH_4 + \tfrac{1}{2}O_2 \rightarrow CH_3OH \quad (4)$$

This direct oxidation by the enzyme, which occurs at ambient temperature and pressure, would serve as an important model for the construction of direct methane oxygenation catalysts that are more robust than the enzyme and can withstand the rigors imposed on it as an industrial catalyst. It is not surprising, therefore, that a reasonable effort has been concentrated on unraveling the structural and dynamic characteristics of the enzyme system capable of effecting the partial oxidation of methane to methanol. The fact that the enzyme system is able to stop the oxidation at the level of methanol is of major significance to catalytic chemists. Despite the fact that for over 85 years chemists have been studying the partial oxidation of methane to methanol using both heterogeneous and homogeneous systems, there have been very few success stories (Gesser et al., 1985). All systems studies so far have indicated that other oxidation products besides methanol are formed (e.g., CO, CO_2, CH_2O,

HCOOH, and even soot), thereby reducing the selectivity of the process (100% selectivity would mean that only methanol would be produced from methane and oxygen). Thus the selectivity of the biological system is 100% since only methanol is normally produced from methane. Exceptionally, formaldehyde may be produced, but only when the methane concentration is vanishingly small and the methanol concentration is sufficiently high to be converted to formaldehyde (the K_m for methanol is around 1 mM (Colby *et al.*, 1977)). Clearly, to understand how the direct oxidation of methane to methanol can be effected under controlled conditions at ambient temperature and pressure still remains one of the greatest unsolved problems in oxidation chemistry (Hill, 1989) and yet bacteria have been able to effect this chemistry *via* MMO, or an enzyme quite similar, for at least 2 or 3 billion years, which would correspond to the time when oxygenic photosynthesis is presumed to have evolved. Understanding the biochemical process therefore ought to yield significant insights into how the controlled oxidation of a reaction with a high intrinsic bond strength (104 kcal/mole) can be effected.

3. PARTICULATE METHANE MONOOXYGENASE

Ribbons and Michalover (1970) were the first to report on a cell-free methane-oxidizing activity that was associated with the particulate fractions of extracts from *M. capsulatus*. Particulate activity was also demonstrated in *Methylomonas methanica* by Ferenci (1974) and Colby *et al.* (1975), who showed that carbon monoxide or bromomethane would serve as useful assay substrates for the enzyme. Furthermore, Hubley *et al.* (1975) and Takeda *et al.* (1976) had indicated that copper ions may play a role in methane oxidation. The latter authors also observed that over 80% of the copper in the cell extracts of the methane-oxidizing *M. margaritae* was associated with the particulate fractions of the cell. Shortly thereafter, Tonge *et al.* (1975, 1977) reported that the methane-oxidizing complex could be released from membrane fractions of *Methylosinus trichosporium* OB3b by treatment with phospholipase C or sonication. Three proteins were reported to be involved in methane oxidation, two of which contained copper. Despite the relative simplicity of the methods used to release and purify the proteins from the membrane, many laboratories, including their own, have been unsuccessful in repeating this work and so it will not be considered further. Likewise, two other reports of purification of membrane-associated MMOs have appeared in the literature, but neither has been verified in other laboratories to date. The first, and most extensive, was by Akent'eva and Gvozdev (1988), who purified two protein fractions from membranes of *M. capsulatus* (strain M) using NaCl and Na$_2$SO$_4$ extraction of unwanted proteins, followed by solubilization with 5% sodium deoxycholate. The proteins were then purified on DEAE cellulose and resolved into the hydroxylase (200–240 kDa; six subunits) and a NADH reductase (180kDa; four subunits). The enzyme (it was not clear which component was referred to) was shown to contain 4g atoms of nonheme iron and 1g atom of copper. In its purified form the hydroxylase loses most of its activity at 4°C and 77°K. The reductase was completely inactivated after a few days at 77°K.

The second report concerned solubilization of the enzyme from *M. capsulatus* (Bath) in which only a partial purification could be effected since activity was lost if attempts were made to purify the solubilized proteins (Smith and Dalton, 1989). Many detergents were tested, including those that were reported to be successful in solubilizing MMO from other organisms. Only dodecyl-β-D-maltoside was able to release particulate MMO in a form that could be subsequently activated by treatment first with Bio-Beads to remove the detergent, followed by the addition of lecithin to recover enzyme activity. Subsequent treatments resulted in an irrecoverable inactivation of the enzyme. The solubilized pMMO comprised of polypeptides of molecular masses 49 kDa, 23 kDa, and 22 kDa and contained 3 nmol copper per milligram protein; copper ions were required for maximum activity. All the data so far acquired for pMMO strongly suggest that it is a copper-containing form of the enzyme that is able to oxidize a limited number of alkanes and alkenes. Neither alicyclic, heterocyclic, nor aromatic compounds appear to be oxidized by pMMO.

Curiously, when the system for *M. trichosporium* OB3b was studied in this author's laboratory, the enzyme was soluble and no evidence for the particulate form of the enzyme was found (Stirling and Dalton, 1979). This was consistent with the earlier observation that the enzyme system from *M. capsulatus* (Bath) could be obtained in a soluble form also (Colby and Dalton, 1976). Although it was not known at the time, it became clear that the environmental growth conditions were responsible for dictating which type of enzyme was present in the cell. Stanley *et al.* (1983) reported that at low biomass concentrations where the copper:biomass ratio was high, the pMMO predominated; at high biomass concentrations where the copper:biomass ratio was low, the sMMO form predominated. In that paper the authors also reported that the enzyme from *M. trichosporium* OB3b could also be switched from a particulate to a soluble form by reducing the copper availability to the cell and that in *Methylomonas albus* BG8 only the particulate form of the enzyme could be formed, irrespective of growth conditions.

It thus appears that copper ions are important in regulating the intracellular location of the MMO enzyme complex, although Scott *et al.*, (1981) did observe substantial pMMO activity in cultures grown with limiting oxygen concentrations, whereas cells grown under nitrate limitation (higher biomass concentrations) contained soluble MMO. These observations would still be consistent with the copper:biomass theory advanced above if we assume that the copper level in the growth medium was the same in both instances.

4. SOLUBLE METHANE MONOOXYGENASE

4.1. The Three Protein Components

The first report of a soluble MMO appeared in 1976 (Colby and Dalton), in which it was demonstrated that the supernatant from a cell extract of *M. capsulatus* (Bath) centrifuged at 160,000g for 1 hr was able to effect an NAD(P)H- and O_2-dependent oxidation of methane or bromomethane. Furthermore, the enzyme could be resolved into two fractions by DEAE cellulose chromatography, both of which were necessary for reconstitution of activity. In subsequent publications it was clearly established that the soluble MMO complex was comprised of three distinct proteins—an NAD(P)H-dependent oxidoreductase (protein C), a regulatory protein (protein B), and a hydroxylase (protein A). It is now quite clear that protein A is the hydroxylase where methane and oxygen interact to form methanol (Fig. 1). Protein C is reduced by NADH and transfers electrons, one at a

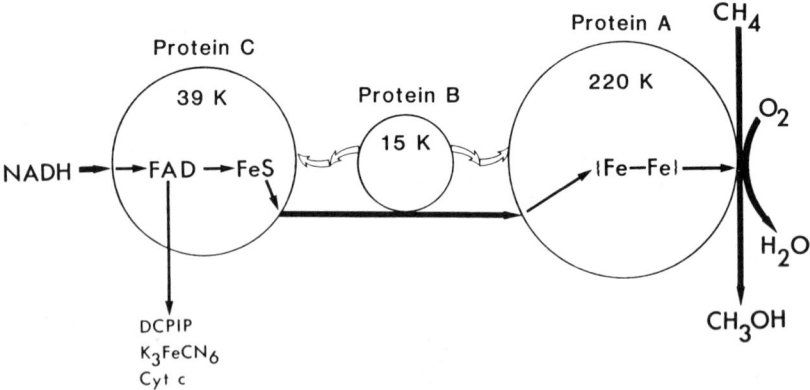

Figure 1. Electron transfer pathway between the three protein components of the MMO complex. The exact location of protein B is unknown but probably associates directly with the hydroxylase protein A.

time, to the hydroxylase. The role of protein B was somewhat enigmatic for some time because it was absolutely necessary for activity under normal conditions but did not contain any prosthetic groups. It is now known to act as a regulator of enzyme activity, converting the enzyme from an oxidase (in the absence of methane) to a hydroxylase (in the presence of methane). Protein C (~40 kDa) contains an FAD and Fe_2S_2 cluster (Colby and Dalton, 1979), which is readily reduced by $NADH_2$ (Lund and Dalton, 1985). Electrons are first transferred to the FAD center, which is fully reduced by the addition of two electrons, whence it is able to effect intramolecular electron transfer of one electron to the Fe_2S_2 center very rapidly with a second-order rate constant of 3.6×10^6 $M^{-1} \cdot sec^{-1}$ at 18°C (Green and Dalton, 1989b) (Fig. 2). Removal of the Fe_2S_2 center of the protein by treatment with the mercurial mersalyl destroyed its ability to act as an NADH acceptor:reductase or to couple to the hydroxylase protein A to form functionally active MMO (Lund et al., 1985). Reconstitution of the

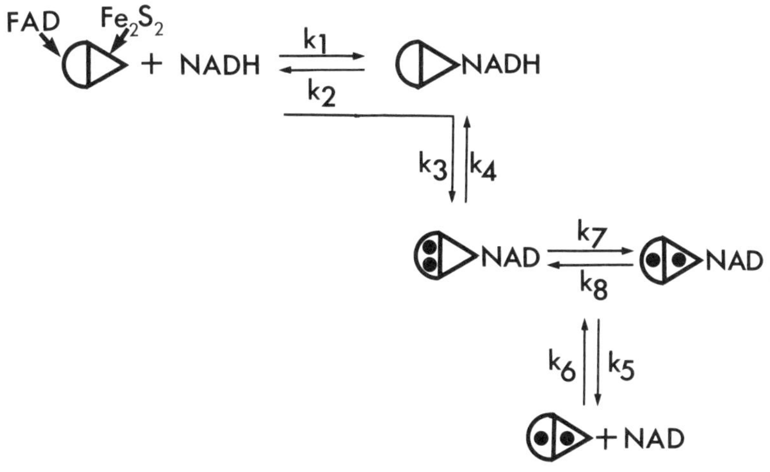

Rate constants at 18°C:

$k_1 = 40 \times 10^6$ $M^{-1} \cdot s^{-1}$ $k_5 = 76$ s^{-1} $k_{diss} = 80 \times 10^{-3}$ M
$k_2 =$ approx 0 s^{-1} $k_6 = 1 \times 10^3$ $M^{-1} \cdot s^{-1}$
$k_3 = 2.9 \times 10^6$ $M^{-1} \cdot s^{-1}$ $k_7 = 3.6 \times 10^6$ $M^{-1} \cdot s^{-1}$
$k_4 =$ approx 0 s^{-1} $k_8 = 1$ s^{-1}

Figure 2. Interaction of protein C (the reductase) with NADH as studied by stopped flow. The rate constants are measured at 18°C.

iron-sulfur center with 2-mercaptoethanol without prior removal of the mercurial restored full activities (sometimes even threefold higher). If the mercurial was removed prior to reconstitution, then the iron center could not be reconstituted but the FAD center remained intact. This apo-Fe_2S_2 protein C now possessed NADH acceptor:reductase activity but was inactive in the MMO assay. If the FAD center was now also removed by treatment with urea, all activities were abolished. Addition of FAD to the apo-Fe_2S_2 protein C would restore the NADH acceptor:reductase activity. These experiments clearly indicated that electron transfer within protein C followed the course:

$$NADH \rightarrow \underset{\substack{\mid \\ \text{acceptor:reductase} \\ \text{activity}}}{FAD} \rightarrow Fe_2S_2 \rightarrow \underset{\substack{\mid \\ \text{MMO activity}}}{\text{protein A}}$$

Electron transfer from the Fe_2S_2 center of protein C to protein A is slower than the reduction of protein C by NADH ($2.9 \times 10^6 M^{-1} \cdot sec^{-1}$), with a second-order rate constant of $6.5 \times 10^5 M^{-1} \cdot sec^{-1}$. The role of protein B has now been established as a regulator of activity. In the absence of a hydroxylatable substrate, it is able to interact with proteins A and C to prevent the wasteful oxidation of NADH that would otherwise occur via proteins A and C (Green and Dalton, 1985). Such a mechanism, of course, would make sense to an organism that was essentially NADH-limited for growth. There is clearly no need to wastefully oxidize NADH in an oxidase reaction that is not coupled to product formation. When methane is present in the environment (and this can fluctuate widely), then coupling between NADH oxidation and methane oxidation can readily occur in the hydroxylase reaction. In the absence of methane, the rate of electron transfer from protein C to protein A in the presence of protein B is 0.35 sec^{-1}, which rises to 5.8 sec^{-1} in the presence of saturating concentrations of methane (Fig. 3). In the presence of protein B, therefore, electrons are readily transferred to the hydroxylase when methane is present but shut down when methane is absent. Using a coupled enzyme assay in the stopped flow system, it has been possible to trap the methanol produced and measure the first-order rate constant of 0.26 sec^{-1} for the reaction at 18°C. This corresponded to a specific activity of 250 nmoles methanol produced per minute per milligram protein.

Purification of the reductase component from soluble extracts of *M. trichosporium* OB3b (Fox et al., 1989) and *Methylobacterium* sp. CRL26 (Prince and Patel, 1986) showed similar features in the reductase purified from *M. capsulatus* (Bath).

Protein B purified from *M. trichosporium* OB3b (Fox et al., 1989) was

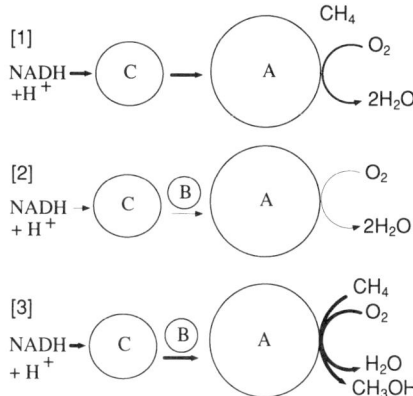

Figure 3. Role of protein B in regulation of electron transfer through the MMO complex. [1] Methane is not oxidized in the absence of B but dioxygen is reduced exclusively to water. [2] In the presence of B but the absence of methane, very little electron transfer is evident and dioxygen is reduced to water. [3] The fully functional complex. Thickness of arrows indicates relative flux through the system.

also physically similar to the protein B form *M. capsulatus* (Bath). Curiously, protein B has not been identified in *Methylobacterium* sp. CRL-26 as a discrete entity, although it should be noted that protein A from *M. capsulatus* (Bath) will bind very tightly to protein B, and unless a chaotrophic agent is present, it may not be readily separated from it. It is possible, since the reductase and hydroxylase from the CRL strain are catalytically competent in the MMO reaction, that protein B is very tightly bound to protein A, with only very small amounts of protein B being necessary to catalyze the reaction.

4.2. The Hydroxylase

The hydroxylase component (protein A) has been purified from a number of methanotrophs, including *M. capsulatus* (Bath) (Woodland and Dalton, 1984), *Methylobacterium* sp. strain CRL-26 (Patel and Savas, 1987), *M. trichosporium* OB3b (Fox and Lipscomb, 1988), and *Methylosinus sporium* 5 (Pilkington and Dalton, 1991). Each protein has a molecular mass of between 210 and 240 kDa, with subunits around 54 kDa, 40 kDa, and 20 kDa in an α_2, β_2, γ_2 arrangement (Table I). As isolated, the preparation from *M. trichosporium* OB3b seems to differ from the others in that it was reported to contain between 4 and 9 moles Fe/mole protein (Fox and Lipscomb, 1988).

Subsequently, preparations were then obtained that were reported to contain between 1.9 and 7.6 moles Fe/mole protein in which there was some correlation between the amount of iron in the protein and its specific activity. Preparations with less than 4.3 moles Fe/mole protein had low specific activities (less than 300 nmoles propylene oxide produced per

TABLE I. Characteristics of Purified Soluble Methane Monooxygenases

Organism	*Methylococcus capsulatus* (Bath)	*Methylobacterium* CRL-26	*Methylosinus trichosporium* (OB3b)	*Methylosinus sporium* strain 5
Classification	Type X	Type 2	Type 2	Type 2
Location	Soluble	Soluble	Soluble	Soluble
Number of component proteins	Three: A Hydroxylase, B Regulatory, C Acceptor reductase	Two: A hydroxylase, C Acceptor reductase	Three: A Hydroxylase, B Regulatory, C Acceptor reductase	Three: A Hydroxylase, B Regulatory, C Acceptor reductase
Component A M_r	210000	220000	245000	220000
Number of subunits	Three	Three	Three	Three
Subunit M_r	54000 42000 17000	55000 40000 20000	54400 43000 22700	56000 40000 20000
Structure	$\alpha_2\beta_2\gamma_2$	$\alpha_2\beta_2\gamma_2$	$\alpha_2\beta_2\gamma_2$	$\alpha_2\beta_2\gamma_2$
Metal content (mol/mol protein)	Fe 2.3	Fe 2.8	Fe 4.3	Fe 2.1
Component B M_r	17000		15000	
Number of subunits	one		one	
Metal content	none		none	
Component C M_r	44600	40000	38300	40000
Number of subunits	One	One	One	One
Prosthetic groups	1 FAD + 1 Fe_2S_2	1 FAD + 1 Fe_2S_2	1 FAD + 1 Fe_2S_2	1 FAD + 1 Fe_2S_2
Catalytic activity	NADH:acceptor reductase	NADH:acceptor reductase	NADH:acceptor reductase	NADH:acceptor reductase

minute per milligram protein), whereas preparations with 4.3 moles iron or greater had specific activities up to 1700 m units/mg. Samples used for Mossbauer for electron paramagnetic resonance studies by Lipscomb's group curiously only appear to contain around 2.3 moles Fe/mole protein and have specific activities around 500.

Using purified hydroxylase protein A from *M. capsulatus* (Bath), Woodland *et al.* (1986) were the first to provide e.p.r. evidence to suggest that the two iron atoms within the protein appeared to coupled via a μ-oxo bridge. This was an unprecedented finding in an oxygenase since similar structures had only previously been reported in hemerythrin (an oxygen-carrying protein with no catalytic activity), purple acid phosphatase, and ribonucleotide reductase. As isolated, the protein gave rise to a small e.p.r. signal at $g = 4.3$, which was presumed to be due to high-spin ferric iron in a rhombic environment and a free radical signal.

The iron atoms were redox active and could exist in one of three states: oxidized (FEIII/FEIII), half-reduced (FEIII/Fe II), or fully reduced (FEII/FEII). Only the half-reduced (or mixed valence) state was e.p.r. active to give rise to a relatively axial e.p.r. spectrum with principal g values of $g_z = 1.78$, $g_y = 1.88$, and $g_x = 1.95$, the g_{ave} being 1.87. The shape and amplitude of the $g = 1.95$ signal changed little when reacted with oxygen, suggesting that the mixed-valence state did not interact with oxygen.

That a μ-oxo binuclear-type bridge structure was present in the hydroxylase was definitely confirmed by EXAFS studies for the *Methylococcus* protein (Ericson *et al.*, 1988) and the *Methylobacterium* protein (Prince *et al.*, 1988). Curve fitting of the mixed-valence form for the Bath protein indicated that the first coordination shell was composed of approximately six nearest N/O neighbors at an average distance of 2.05 Å. This has recently been redefined to be closer to 2.11 Å. No short Fe—O distance could be fitted to the data, and it was concluded, based also on comparison with characterized model complexes, that the bridge was more likely to be μ-hydroxo or -alkoxy rather than μ-oxo. The data for the CRL-26 protein was similar but indicated an Fe—Fe distance of 3.05 Å as opposed to 3.41 Å in the Bath protein.

Proof that the iron atoms were involved in catalysis was obtained by the Warwick group in the same year (Green and Dalton, 1988). They were able to remove the iron center from the hydroxylase protein A by subjecting the protein to three freeze/thaw cycles or by dialysis against 8-hydroxyquinoline. This process reduced the iron content of the protein from 2.4 moles/mole protein to 0.26 and reduced its specific activity by 90%. The e.p.r. spectrum of the iron-depleted protein in the mixed-valence state was also dramatically reduced. In a separate experiment, reduction of the iron content to 0.66 moles/mole protein completely abolished enzyme activity. Reconstitution of the iron center was only possible by incubation of the iron-depleted protein with chelated iron (as Fe-EDTA) in the presence of dithiothreitol. Typically, the reconstituted protein was three times more active than the original protein, and under some circumstances activities of 1400 m units/mg protein were obtained (a 10-fold enhancement of the original sample). The reconstituted protein also had its iron content restored to its original form, clearly strongly implicating the iron center in enzyme catalysis. In some circumstances, the iron content of the reconstituted protein was 50% higher (3.3 moles Fe/mole protein) than the original protein, which may indicate that it has an important role in catalysis since the protein also possessed an enhanced activity. These results should be considered along with the observations of Fox *et al.* (1989), who observed preparations with 7.6 moles Fe/mole protein, which had an activity one-third of the maximum observed in protein with 4.3 moles Fe/mole protein: presumably adventitious iron which is non-specifically bound to the protein may be involved in the former case.

Only iron could be reintroduced functionally into the iron-depleted protein (Green and Dalton, 1988); neither Ni^{2+}, Co^{2+}, Zn^{2+}, Mo^{6+}, nor V^{3+} would give rise to functionally active protein.

When total RNA from cells actively expressing MMO was incubated in the *in vitro* wheat germ translation system with radioactive methionine, the ^{35}S-labeled protein that was produced had the same molecular mass as the native protein labeled the same way *in vivo*, indicating that the hydroxylase did not undergo any posttranslational processing. Furthermore, when run on nondenaturing gels, the subunits self-assembled to a protein of 220 kDa molecular mass with no MMO activity, but showed MMO activity when Fe-EDTA and dithiothreitol was added, albeit at about 10% the activity of the native protein.

The catalytic role of iron in the hydroxylase is therefore in little doubt. To effectively study the nature of the iron species, however, requires the use of magnetic spectroscopic techniques, e.p.r. in particular. As stated above, only the FEIII/FEII mixed-valence state was observed to be e.p.r. active as far as the $g_{ave} = 1.87$ was concerned. Generation of this signal has been achieved in a variety of ways, using either proteins B and C with NADH or sodium dithionite (Woodland *et al.*, 1986), dithionite and phenazine methosulfate (Prince *et al.*, 1988; Fox *et al.*, 1989), or even with photoreduction in the x-ray beam (Ericson *et al.*, 1988). No reliable method has been developed to reproducibly obtain the mixed-valence signal as the majority species in any preparation to date. However, it was noted by Fox *et al.* (1988) that during reductive titration of the protein, the $g_{ave} = 1.85$ signal in *M. trichosporium* OB3b hydroxylase increased on partial reduction and decreased on full reduction as expected, but that a signal at $g = 15$ [later corrected to $g = 16$ by Hendrich et al. (1990)] appeared concomitantly with the disappearance of the $g_{ave} = 1.85$ signal. Addition of oxygen to this resulted in the disappearance of the $g = 15$ signal with no $g_{ave} = 1.85$ signal appearing, suggesting that the fully reduced form reacts with O_2. These data are consistent with the earlier observations (Woodland et al., 1986) on the noninteraction of oxygen with the mixed-valence state, but go further in that the $g = 15$ signal can now be used as a measure of the FEII/FEII form of the protein. Similar observations were also made by Bentsen et al. (1989) in which a broad $g = 12.3$ signal was observed on full reduction of the hydroxylase protein from M. capsulatus (Bath).

Recent analysis of the $g = 15$ signal gives the value closer to $g = 16$ (Hendrich et al., 1990), and analysis of the data suggests that the $g = 16$ resonance represents the majority (between 70% and 90% of the iron in the sample) of iron in the diferrous hydroxylase. Simulations of the data indicated that the iron was found in a cluster containing two high-spin ferrous ions that were ferromagnetically coupled. In the oxidized (FeIII/FEIII) and mixed valence (FEIII/FEII) states there is *anti*ferromagnetic coupling between the two iron atoms. Two coupling schemes, one with strong and one with weak ferromagnetic coupling, were considered to be compatible with the data, but further information using other spectroscopic techniques will be required to unravel the complexities of the system.

Single turnover experiments have been used to show that chemically reduced hydroxylase is able to effect the oxidation of propane or propene, two substrates of MMO (Fox et al., 1988). The protein was reduced with sodium dithionite and phenazine methosulfate (for the mixed-valency form) or dithionite with proflavin and methyl viologen (for the fully reduced form). In the absence of protein B or protein C, propan-1-ol or epoxypropane was produced, indicating that proteins B and C were not necessary in these experiments and that so long as the hydroxylase could be kept in a reduced form, product formation would occur. Mixed-valency hydroxylase was also effective, but only at 8% of the fully reduced form.

The role of protein B in M. trichosporium in multiple turnover experiments appeared to be different from that observed for M. capsulatus. In the former case it was not possible to shut down electron transfer to the hydroxylase in the absence of substrate; indeed, addition of protein B appeared to stimulate O_2 utilization (O_2 utilization by the hydroxylase will occur as the oxidase reaction of MMO in the absence of substrate, *vide supra*). Furthermore, in the M. trichosporium case (Fox et al., 1988), the ratio of NADH oxidized:O_2 reduced remained at 1, regardless of the presence or absence of substrate. Since oxygen was not incorporated into the product in the absence of substrate to maintain a stoichiometry of NADH:O_2 of

1, then a 2e⁻ reduction of O_2 must occur, which would produce hydrogen peroxide. This was not detected by Fox *et al.*, (1988), nor indeed was peroxide detected in the *M. capsulatus* system, where an NADH:O_2 ratio of 2 was observed in the uncoupled reaction (Green and Dalton, 1985), which indicates that only water is produced—a novel finding in uncoupled oxidative prokaryotic systems. The puzzling lack of peroxide detection in the *M. trichosporium* system was rationalized by assuming that 1e⁻ or 2e⁻ reduction of O_2 would lead to superoxide or peroxide, which could cause the observed inactivation of the enzyme and not be detectable in solution.

Pre-steady-state kinetics on the functioning complex (Green and Dalton, 1989b) indicated that at equimolar concentrations of A, B, and C, electron transfer between C and A was about 11% of the rate in the absence of B, thus confirming the steady-state kinetic data obtained previously for *M. capsulatus* (Green and Dalton, 1985).

The system *Methylobacterium* sp. strain CRL-26 has also been purified (Patel and Savas, 1987), but was shown not to contain any protein resembling protein B from the other organisms. Specific activity and Fe content of the hydroxylase were similar to those reported for *M. capsulatus* (Bath).

5. KINETIC MECHANISM OF THE MMO REACTION

Both steady-state and pre-steady-state kinetics have helped establish the overall kinetic pathway of intra- and intermolecular electron transfer coupled to methane oxidation. The enzyme complex was found to follow a concerted-substitution mechanism (Green and Dalton, 1986). The sequence of events during the catalytic cycle closely resembles the P450 cycle in which the substrate (methane in this case) binds first to the enzyme followed by NADH to give the first ternary complex. The reduced enzyme/methane complex then binds O_2 to give a second ternary complex, which ultimately breaks down to release methanol and water (Fig. 4). The enzyme system has a relatively high affinity for methane (K_m = 3 µM), which is a lower value than 160 µM reported previously for cell extracts (Colby *et al.*, 1977). The K_m for O_2 is also low (16.8 µM), but these values are both higher than the values reported by Joergensen (1985) for a methanotroph (strain OU-4-1) which gave K_m values for methane of 1 µM and O_2 of 0.14 µM as measured by mass spectrometry.

Pre-steady-state kinetic analysis has indicated that intra- and intermolecular electron transfer reactions are all higher than the turnover rate of the enzyme (Green and Dalton, 1989b). Kinetic isotope studies using [$^2H_4$] methane gave a value of 11.8, indicating that C–H bond breakage is the rate-limiting step in enzymic methane oxidation. This value is higher

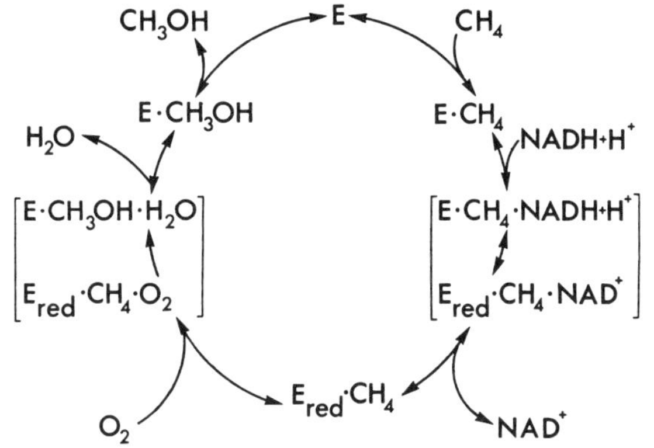

Figure 4. Catalytic cycle of methane monooxygenase showing the presence of two ternary complexes. Dioxygen is bound to the reduced enzyme/methane complex ultimately to release water, methanol, and reoxidized enzyme. (From Green and Dalton, 1986.)

than that reported by Gvozdev *et al.*, (1984) for the particulate enzyme from *M. capsulatus* (strain M) of 5, but comparable to that for *Methylomonas rubrum* reported by Belova *et al.*, (1976) of 12.5. Studies of individual rate constants during the catalytic cycle reveal several interesting features (Fig. 5) (Green and Dalton, 1989b; Green and Dalton, unpublished). The dissociation constant for methane binding to the enzyme (k_2/k_1) is 0.5×10^{-6} M^{-1}, thus confirming the low K_m values determined independently. Second, the second-order rate constant for oxygen binding of 3×10^7 M^{-1}sec^{-1} is close to diffusion limitation and comparable to the binding of oxygen to hemerythin (Wilkins and Wilkins, 1987). The K_{diss} for the enzyme–methanol complex was 4.2×10^{-3} M (Green and Dalton, 1989b), suggesting that product release from the enzyme is not rate limiting and that C–H bond breakage is therefore the most probably rate limiting.

6. SUBSTRATE SPECIFICITY—MECHANISTIC IMPLICATIONS

6.1. Oxygen or Hydrocarbon Activation?

The data presented so far indicate that at the active site of the hydroxylase, a binuclear iron cluster is probably responsible for activation of methane. Exactly how that occurs is the next important question. For example, it is not known whether methane *per se* is bound to the metal and then allowed to react with oxygen, or whether oxygen is activated, which then interacts with methane. Several well-known mechanisms have been

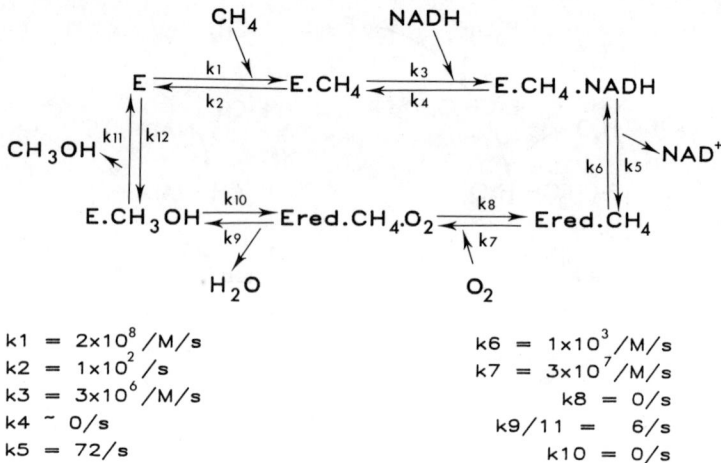

Figure 5. Rate constants of the complete catalytic cycle as measured by stopped flow at 18°C.

studied for many years for hydrocarbon functionalization and each relies on oxygen activation to effect C—H bond activation. The Fenton reagent (essentially ferrous iron and hydrogen peroxide) leads to the production of hydroxyl radicals and is able to oxidize many organic compounds. Clearly, such a reagent would be quite harmful if present in biological systems and as such would not be a good model. However, if the active species is a high-valent metal–oxy compound, then this may be more manageable (see Fee, 1980). In both instances they are powerful oxidants capable of hydrogen abstraction from saturated hydrocarbons. There have been a number of derivatives of this system since it was first discovered in 1893, and all appear to rely on generation of the superoxide anion, which acts as an effective reductant for the high-valent metal species once it has abstracted the hydrogen atom from the substrate.

The other well-studied system is the cytochrome P450 system and the models designed to mimic its ability to activate saturated hydrocarbons. Here again, there is an overwhelming body of evidence that radical chemistry is involved in the activation of hydrocarbons, which is mediated via a high-valent metal–oxo species (see Ortiz de Montellano, 1986). The elegant studies by Ortiz de Montellano and his colleagues on the P450 enzyme and by Groves (McMurray and Groves, 1986) and Mansuy (Mansuy and Battioni, 1989) on models for P450 systems have provided valuable information on the mechanism of action of P450. In particular, they have led to the conclusion that for alkane hydroxylation, a high-valent iron–oxo species, formed by direct interaction between dioxygen and a high-spin ferrous form of the enzyme, abstracts a hydrogen atom from the substrate to yield a carbon radical and an iron-bound hydroxyl radical, which recombine to form the alcohol. Although the exact nature of the high-valent species has not been verified, the circumstantial evidence for its existence is good. This system may provide valuable clues to the nature of the mechanism of action of MMO.

6.2. Substrate Oxidation

In 1977 Colby *et al.*, published an important paper showing that crude preparations of soluble MMO from *M. capsulatus* (Bath) were extremely broad and were able to insert oxygen into alkanes, haloalkanes, alkenes, ethers, alicyclic, aromatic, and heterocyclic compounds. Even methanol, the product of the MMO reaction from methane, was also a substrate for the enzyme. The experiments reported in that paper were performed on each substrate using five controls to ensure that the reaction studied was catalysed by MMO and was not due to the activities of other enzymes in the preparation. Subsequently a flurry of papers appeared in the literature, extending these observations to include a wider variety of substrates, but most of these papers concentrated on whole-organism biotransformations in which it is believed that the MMO enzyme is responsible for the particular oxygenation reaction observed. Unless adequate controls are executed in each reaction, one cannot always be certain that the reaction under study is catalyzed by MMO. Consequently, only those reactions catalyzed by purified MMO preparations or those in which adequate controls have been performed will be considered valid for mechanistic studies.

Oxidation of straight-chain and branched n-alkanes can give useful information on the mechanistic nature of the hydroxylation process. Briefly, there are two possible mechanisms we can consider for introduction of the activated oxygen species into the substrate. The first is a concerted mechanism in which direct insertion of oxygen into a C—H bond gives rise to only one transition state and should proceed with retention of configuration. The second is a nonconcerted mechanism in which an intermediate species is formed following initial abstraction of hydrogen, a proton, or a hydride ion giving two or more transition states with little retention of configuration. If a nonconcerted radical-type mechanism operates, then radical chemistry predicts that saturated hydrocarbons would be attacked with a selectivity of tertiary > secondary > primary.

The oxidation of straight-chain n-alkanes (C_5–C_7) by purified MMO has indicated that oxidations at the secondary position are favored (there are, of course, no tertiary positions available for attack). Substrates that contain all three types of hydrogen atoms, such as 2-methylpropane, gave 70% tertiary and 30% primary products (there are no secondary positions available), whereas 2,3-dimethylpentane gave exclusively attack at the secondary position. Presumably, steric hindrance played a major role in determining the position of attack in the latter case (Green and Dalton, 1989a). Of particular interest was the report that adamantane was attacked with a C_2/C_3 selectivity of 1.0 (in which C_2 represents the total amount of secondary products and C_3 the total amount of tertiary products). If there was no selectivity, the ratio would be 3 (i.e., 12/4), so that a ratio of 1 would indicate selectivity for the tertiary position, which is clearly indicative of radical-type attack. Selectivity for adamantane by the P450 systems is generally around 0.15, but for the iron powder–pyridine Gif III system, this ratio is around 1.15 (Barton *et al.*, 1990), which has led the Texas group to draw close parallels between their alkane-functionalizing system and MMO. Indeed, there are a significant number of similarities between the two systems when substrate specificities are concerned (Fig. 6). Unfortunately, the Gif system is unable to epoxidate alkenes—a feature that is characteristic of many methane monooxygenase systems—nor is it able to activate methane.

A wide variety of substrates were tested by Green and Dalton (1989a) of which the following observations strongly indicated that a nonconcerted mechanism was operational during substrate oxidation: the loss of stereochemistry on oxidation of *cis*-dimethyl cyclohexanes; allylic rearrangements observed in the oxidation of β-pinene and methylene cyclohexane; the dissociation of the cyclopropyl ring during hydroxylation; the dissociation of electron transfer from oxygen transfer in quadricyclane oxidation; and the large kinetic isotope effect observed for methane. Further substantive evidence that a hydrogen-abstracting species could form a radical came from studies by Frey's group in Wisconsin (Ruzicka *et al.*, 1990), who studied the oxygenation of dimethyl cyclopropane by a purified system from *M. trichosporium*. The various products formed [1-methylcyclobutanol, (1-methylcyclopropyl) methanol, and 3-methyl-3-buten-1-ol] could be rationalized by assuming that a radical intermediate is first formed by hydrogen abstraction. This radical could then either rearrange ultimately to give the eneol or undergo a further 1e oxidation to give the carbocation that would ring-expand (to give cyclobutanol) or undergo conventional oxidation to give the cyclopropylmethanol.

Direct evidence for the involvement of carbon-centered radicals has recently been obtained in the *M. capsulatus* system using two different spin traps (Deighton *et al.*, 1991). Methane, methanol and acetonitrile which are all MMO substrates were trapped as their radicals during enzyme turnover and identified by their characteristic hyperfine splitting constants in their electron paramagnetic resonance spectra.

Both dichloroethylene (DCE) and trichloroethylene (TCE) were shown unequivocally to be MMO substrates (Green and Dalton, 1989a). In the case of DCE, rapid enzyme inactivation occurred, suggesting a radical intermediate was formed in a manner similar to that observed in the P450 system. The radical could then cause inactivation through alkylation or lose an electron to form a carbocation and ultimately monochloroacetic acid. TCE (an important environmental pollutant) is also readily oxidized by MMO, presumably via TCE oxide and thence formate, carbon monoxide, and HCl. It also caused enzyme inactivation, but at a slower rate than DCE (Green and Dalton, 1989a). Subsequent studies by Fox *et al.* (1990) investigated the oxidation of haloalkanes in greater detail and observed essentially similar effects in that the data were consistent with an intermediate being formed.

Finally, oxidation of monohalobenzenes has been useful in determining that hydroxylation proceeds *via* a charged intermediate and also lends support to the view that a P450-type mechanism probably operates in the MMO reaction even though the iron present at the active site is in a totally different environment. If we assume, therefore, that MMO oxidizes substrates via a two-step, nonconcerted mechanism in which hydrogen abstraction from the alkane substrate precedes hydroxylation via radical or carbocation intermediates, then it is possible to propose a mechanism that accounts for the data presented above.

6.3. Mechanism of Action of MMO

The prerequisite for hydrogen atom extraction in the mechanism implies the generation of a strong electrophile similar to the oxene species proposed for cytochrome P450 (Hamilton, 1974). Generation of this electrophile would require cleavage of the bound peroxy species (from dioxygen bound to ferrous iron at the active site), which could proceed via heterolytic or homolytic cleavage such that the remaining oxygen atom contains the two oxidation equivalents associated with the peroxide. This hypothetical high-valent electrophilic species needs to be stabilized in some manner since the electron inventory for such a species would indicate that the oxygen has six electrons if the iron was in the ferric state (formally $[Fe^{3+}O]$). In cytochrome P450, the full octet of electrons on oxygen could be achieved by donation of one electron from the iron to give an $Fe^{4+}O^-$ species with seven electrons on the oxygen or two electrons from the iron to give $[Fe^{5+}O^{2-}]$ species, but these complexes are not known to exist formally (Ortiz de Montellano, 1986). Since the exact nature of the

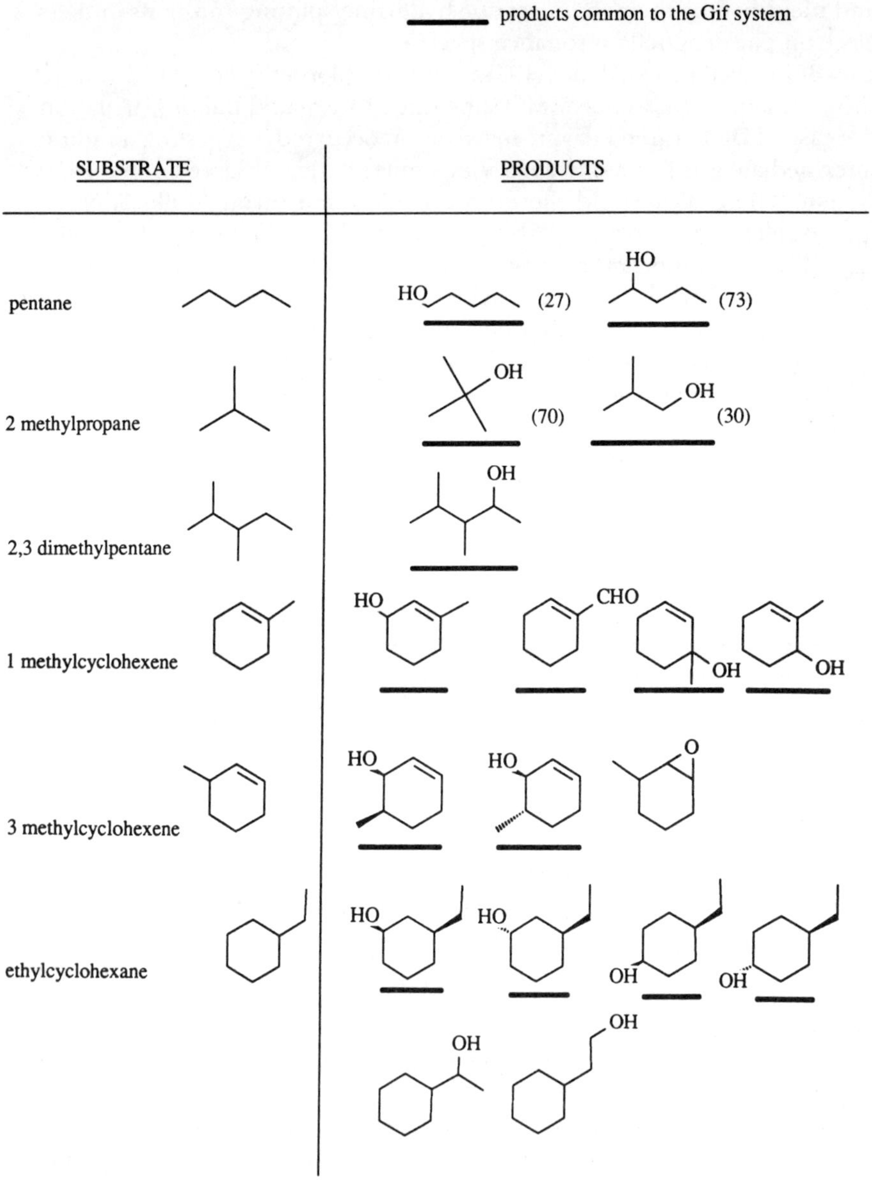

Figure 6. Comparison of substrate specificities between MMO and the Gif systems.

METHANE OXIDATION BY METHANOTROPHS 105

———— products common to the Gif system

SUBSTRATE	PRODUCTS
cyclohexene	cyclohex-2-en-1-ol (underlined); cyclohexene oxide; cyclohexanone
1 methylcyclohexane	2-methylcyclohexanol (OH); cis/trans 3-methylcyclohexanol (underlined); 4-methylcyclohexanol; (1-methylcyclohexyl)methanol; 1-methylcyclohexan-1-ol
cis 1,4 dimethyl cyclohexane	2-hydroxy derivative; 3-hydroxy derivative; (hydroxymethyl) derivative (underlined)
trans 1,4 dimethyl cyclohexane	(hydroxymethyl) derivative (underlined); 2-hydroxy derivative
adamantane	1-adamantanol (50) (underlined); 2-adamantanol (50) (underlined)
β-pinene	(hydroxymethyl) product (72) (underlined); epoxide (28)

active oxidant is unknown, a certain amount of speculation is permissible. Completion of the octet and hence stabilization could come from the porphyrin ring in cytochrome P450 to generate a porphyrin radical cation on an $Fe^{4+}O^-$ complex. In the case of MMO, stabilization could arise from the binuclear iron cluster (Shilov, 1984; Green and Dalton, 1989a) through delocalization of electron density (Fox *et al.*, 1990).

The data accumulated for MMO are summarized in Figure 7. In the native state, the iron species at the active site of the enzyme is present as a pair of high-spin ferric irons (1). Methane is presumed to bind to this oxidized state of the enzyme at a site close to the binuclear center, presumably in some hydrophobic pocket in which displacement of water by methane would provide the entropy for binding (2). To date there is no evidence that methane binds to the iron species directly. A recent revision of the mechanism of the Gif-type system by Barton *et al.* (1990), in which it was presumed that a binuclear μ-oxo bridge iron species is also present in that system, includes the intermediacy of a weak Fe—C bond from the high-valent Fe^v=O species. There is no direct evidence for the existence of the Fe—C bond in that system to date. Kinetic evidence (Green and Dalton, 1986, 1988) indicates that reduction of the protein by NADH follows methane binding. Since electrons are delivered to the hydroxylase one at a time from protein C, then one electron transfer gives rise to the mixed-valence form (3), which is now e.p.r. active, followed by a second electron to give the fully reduced, but e.p.r. silent, form (for the g_{ave} = 1.87 signal), but active for the g = 16 form (Fox *et al.*, 1988; Fox *et al.*, 1989). Dioxygen binding to the ferrous form proceeds with a second-order rate constant of $3 \times 10^7 M^{-1}sec^{-1}$ (Green and Dalton, 1989b), implying that oxygen binds directly to the iron without having to displace even a weakly held ligand in a manner similar to that observed for the μ-oxo-bridged, oxygen-carrying protein hemerythrin (Wilkins and Wilkins, 1987). Presumably one of the iron atoms in the binuclear site is penta-coordinate to allow for the binding of the incoming dioxygen. Stabilization of the peroxide may be achieved through hydrogen bonding between the hydroperoxy group and the μ-oxo bridge (5). Two possible fates of the peroxide can then be envisaged. Homolysis of the peroxo-oxygen bond may occur, catalyzed by associated protons, to produce the hydroxyl radical that would then abstract hydrogen from the methane molecule in a Fenton-like reaction to form a methyl radical and water. The methyl radical would then collapse with the resultant $Fe^{4+}O^-$ species and a proton to form methanol. Alternatively, heterolytic cleavage of the peroxide with a proton would generate water and an oxene species (6) (formally $Fe^{5+}O^{2-}$). The hydrogen-abstracting species would, of course, be highly electrophilic; so to stabilize this, in the absence of an electron-rich heme pocket as found in

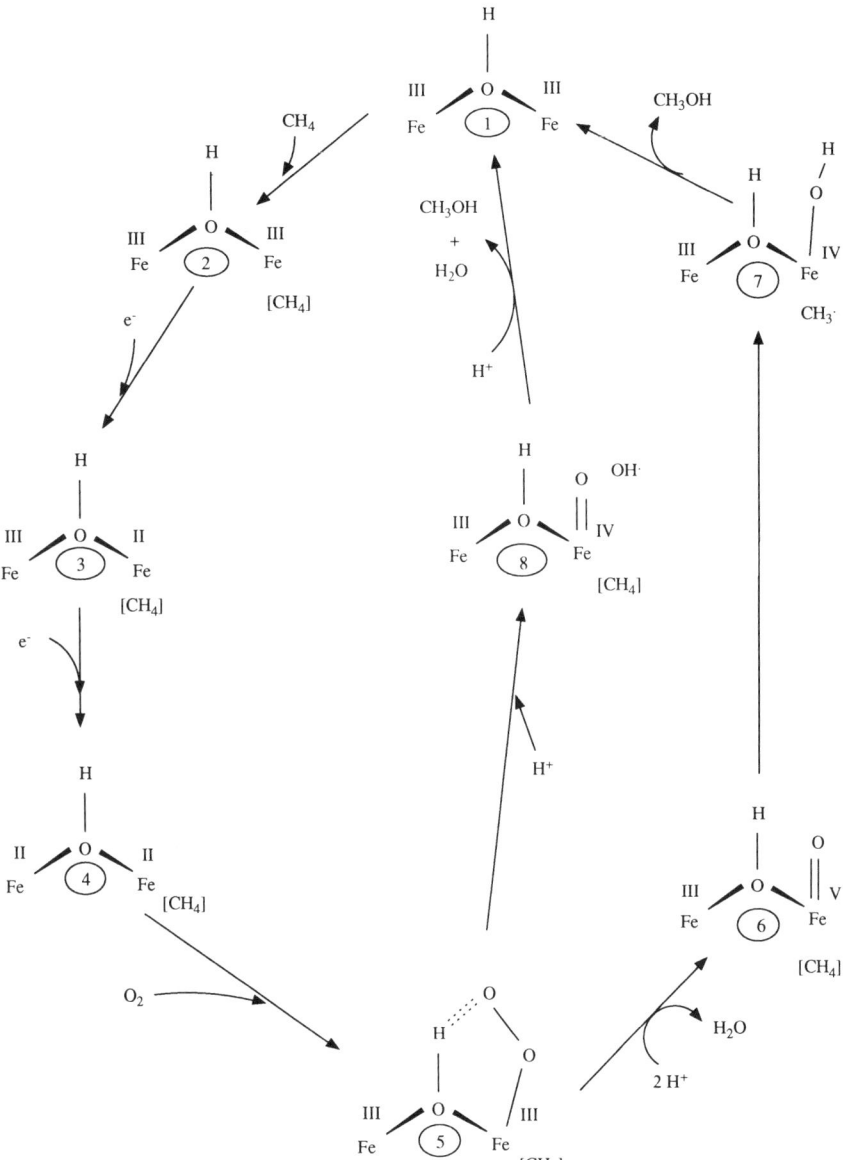

Figure 7. Proposed mechanism for the status of the iron species at the active site of MMO during catalysis.

cytochrome P450, it would be necessary to delocalize electron density from the immediate environment. This, one could presume, is the function of the second iron in the binuclear species. It would be able to channel electrons into the vacant shell (orbital) of the oxygen to transiently complete the octet. Thus a tautomer between Fe^{4+}—OH–Fe^{4+}–O^- and Fe^{3+}–OH–Fe^{5+}—$O^=$ may exist to stabilize the oxene. The oxene species (6) would then abstract hydrogen from methane to form a transient carbon radical in close proximity with an iron-bound hydroxyl radical. Radical recombination via a Groves and McCluskey (1976) oxygen rebound mechanism would produce the alcohol.

Radical-trapping experiments have recently shown that only carbon-centered radicals are formed during catalysis with no evidence for hydroxyl radical formation (Deighton *et al.*, 1991), thus favoring the heterolytic route for peroxide cleavage. Moreover, hydrogen peroxide *per se* has also been shown to directly supply both electrons and oxygen to the hydroxylase protein, thus obviating the need for proteins B and C, NADH, and oxygen (Andersson *et al.*, 1991). This important demonstration of a peroxide shunt reaction brings MMO very close, mechanistically, to the cytochrome P450 system which can also be driven by peroxide (Mansuy and Battioni, 1989).

7. THE PHYSICAL NATURE OF THE ACTIVE SITE

Two important developments in the last few years have brought us closer to defining the exact nature of the active site of MMO.

The first is the cloning of the structural genes for MMO (see Chapter 4), which has now given a complete amino acid sequence for the reductase, hydroxylase, and protein B from *M. capsulatus* (Bath) and *M. trichosporium* (OB3b) (Mullens and Dalton, 1987; Stainthorpe *et al.*, 1989, 1990; Pilkington *et al.*, 1990; Cardy *et al.*, 1991). Structural homology between the hydroxylase proteins from these two organisms is remarkably high, with over 90% sequence identity in the α subunit of the hydroxylase. The second is a recent report (Nordlund *et al.*, 1990) on the x-ray crystal structure of the B2 subunit of *Escherichia coli* ribonucleotide reductase, which has identified those ligands associated with the binuclear iron-oxo center of this protein. The complete amino acid sequence alignment of 11 of the known sequences of this protein from different sources was well known, with only 18 residues out of 375 being conserved between these proteins (B-M. Sjöberg, personal communication), which include residues of the iron ligands, the radical tyrosine, and a few residues at the C terminus essential for binding to the B1 subunit. Of immediate interest are

METHANE OXIDATION BY METHANOTROPHS 109

	glu	his	tyr	glu	his		
	115	118	122	238	241		
──── E	T I H	S R S Y T H	────	R D E A L H	──	*E. coli*	38kDa
──── E	V V H	S R V Y N I	────	R D E A V H	──	HSV2	38kDa
──── E	N V H	G E T Y A N	────	R D E L L H	──	EBV	34kDa
──── E	N I H	S E M Y S L	────	R D E G L H	──	Clam	42kDa
──── E	T I H	S R S Y T H	────	R D E Q L H	──	T4	
──── E	N I H	S E M Y S L	────	R D E G L H	──	Mouse	
──── E	N I H	S E T Y S L	────	R D E G L H	──	Yeast	
──── E	N I H	S E M Y S L	────	R D E G L H	──	Vaccinia	
──── E	V V H	A R V Y S Q	────	R D E A I H	──	Varicella	
──── E	I R H	T H T H Q C	────	T D E L R H	──	OB3b	60.5kDa
──── E	I R H	T H T H Q C	────	T D E L R H	──	M.c.	60.5kDa
	144	147	151	242	246		
	glu	his	his	asp	his		

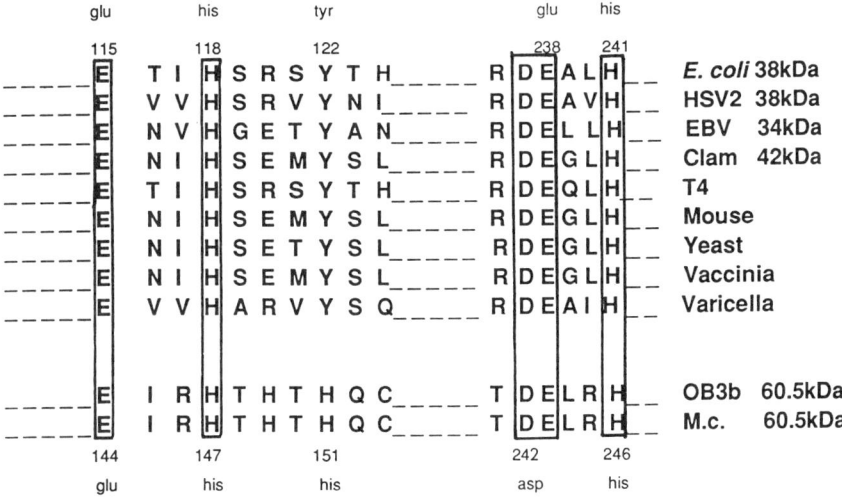

Figure 8. Sequence homology between the Fe-binding sites in the B2 protein of ribonucleotide reductase from nine different sources and the α subunit of protein A of MMO from *M. capsulatus* (Bath) and *M. trichosporium* OB3b. (After Nordlund *et al.*, 1990.)

those ligands involved in binding the binuclear iron center to the protein. The B2 protein exists as a dimer, with one binuclear iron center in each subunit. At 2.2 Å, resolution geometry at the iron center can be reasonably well described. Figure 8 shows the arrangement of ligands of iron in the B2 protein. Of particular interest is the finding that each iron atom has water ligated to it and that apart from this and the oxo bridge, there are only six ligands in total associated with the center. One is a bridging carboxylate (Gln 115), one is bidentate (Asp 84), and the others are His (118, 241) and Gln (204, 238). Inspection of the amino acid sequences from nine different sources of the B2 protein indicates these amino acids are completely conserved in each (residue 84 is also identical in each but not shown here). Alignment of the sequences of the a subunit of the hydroxylase protein A from *M. capsulatus* (Bath) and *M. trichosporium* (OB3b) shows two structural regions where perfect complementarity can be made with the sequences for the iron-binding site of the B2 protein. Earlier studies (Prior and Dalton, 1985) had shown that it is quite likely that the active site in the hydroxylase resides in the a subunit. Treatment of the MMO complex with radioactive acetylene (a suicide substitute for MMO) revealed strong binding of the resultant ketene with ligands closely associated with the a subunit of protein A, indicating that substrate activation by the iron-oxene species is closely associated with that region of the protein. Thus, in the first structural region in the B2 protein, Gln 115 and His 118 could be aligned with Gln 144 and His 147 in MMO a, which is upstream of the second iron-binding region at Glu 238 and His 241. Here again, close sequence homology can be found in the MMO a proteins corresponding to Glu 244 and His 241. These are the only two regions in the MMO a protein that contain Glu and His residues separated by two other residues, and it is therefore tempting to speculate that these residues are involved in ligation to the binuclear iron center. Studies in the author's laboratory are currently underway to determine which, if any, of these histidine residues are involved at the iron coordination site.

ACKNOWLEDGMENTS. I would like to thank all my postdocs and students whose efforts have contributed significantly to our progress in this area. British Petroleum, British Gas, SERC, and Gas Research Institute have all made financial contributions to our work, and their generosity is gratefully acknowledged.

REFERENCES

Akent'eva, N. F., and Gvozdev, R. I., 1988, Purification and physicochemical properties of methane monooxygenase from membrane structures of *Methylococcus capsulatus, Biokhimiya* **53**:91–96.

Andersson, K. K., Froland, W. A., Lee, S.-K., and Lipscomb, J. D., 1991, Dioxygen independent oxygenation of hydrocarbons by methane monooxygenase hydroxylase component, *New. J. Chem.* **15**:411–415.

Barton, D. H. R., Csuhai, E., Doller, D., Ozbalik, N., and Balavoine, G., 1990, Mechanism of the selective functionalization of saturated hydrocarbons by Gif systems: relationship with methane monooxygenase, *Proc. Natl. Acad. Sci. USA* **87**:3401–3404.

Belova, V. S., Gvozdev, R. I., Malashenko, Y. R., Sadkov, A. P., and Yurchenko, V. V., 1976, Isotopic effect in enzymatic oxidation of methane, *Biokhimiya* **41**:1903–1904.

Bentsen, J. G., Lippard, S. J., De Witt, J., Hedman, B., Ericson, A., Hodgson, K. O., Green, J., and Dalton, H., 1989, ESR and x-ray absorption spectroscopy of protein A of soluble methane monooxygenases. Poster presentation at Metals in Biology Gordon Conference, Ventura, CA, Jan. 22–27, 1989.

Best, D. J., and Higgins, I. J., 1981, Methane oxidizing activity and membrane morphology in a methanol-grown obligate methanoltroph *Methylosinus trichosporium* OB3b, *J. Gen. Microbiol.* **125**:73–84.

Cardy, D. L. N., Laidler, V., Salmond, G. P. C., and Murrell, J. C., 1991, Molecular analysis of the methane monooxygenase (MMO) gene cluster of *Methylosinus trichosporium* OB3b, *Mol. Microbiol.* **5**:335–342.

Colby, J., and Dalton, H., 1976, Some properties of a soluble methane monooxygenase from *Methylococcus capsulatus* strain Bath, *Biochem. J.* **157**:495–497.

Colby, J., and Dalton, H., 1979, Characterization of the second prosthetic group of the flavoenzyme NADH-acceptor reductase (component C) of the methane monooxygenase from *Methylococcus capsulatus* (Bath), *Biochem. J.* **177**:903–908.

Colby, J., Dalton, H., and Whittenbury, R., 1975, An improved assay for bacterial methane monooxygenase: some properties of the enzyme from *Methylomonas methanica, Biochem. J.* **151**:459–462.

Colby, J., Stirling, D. I., and Dalton, H., 1977, The soluble methane monooxygenase of *Methylococcus capsulatus* (Bath): its ability to oxygenate n-alkanes, n-alkenes, ether, and alicyclic, aromatic and heterocyclic compounds, *Biochem. J.* **165**:395–402.

Davies, S. L., and Whittenbury, R., 1970, Fine structure of methane- and other hydrocarbon-utilizing bacteria, *J. Gen. Microbiol.* **61**:227–232.

De Boer, W. E., and Hazeu, W., 1972, Observations on the fine structure of a methane-oxidizing bacterium, *Antonie van Leeuw. J. Microbiol. Serol.* **38**:33–47.

Deighton, N., Podmore, I. D., Symons, M. C. R., Wilkins, P. C., and Dalton, H., 1991, Substrate radical intermediates are involved in the soluble methane monooxygenase catalyzed oxidations of methane, methanol, and acetonitrile, *J. Chem. Soc. Chem. Commun.* 1086–1088.

Ericson, A., Hedman, B., Hodgson, K. O., Green, J., Dalton, H., Bentsen, J. G., Beer, R. H., and Lippard, S. J., 1988, Structural characterization by EXAFS spectroscopy of the binuclear iron centre in component A of methane monooxygenase from *Methylococcus capsulatus* (Bath), *J. Am. Chem. Soc.* **110**:2330–2332.

Fee, J. A., 1980, Superoxide, superoxide dismutases and oxygen toxicity, in: *Metal Ion Activation of Dioxygen* (T. G. Spiro, ed.) J. Wiley, New York, pp. 209–237.

Ferenci, T., 1974, Carbon monoxide-stimulated respiration in methane-utilizing bacteria, *FEBS Lett.* **41**:94–98.

Fox, B. G., and Lipscomb, J. D., 1988, Purification of a high specific activity methane monooxygenase hydroxylase component from a type II methanotroph, *Biochem. Biophys. Res. Commun.* **154**:165–170.

Fox, B. G., Surerus, K. K., Munck, E., and Lipscomb, J. D., 1988, Evidence for a μ-oxo-bridged binuclear iron cluster in the hydroxylase component of methane monooxygenase, *J. Biol. Chem.* **263**:10553–10556.

Fox, B. G., Froland, W. A, Dege, J. E., and Lipscomb, J. D., 1989, Methane monooxygenase from *Methylosinus trichosporium* OB3b, *J. Biol. Chem.* **264**:10023–10033.

Fox, B. G., Borneman, J. G., Wackett, L. P., and Lipscomb, J. D., 1990, Haloalkene oxidation by the soluble methane monooxygenase from *M.trichosporium* OB3b: mechanistic and environmental implications, *Biochemistry* **29**:6419–6427.

Gesser, H., Hunter, N. R., and Drakash, C. B., 1985, The direct conversion of methane to methanol by controlled oxidation, *Chem. Rev.* **85**:235–244.

Green, J., and Dalton, H., 1985, Protein B of soluble methane monooxygenase from *Methylococcus capsulatus* (Bath), *J. Biol. Chem.* **260**:15795–15801.

Green, J., and Dalton, H., 1986, Steady-state kinetic analysis of soluble methane monooxygenase from *Methylococcus capsulatus* (Bath), *Biochem. J.* **236**:155–162.

Green, J., and Dalton, H., 1988, The biosynthesis and assembly of protein A of soluble methane monooxygenase of *Methylococcus capsulatus* (Bath), *J. Biol. Chem.* **263**:17561–17565.

Green, J., and Dalton, H., 1989a, Substrate specificity of soluble methane monooxygenase: mechanistic implications, *J. Biol. Chem.* **264**:17698–17703.

Green, J., and Dalton, H., 1989b, A stopped-flow kinetic study of soluble methane monooxygenase from *Methylococcus capsulatus* (Bath), *Biochem. J.*, **259**:167–172.

Groves, J. T., and McCluskey, G. A., 1976, Aliphatic hydroxylation *via* oxygen rebound: oxygen transfer catalysed by iron, *J. Am. Chem. Soc.* **98**:859–861.

Gvozdev, R. I., Shushenacheva, E. V., Pylyashenko-Novochatnii, A. I., and Belova, V. S., 1984, Investigation of enzymatic oxidation by methane monooxygenase of *Methylococcus capsulatus* strain M, membrane structure, *Oxidation Commun.* **7**:249–266.

Hamilton, G. A., 1974, Chemical models and mechanisms for oxygenases, in: *Molecular Mechanisms of Oxygen Activation* (O. Hayaishi, ed.), Academic Press, New York, pp. 405–448.

Hendrich, M. P., Munck, E., Fox, B. G., and Lipscomb, J. D., 1990, Integer-spin epr studies of the fully reduced methane monooxygenase hydroxylase component, *J. Am. Chem. Soc.* **112**:5861–5865.

Hill, C. L., 1989, Catalytic oxygenation of unactivated C–H bonds: superior oxo transfer catalysts and the "organic metalloporphyrin," in: *Activation and Functionalization of Alkanes* (C. L. Hill, ed.), Wiley, New York, pp. 243–279.

Hubley, J. H., Thomson, A. W., and Wilkinson, J. F., 1975, Specific inhibitors of methane oxidation in *Methylosinus trichosporium*, *Arch. Microbiol.* **102**:199–202.

Joergensen, L., 1985, The reaction of methane monooxygenase studied by membrane-inlet mass spectrometry in whole cells of methanotrophic bacteria, in: *Gas Enzymology* (H. Degn, R. P. Cox, and H. Toftlund, eds.), D. Reidel, Dordrecht, pp. 187–200.

Linton, J. D., and Vokes, J., 1978, Growth of the methane-utilizing bacterium *Methylococcus* NC1B 11083 in mineral salts medium with methanol as the sole source of carbon, *FEMS Lett.* **4**:125–128.

Lund, J., and Dalton, H., 1985, Further characterization of the FAD and Fe_2S_2 redox centres of component C, the NADH: acceptor reductase of the soluble methane monooxygenase of *Methylococcus capsulatus* (Bath), *Eur. J. Biochem.* **147:**291–296.

Lund, J., Woodland, M. P., and Dalton, H., 1985, Electron transfer reactions in the soluble methane monooxygenase of *Methylococcus capsulatus* (Bath), *Eur. J. Biochem.* **147:**297–305.

Mansuy, D., and Battioni, P., 1989, Alkane functionalization by cytochromes P-450 and by model systems using O_2 or H_2O_2, in: *Activation and Functionalization of Alkanes* (C. L. Hill, ed.), John Wiley and Sons, New York, pp. 195–218.

McMurray, T. J., and Groves, J. T., 1986, Metalloporphyrin models for cytochrome P-450, in: *Cytochrome P450, Structure, Mechanism, and Biochemistry* (P. R. Ortiz de Montellano, ed.), Plenum Press, New York, pp. 1–28.

Mullens, I. A., and Dalton, H., 1987, Cloning of the gamma-subunit methane monooxygenase from *Methylococcus capsulatus*, *Bio/Technology* **5:**490–493.

Nordlund, P., Sjöberg, B-M., and Eklund, H., 1990, The three-dimensional structure of the free radical protein of ribonucleotide reductase, *Nature* **345:**593–598.

Ortiz de Montellano, P. R., 1986, Oxygen activation and transfer, in: *Cytochrome P450, Structure, Mechanism and Biochemistry* (P. R. Ortiz de Montellano, ed.), Plenum Press, New York, pp. 217–271.

Patel, R. N., and Savas, J. C., 1987, Purification and properties of the hydroxylase component of methane monooxygenase, *J. Bacteriol.* **169:**2313–2317.

Pilkington, S. J., Salmond, G. P. C., Murrell, J. C., and Dalton, H., 1990, Identification of the gene encoding the regulatory protein B of soluble methane monooxygenase, *FEMS Lett.* **72:**345–348.

Pilkington, S. J., and Dalton, H., 1991, Purification and characterization of the soluble methane monooxygenase from *Methylosinus sporium* 5 demonstrates the highly conserved nature of this enzyme in methanotrophs, *FEMS Lett.* **78:**103–108.

Prince, R. C., and Patel, R. N., 1986, Redox properties of the flavoprotein of methane monooxygenase, *FEMS Lett.* **203:**127–130.

Prince, R. C., George, G. N., Savas, J. C., Cramer, S. P., and Patel, R. N., 1988, Spectroscopic properties of the hydroxylase of methane monooxygenase, *Biochem. Biophys. Acta* **952:**220–229.

Prior, S. D., and Dalton, H., 1985, The effect of copper ions on membrane content and methane monooxygenase activity in methanol-grown cells of *Methylococcus capsulatus* (Bath), *J. Gen. Microbiol.* **131:**155–163.

Proctor, H. M., Norris, J. R., and Ribbons, D. W., 1969, Fine structure of methane-utilizing bacteria, *J. Appl. Bacteriol.* **32:**118–121.

Ribbons, D. W., and Michalover, J. L., 1970, Methane oxidation by cell-free extracts of *Methylococcus capsulatus*, *FEBS Lett.* **11:**41–44.

Ruzicka, F., Huang, D-S., Donnelly, M. I., and Frey, P. A., 1990, Methane monooxygenase catalysed oxygenation of 1,1-dimethylcyclopropane. Evidence for radical and carbocationic intermediates, *Biochemistry* **29:**1696–1700.

Scott, D., Brannan, J., and Higgins, I. J., 1981, The effect of growth conditions on intracytoplasmic membranes and methane monooxygenase activities in *Methylosinus trichosporium* OB3b, *J. Gen. Microbiol.* **125:**63–72.

Shilov, A. E., 1984, *Activation of Saturated Hydrocarbons by Transition Metal Complexes*, D. Reidel, Dordrecht.

Smith, D. D. S., and Dalton, H., 1989, Solubilization of methane monooxygenase from *Methylococcus capsulatus* (Bath), *Eur. J. Biochem.* **182:**667–671.

Smith, U., Ribbons, D. W., and Smith, D. S., 1970, The fine structure of *Methylococcus capsulatus, Tissue Cell* **2:**512–520.

Stainthorpe, A. C., Murrell, J. C., Salmond, G. P. C., Dalton, H., and Lee, V., 1989, Molecular analysis of methane monooxygenase from *Methylococcus capsulatus* (Bath), *Arch. Microbiol.* **152:**154–159.

Stainthorpe, A. C., Lees, V., Salmond, G. P. C., Murrell, J. C., and Dalton, H., 1990, The methane monooxygenase gene cluster of *Methylococcus capsulatus* (Bath) *Gene* **91:**27–34.

Stanley, S. H., Prior, S. D., Leak, D. J., and Dalton, H., 1983, Copper stress underlies the fundamental change in intracellular location of methane monooxygenase in methane-oxidizing organisms: studies in batch and continuous cultures, *Biotechnol. Lett.* **5:**487–492.

Stirling, D. I., and Dalton, H., 1979, Properties of the methane monooxygenase from extracts of the *Methylosinus trichosporium* OB3b and evidence for its similarity to the enzyme from *Methylococcus capsulatus* (Bath), *Eur. J. Biochem.* **96:**205–212.

Takeda, K., and Tanaka, K., 1980, Ultrastructure of intracytoplasmic membranes of *Methanomonas margaritae* cells grown under different conditions, *Antonie van Leeuw. J. Microbiol. Serol.* **46:**15–25.

Takeda, K., Tezuka, C., Fukuoka, S., and Takahara, Y., 1976, Role of copper ions in methane oxidation by *Methanomonas margaritae, J. Ferm. Technol.* **54:**557–562.

Tonge, G. M., Harrison, D. E. F., Knowles, C. J., and Higgins, I. J., 1975, Properties and partial purification of the methane-oxidizing enzyme system from *Methylosinus trichosporium, FEBS Lett.* **58:**293–299.

Tonge, G. M., Harrison, D. E. F., and Higgins, I. J., 1977, Purification and properties of the methane monooxygenase enzyme system from *Methylosinus trichosporium* OB3b, *Biochem. J.* **161:**333–344.

Wilkins, P. C., and Wilkins, R. G., 1987, The coordination chemistry of the binuclear iron site in hemerythrin, *Coord. Chem. Rev.* **79:**195–214.

Woodland, M. P., and Dalton, H., 1984, Purification and characterization of component A of the methane monooxygenase from *Methylococcus capsulatus* (Bath), *J. Biol. Chem.* **259:**53–60.

Woodland, M. P., Patil, D. S., Cammack, R., and Dalton, H., 1986, ESR studies of protein A of the soluble methane monooxygenase from *Methylococcus capsulatus* (Bath), *Biochim. Biophys. Acta* **873:**237–242.

The Genetics and Molecular Biology of Obligate Methane-Oxidizing Bacteria

4

J. COLIN MURRELL

1. INTRODUCTION

A considerable amount of information on the physiology and biochemistry of obligate methane-oxidizing bacteria (methanotrophs) has emerged over the last 20 or so years (Anthony, 1986; Chapter 3, this volume), particularly regarding the metabolism of methane into cell carbon and also nitrogen assimilation. In contrast, the genetics of these organisms is far less well developed, which is in part due to the relatively slow growth of methanotrophs on solid media. To carry out genetic studies on microorganisms, it is necessary to develop the capabilities for mutant isolation and gene transfer. Traditionally, methanotrophs have proved refractory to most forms of conventional genetic analysis. However, with the advent of gene-cloning technology and the development of numerous broad-host-range cloning vectors, a number of significant advances are being made in the molecular genetics of methanotrophs. In this chapter, I will review the current literature on the mutagenesis, plasmid, bacteriophage, and gene transfer studies carried out on methanotrophs and then concentrate on the more recent developments in molecular genetics related to C_1-specific genes, such as those encoding for methanol dehydrogenase and methane monooxygenase, and also the molecular biology of nitrogen assimilation genes, which has served as a model system in which to develop genetic techniques for this unique group of bacteria.

J. COLIN MURRELL • Department of Biological Sciences, University of Warwick, Coventry CV4 7AL, England.

Methane and Methanol Utilizers, edited by J. Colin Murrell and Howard Dalton. Plenum Press, New York, 1992.

2. MUTAGENESIS

It has generally been thought that genetic studies of methanotrophs have been hampered by the slow growth of these organisms on solid media. However, most of the original isolates of Whittenbury and his co-workers (Whittenbury *et al.*, 1970) will exhibit good turbidity in batch liquid cultures after 2–3 days and form good single isolated colonies on nitrate mineral salts agar plates in 4–5 days when regularly subcultured (Whittenbury and Dalton, 1981). Probably the major limitation to mutagenesis studies is their limited substrate capabilities, i.e., growing only on methane as sole source of carbon and energy. However, many methanotrophs are capable of good growth on methanol, although for certain strains it may be necessary for a period of adaptation to methanol, usually carried out by gradually adapting from an initial methane/air grown culture to one containing an increasing concentration of methanol. This has been done successfully with *Methylocystis parvus* OBBP (Hou *et al.*, 1978) and *Methylosinus trichosporium* OB3b (Best and Higgins, 1981) such that the organisms will readily grow on 1% v/v high-purity methanol. The molecular basis for this "adaptation" to growth on methanol is unclear; however, it appears that one of the most convenient ways of quickly achieving methanol tolerance in a methanotroph is to gradually increase the concentration of methanol flowing into a methane-limited continuous culture of a methanotroph and regularly plating onto methanol-containing agar plates to check for methanol tolerance and growth (Murrell, unpublished).

2.1. Conventional Mutagenesis

One of the first mutagenesis studies on methanotrophs was by Harwood *et al.* (1972), who investigated the frequency of spontaneous mutation in *Methylococcus capsulatus* (Foster and Davis strain). Mutants resistant to antibiotics, amino acid analogs, and other compounds were obtained at frequencies similar to those found in other bacteria (e.g., ≤1 in 1×10^8). However, attempts to increase the frequency of spontaneous mutation using mutagens such as nitrosoguanidine, ethyl methane-sulfonate, and ultraviolet light were unsuccessful. Subsequent studies by these workers showed that it was very difficult to mutate both *Methylococcus* and *Methylomonas* strains (Williams *et al.*, 1977). Apart from spontaneous drug-resistant mutants, the only mutant they isolated was one requiring *p*-amino benzoic acid for growth, and even this was a leaky mutant. One other report worth noting was that of Williams and Shimmin (1978), who showed that it was possible to induce filamentation in methanotrophs using ultraviolet light and that the sensitivity of *Methylosinus* sp. and *Me-*

thylobacter sp. to DNA-damaging agents is increased when in the filamentation state. Related studies suggested that methanotrophs may lack the error-prone SOS repair processes characterized in *Escherichia coli*, although there is no direct evidence for this and no detailed studies following up these observations have been forthcoming (Williams and Bainbridge, 1976).

A suggestion to account for the difficulties in isolating amino acid auxotrophs of methanotrophs has been the sensitivity of certain methanotrophs to amino acids (Bainbridge, 1983; Lidstrom *et al.*, 1984). In a detailed study on the assimilation, toxicity, and uptake of exogenous amino acids by *M. capsulatus* (Foster and Davis strain), Eccleston and Kelly (1972a,b) showed that of 21 amino acids tested only threonine, phenylalanine, histidine, tyrosine, and homoserine inhibited exponential growth of *M. capsulatus* when supplied exogenously at 1.0 mM. Their incorporation experiments using radiolabeled amino acids demonstrated a broad-specificity, common-amino-acid transport system in this microorganism. It has also been shown that a number of amino acids will serve as nitrogen sources for methane oxidizers (Murrell, 1981) and therefore the failure to isolate amino acid auxotrophs may not be a problem of either toxicity or transport of amino acids into the cell.

2.2. Dichloromethane Mutagenesis

Despite the difficulties outlined in the previous section, it has proved possible to isolate methanotrophs that are defective in growth on methane. This has been achieved for *Methylomonas albus* BG8 (McPheat *et al.*, 1987a) and for *Methylosinus trichosporium* OB3b (Nicolaidis and Sargent, 1987). The selection of methane monooxygenase (MMO) deficient mutants relies on the fact that MMOs have a broader substrate specificity, with the soluble form of the enzyme being more catholic in its properties than the particulate (membrane-bound) form of the enzyme (see Chapter 3). Included in the long list of substrates for this enzyme is dichloromethane, which is cooxidized by MMO to carbon monoxide, a potentially toxic product. Thus growing methanotrophs in the presence of dichloromethane creates potentially lethal conditions for all "wild type" cells. Methanol-adapted *M. albus* and *M. trichosporium* were incubated on methanol-containing plates in an atmosphere of dichloromethane in Tupperware boxes. This procedure gave rise to dichloromethane (DCM)-resistant colonies at frequencies in the order of 10^{-5} to 10^{-4}. When analyzed, at least some of these DCM-resistant mutants were unable to grow on methane. Reversion rates (to growth on methane) were approximately 1×10^{-8} (for *M. albus*), indicating the stable nature of these mutants.

The failure of DCM-resistant mutants of *M. albus* to oxidize propylene oxide in a standard whole-cell MMO assay (Leak and Dalton, 1983) indicated that this procedure was successful for the isolation of MMO mutants. As *M. albus* only contains a particulate MMO and DCM-resistant mutants grew on methanol, it was assumed that mutations lay within the structural genes for subunits of particulate MMO (McPheat *et al.*, 1987a). The situation is more complex with the DCM-resistant, MMO⁻ mutants of *M. trichosporium* OB3b because the organism has the ability to express either a soluble or particulate MMO depending on the copper-to-biomass ratio "perceived" by the organism (Burrows *et al.*, 1984). Since the nitrate mineral salts growth medium used by Nicolaidis and Sargent (1987) contained 0.2 µg/liter, it is likely that their *M. trichosporium* would be expressing only particulate MMO and not soluble MMO, so perhaps mutants still contained a functional soluble MMO.

It is clear that resistance to DCM is a powerful tool for the isolation of MMO-deficient mutants. When analyzing mutants of methanotrophs known to possess two forms of MMO, attention must be given to the growth conditions and expression of soluble or particulate MMO. A quick, reliable plate assay for determining whether methanotroph colonies are expressing the soluble form of the enzyme may be one that relies on the conversion of naphthalene to 1-naphthol and 2-naphthol by soluble MMO but not particulate MMO. The naphthols thus formed may then be reacted with tetrazotized *O*-dianisidine to form purple diazo dyes with large molar absorptivities (Brusseau *et al.*, 1990). The possibility of adapting this assay to differentiate between soluble and particulate MMO⁻ mutants of *M. trichosporium* OB3b on nitrate mineral salts agar plates is currently being investigated (Murrell *et al.*, unpublished).

3. GENE TRANSFER SYSTEMS

Another major hurdle to the successful development of genetic systems for the obligate methanotrophs has been the lack of suitable means of introducing either foreign or homologous DNA into these bacteria. The quest for such a technique has generated some interest in plasmid and bacteriophage biology of methanotrophs, and these will be reviewed in the following sections. The techniques that have been developed, together with recent advances in methanotroph genetics, are considered.

3.1. Plasmids in Methanotrophs

One of the first reports on the examination of methane oxidizers for the presence of plasmids was by Warner *et al.* (1977), who used a lysozyme/detergent treatment to lyse the bacteria, a mechanical shearing step, and then a cesium chloride ethidium bromide centrifugation step to separate any plasmids away from chromosomal DNA. Although successful in identifying plasmids in the methanotroph *Pseudomonas* AM1 (now *M. extorquens* AM1), they could not detect plasmids in *M. trichosporium*, *M. capsulatus*, or *Methylomonas methanica*. A later study by Lidstrom and Wopat (1984) showed that of 10 strains of methanotroph examined, only *M. capsulatus* (Bath) did not contain plasmid DNA. *M. trichosporium* strains, for example, contained at least three plasmids of approximate sizes 186, 159, and 154 kb. Representatives of *Methylomonas, Methylobacter,* and *Methylocystis* also contained large plasmids. (See Table I.) Some cross-hybridization with plasmid DNA from *Methylosinus sporium* 5 and *M. trichosporium* plasmid DNA was observed although, despite examination of antibiotic and heavy metal resistance patterns among all strains, the function of these plasmids remained cryptic. The authors also pointed out the difficulties in detecting these plasmids in methanotrophs, and a number of different techniques were employed during the study.

Although a rudimentary restriction map of the 55-kb plasmid from *M. albus* BG8 is available (Lidstrom and Wopat, 1984), difficulties in isolating these plasmids, their large size, and the absence of a selectable phenotype may hamper their immediate use as cloning vectors for these organisms. However, their presence in methanotrophs is an important factor in studies on the genetics of these organisms because they represent significant

TABLE I. Plasmids Detected in Obligate Methanotrophs

Strain	Approximate size (kb)		
Type I			
Methylomonas albus BG8	55		
Methylobacter capsulatus Y	94		
Methylomonas methanica S1	186		
Type X			
Methylococcus capsulatus (Bath)	N.D.[a]		
Type II			
Methylosinus trichosporium OB3b	186,	159,	145
OB3bH	186,	159,	145
OB5b	186,	159,	145
Methylosinus sporium 5	170,	108	
Methylocystis parvus OBBP	186,	159	
Methylocystis POC	176,	152,	75

[a]N. D., not detected.

potential coding regions, their replicon may be of use in development of cloning vectors, and their incompatibility with "foreign" plasmids may prove problematical.

The above study by Lidstrom and Wopat was prompted by their discovery that the apparent facultative methanotroph *Methylobacterium ethanolicum* (Lynch, 1980) contained plasmids. Upon closer examination, this culture was found to consist of a syntrophic relationship between an obligate methane oxidizer (containing the plasmids) and a facultative methanol-oxidizer (Lidstrom-O'Connor *et al.*, 1983). At that time, it was postulated that methane oxidation was a plasmid-borne trait (Hanson, 1980; O'Connor, 1981; Haber *et al.*, 1983), although with the findings of Lidstrom and colleagues (see above) and the recent cloning of soluble methane monooxygenase genes (see Section 4.2.2.), it is now generally thought that the absolutely essential function of methane oxidation is not plasmid-encoded in methanotrophs. This view is further strengthened by the observation of McPheat *et al.* (1987a) that the loss of methane oxidation in the dichloromethane-resistant mutants of *M. albus* BG8 was not accompanied by the loss of the indigenous 55 kb plasmid.

3.2. Bacteriophages of Methanotrophs

There have been a number of reports on the isolation of bacteriophages of methanotrophic bacteria (Tyutikov *et al.*, 1976, 1980, 1983; Wünsche *et al.*, 1977; Tikhonenko *et al.*, 1982). Apparent from these reports is the ease with which it is possible to isolate such bacteriophages from an extremely wide range of environments, including oil and gas installation waters, groundwater, fermenters, soils, rumen of cattle, and fish. Particularly prevalent were bacteriophages specific for *Methylosinus* and *Methylocystis* species. A variety of characteristics of such bacteriophages, including plaque morphology, ultrastructure, lytic spectrum (usually each bacteriophage had a very limited host range), antigenic properties, nucleic acid content, and restriction endonuclease analysis, were reported by the above workers.

From these studies it is clear that temperate bacteriophages of methanotrophs are widespread in nature but little attempt has been made to use them for transduction. Their apparent ubiquitous nature should at least be noted with respect to potential problems in industrial fermentations with C_1 utilizers.

3.3. Conjugation Systems

One of the first reports of conjugal transfer of plasmids to a methanotroph was by Warner *et al.*, (1980), who demonstrated the successful filter matings of *Pseudomonas aeruginosa* PAO8 containing R68.45 with *M. trichosporium* OB3b. Kanamycin resistance was used as the selective marker and the frequency of transfer to *M. trichosporium* bacteria was given as 10^{-2} to 10^{-3} (per donor). It was interesting to note that plate matings between *M. trichosporium* and *P. aeruginosa* were unsuccessful; also, Warner and colleagues failed to transfer kanamycin or carbenicillin resistance from (R68.45) to *Methylococcus* sp. NC1B 11083 by either plate mating or patch mating. They attributed their lack of success to the possibilities of inadequate uptake, inadequate expression and replication, or restriction of foreign DNA in this organism and were unable to demonstrate chromosome mobilization in the methanotrophs.

Around this time, considerable success was being achieved in the laboratories of Holloway, Goodwin and Hanson, and others in the conjugal transfer of plasmids into facultative methanol utilizers such as *Methylobacterium organophilum*, *Methylobacterium extorquens* AM1, and the obligate methanol utilizer *Methylophilus methylotrophus* AS1. Several studies reported mobilization of chromosomal markers between different strains using broad-host-range plasmids (Haber *et al.*, 1983; Tatra and Goodwin, 1984; Holloway, 1981; Holloway *et al.*, 1987; also reviewed in Chapter 6). These studies showed that broad-host-range conjugative and mobilizable plasmid derivatives could be used as cloning vectors in methylotrophs (Chapter 6, this volume).

Lidstrom and colleagues developed a conjugative system for gene transfer in three different methanotrophs (Lidstrom *et al.*, 1984) using the Inc P1 cosmid cloning vector pVK100 (Knauf and Nester, 1982), which was mobilized by the hybrid conjugative plasmid pRK2013 (Figurski and Helinski, 1979). Filter matings allowed transfer of frequencies of 10^{-2} for *M. albus* and *Methylocystis* POC, sufficiently high to allow for direct complementation of mutants. Lower transfer frequencies (10^{-8}) were achieved for *Methylosinus* 6, sufficient only for transferring specific hybrid plasmids into cells (Lidstrom *et al.*, 1984). However, this conjugation system was still of use in directly isolating nitrogen fixation mutants of *Methylosinus* 6 by marker exchange mutagenesis (see Section 4.1.).

Inc W and Inc Q plasmids have also been transferred to the obligate methanotroph *M. albus* BG8 at low frequencies (McPheat *et al.*, 1987b). Using filter matings, the Inc P plasmids R68.45, R751, and derivatives carrying bacteriophage Mu and/or transposons and pS-a (Inc W plasmids) were transferred to *M. albus* from *E. coli* at frequencies of 10^{-7} to 10^{-8}. Higher frequencies of transfer were noted for RP4 and R300B (10^{-3}). The transfer frequencies for Mu-containing plasmids indicated that the Mu-associated suicide phenomenon did not occur in *M. albus*. McPheat *et al.* also investigated the effect of heat shock on the transfer of RP4 and R300B to *M. albus*. However, the relatively low increase (five- to sixfold) in transfer frequency after heat shock suggested that a host restriction system is not the limiting factor in the efficient transfer of plasmids to this methanotroph.

In a similar study, Al-Taho and Warner (1987) successfully transferred plasmid pULB113, a derivative of RP4 into which a deleted fragment of bacteriophage Mu has been inserted, from *E. coli* to *M. trichosporium* OB3b again using filter matings. Transfer frequencies were in the order of 1×10^{-5} per recipient. In the same study, these workers showed evidence for mobilization of chromosomal markers from the methanol utilizer *M. extorquens* leading to restoration of phenotype in *E. coli* auxotrophs; however, these experiments were unsuccessful when repeated with *M. trichosporium* OB3b harboring pULB113. Nevertheless, the potential at least is available for gene mapping and cloning studies on methanotrophs.

With the increase in availability of broad-host-range cloning vectors, it is likely that the successes outlined above will be further improved and developed. Conjugal transfer of plasmids to methanotrophs should be a good enough gene transfer system for most gene manipulation studies, and we await a detailed study that considers factors such as ratio of organism, contact time, filter type, recovery time, nutrient supplementation (e.g., proteose peptone or yeast extract in trace amounts), and choice of antibiotic for plasmid selection. Undoubtedly, the most significant advances in such gene transfer systems will be in the few methanotrophs that are well studied biochemically, e.g., *M. trichosporium* and *M. capsulatus*. Table II lists a number of broad-host-range plasmids that have been used successfully with methanotrophs.

TABLE II. Plasmids Transferred to Methanotrophs by Conjugation

Vector	Markers	Methanotroph	Transfer frequency	Reference
R68.45	Km, Tc, Ap	*M. trichosporium*	10^{-2} to 10^{-3}	Warner *et al.* (1980)
RP4	Ap, Km, Tc	*M. albus*	10^{-3}	McPheat *et al.* (1987b)
R300b	Su, Sm	*M. albus*	10^{-3}	McPheat *et al.* (1987b)
pJB4JI	Gm	*M. albus*	10^{-8}	McPheat *et al.* (1987b)
p-Sa	Km, Su, Cm, Sp	*M. albus*	10^{-7}	McPheat *et al.* (1987b)
pVK100	Tc, Km	*M. albus*, *Methylocystis*	10^{-2}	Lidstrom *et al.* (1984)
pVK100	Tc, Km	*Methylosinus* 6	10^{-8}	Lidstrom *et al.* (1984)
pVK100	Tc, Km	*M. trichosporium*	10^{-3}	Murrel *et al.*, unpublished
pULB113	Km	*M. trichosporium*	10^{-5}	Al-Taho and Warner, (1987)
pGSS33	Ap, Km, Sm, Tc	*M. trichosporium*	10^{-5}	Murrell *et al.*, unpublished

3.4. Transformation Studies and Electroporation

A gene transformation system was described for *M. capsulatus* by Williams and Bainbridge (1971); however, very high concentrations of DNA and long contact times made this a very inefficient transformation system. The problem with the lack of suitable mutants for transformation was alluded to in this study. Since then, few, if any, transformation systems have been described for the obligate methanotrophs, presumably because of the lack of suitable plasmids or genetic markers. Probably the most significant advance in the development of gene transfer systems for bacteria that are refractory to most conventional transformation systems has been the use of electroporation (Wirth *et al.*, 1989; Fiedler and Wirth, 1988; reviewed by Chassey *et al.*, 1988). This method of transformation uses electric current for the generation of membrane distortions that allow the uptake of DNA and has the advantage of being quick and in many cases (e.g., *E. coli*) highly efficient. Suitable apparatus includes the Gene Pulser electroporator from Bio-Rad Laboratories (Richmond, CA). This is a capacitor discharge device producing exponential declining pulses with field strengths of 125–6250 V/cm. Duration of the pulse is controlled by the choice of capacitor used to deliver the pulse. Pulses are delivered to cells in sterile, disposable plastic cuvettes, with electrodes being either 0.2 or 0.4 cm apart. Capacitors of 0.25, 1.0, 3.0, and 25 µF can be selected. Voltage can also be varied between 50 and 2500 V.

A wide variety of gram-negative bacteria have been successfully transformed by electroporation (Wirth *et al.*, 1989), and preliminary results suggest that with the correct choice of broad-host-range plasmid, the method will also work with methanotrophs. *M. capsulatus* (Bath), *M. trichosporium*, and *M. albus* have all been transformed (albeit at low frequencies of 10^{-2} to 10^{-3} transformants per µg DNA) with the broad-host-range cosmid PVK100. In these experiments, 0.2-cm cuvettes were used with cells resuspended in 10% glycerol and Gene Pulser settings of 25 µF, 2.5 kV, and 200 Ω giving a time constant of 3–5 msec, corresponding to a maximum field strength of 12.5 kV/cm (Murrell *et al.*, unpublished). It is surprising that a plasmid as large as pVK100 (23 kb) is transformed in this way, but from preliminary observations it is clear that by careful choice of plasmids, possibly including some in Table III, and optimization of the numerous parameters outlined above, a very efficient transformation system for selected methanotrophs will soon be available.

4. MOLECULAR BIOLOGY OF METHANOTROPHS

Since the obligate methanotrophs have until recently proved refractory to many forms of genetic analysis, it is not surprising that the molec-

TABLE III. Potential Vectors for Methanotrophs

Vector	Useful markers	Properties	Reference
pVK100 series	Tc, Km	Mobilizable derivatives of pRK290, cosmids	Knauf and Nester (1982)
pLAFR1	Tc	Mobilizable cosmid derivative of pRK290	Friedman et al. (1982)
pLA2901 series	Tc, Km	Mobilizable derivatives of pRK290, multiple cloning sites, cosmids	Allen and Hanson (1985)
pKT230, pKT231	Km, Sm	Mobilizable derivatives of RSF1010	Bagdasarian et al., (1981)
pLG221, pLG223	Km or Tc	Can be used to introduce Tn5 or Tn10 into a wide range of gram-negative bacteria	Boulnois et al. (1985)
pDSK509, pDSK519, pRK415	Km	Plasmids based on RSF1010 and pRK404 with additional cloning sites and antibiotic resistance genes; also improved cosmid pLAFR5	Keen et al. (1988)
pGSS33 series	Ap, Cm, Sm, Tc	Series of cloning vectors based on broad-host-range plasmid R300B	Sharpe (1984)
pSUP series	Tc, Km, various	Based on RSF1010 broad-host-range, Tn5 derivatives only	Simon et al. (1983)

ular biology of these organisms has lagged behind our knowledge of their physiology and biochemistry. However, within the last 5 or 6 years, a considerable amount of effort has been directed, by only a few research groups, toward a better understanding of the molecular biology of nitrogen and carbon, particularly methane and methanol metabolism. In at least two groups, the model system chosen to develop methanotroph molecular genetics was nitrogen metabolism, due in part to the availability of numerous heterologous gene probes for identifying and cloning nitrogen fixation genes.

4.1. Nitrogen Metabolism Genes

4.1.1. Nitrogen Fixation Genes

Using plasmid pSA30, containing the nitrogen fixation (*nif*) structural genes K, D, and H of *Klebsiella pneumoniae*, Toukdarian and Lidstrom (1984a) were able to identify homologous DNA fragments in the chromosomal DNA from the type II methanotroph *Methylosinus* sp. strain 6. This heterologous *Klebsiella nif* gene probe also allowed the cloning of a 2.3-kb *nif*-specific DNA fragment from *Methylosinus* into pBR325 to yield pAT600. Subsequent analysis revealed that the interspecific *nif* homology was limited to DNA sequences encoding the *K. pneumoniae nif*D and *nif*H genes. Identification of proteins encoded by pAT600 and pAT601 (insert in reverse orientation) in *E. coli* maxi cells indicated that the cloned 2.3-kb *Methylosinus nif* DNA fragment directed the synthesis of two polypeptides of molecular weights 57 and 34 kDa. Expression was thought to be from vector promoters rather than a cloned *Methylosinus* promoter. Although no DNA sequencing, complementation studies, or Western blotting was carried out to prove the exact identity of the cloned *nif* DNA, an important extension of this work was the molecular construction and characterization of *nif* mutants of *Methylosinus* by a one-step marker exchange step.

The recombinant plasmid pAT600, consisting of pBR325 containing a 2.3-kb *Hind*III fragment of *Methylosinus nif* DNA, was subjected to Tn5 mutagenesis and the Tn5 insertions within the *Methylosinus* DNA were mapped with respect to known restriction sites. A three-way filter mating procedure was used to mobilize pAT600::Tn5 derivatives into *Methylosinus* 6 recipients, with pRK2013 as helper plasmid. Since pBR325 was not maintained in *Methylosinus* 6, resulting kanamycin-resistant (Km^R) colonies arose by homologous recombination between the *nif* region contained in pAT600::Tn5 and the chromosomal *nif* region (summarized in Fig. 1). This was confirmed and all Km^R *Methylosinus* 6 colonies tested after mating

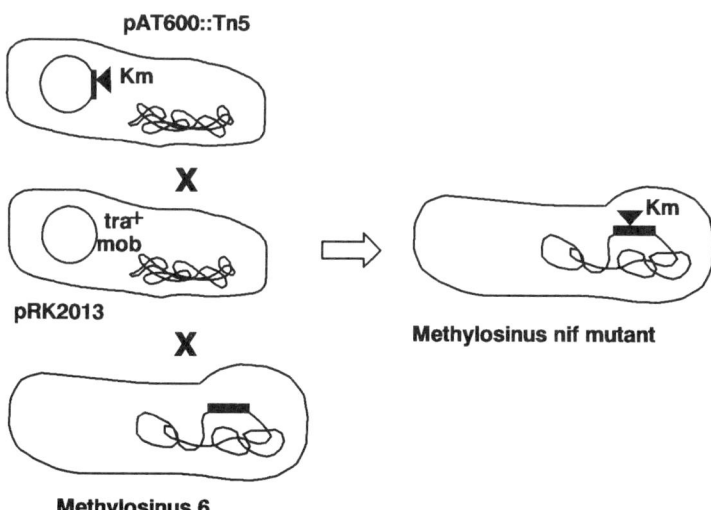

Figure 1. Marker exchange mutagenesis of *nif* genes from *Methylosinus*.

exhibited a *nif⁻* phenotype. They were unable to grow in the absence of a fixed nitrogen source and under *nif*-derepressing growth conditions lacked three *nif*-specific polypeptides when examined by 2-D polyacrylamide gel electrophoresis. Possibilities for these polypeptides were *nif*K and *nif*D gene products (Toukdarian and Lidstrom, 1984b). Although the frequency of transfer of pAT600::Tn5 was very low, the general principle and success of this one-step marker exchange mutagenesis illustrates that once a gene of interest has been cloned, it is possible to generate methanotroph mutants in a highly selective way (see Section 4.2.2.).

Heterologous DNA hybridization has also been used to identify *nif* genes in other methanotrophs. Oakley and Murrell (1988) screened chromosomal DNAs prepared from 13 representative strains of methanotrophs with the *nif*H gene of *K. pneumoniae*. All type II methanotrophs and *M. capsulatus* (Bath) contained *nif*H homologs, supporting the physiological data that all type II and type X methanotrophs are capable of fixing N_2 (Murrell and Dalton, 1983a). Surprisingly, *nif*H homologs were also found in the non-N_2-fixing species *M. methanica*, which was perhaps indicative of loss of N_2-fixing ability by evolutionary sequence diversification.

The homology between DNA sequences from *M. capsulatus* (Bath) and *nif* genes from *K. pneumoniae* was exploited to clone the structural genes from this methanotroph. Probing a cosmid library of *M. capsulatus* (Bath) DNA, constructed in pVK100, revealed a cosmid containing the structural *nif* gene cluster. The 10-kb region containing these genes is shown in Figure 2. As in most other diazotrophs, the structural genes lay in the order *nif*H, D, and K and were chromosomally located. Sequencing of the *nif*H gene has also revealed a high degree of homology with other *nif* genes (Murrell and Oakley, unpublished). Attempts to complement *K. pneumoniae nif* structural gene mutants with this *nif* DNA from *Methylococcus* were unsuccessful. The reason for this is unclear because subsequently, nitrogen metabolism genes from methanotrophs have been used to complement *E. coli* glutamine auxotrophs (see below).

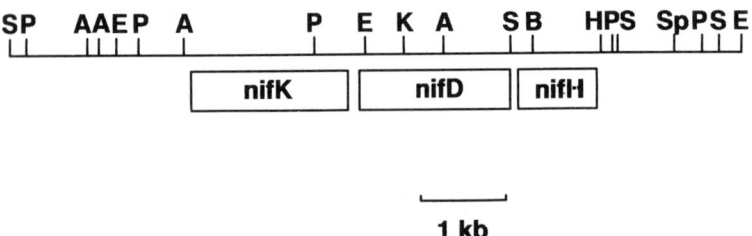

Figure 2. The *nif* gene cluster of *M. capsulatus* (Bath). S, *Sal*I; P, *Pst*I; E, *Eco*RI; A, *Ava*I; K, *Kpn*I; H, *Hind*III; B, *Bgl*II; Sp, *Sph*I.

4.1.2. Ammonia Assimilation Genes

Owing to an abundance of heterologous gene probes from *Klebsiella* and *Azotobacter*, Cardy and Murrell carried out a systematic screening of methanotroph chromosomal DNAs with DNA probes coding for the enzyme glutamine synthetase (*gln*A) and the nitrogen regulatory genes *ntr*A, B, and C. *gln*A homologs were found in the chromosomes of all type I and type II methanotrophs tested, indicating the conserved nature of this gene throughout the species. Similar results were obtained when the *ntr*C and *ntr*A genes of *K. pneumoniae* were used as DNA probes against a number of representative species of methanotrophs (Cardy, 1989). Little or no homologies were observed with *ntr*B probes, which is consistent with reports from other workers studying *ntr* genes in other gram-negative bacteria. The existence of *ntr*A and *ntr*C homologs in all methanotrophs tested suggests that nitrogen metabolism in these organisms is regulated by a global *ntr* system similar to that found in a wide range of bacteria (Merrick, 1988).

Once the presence of putative *ntr* genes had been established within the *M. capsulatus* (Bath) genome, cosmid libraries of this organism were screened with the *Klebsiella ntr*C gene, resulting in the isolation of pCOS1. On analysis, this cosmid was found to contain the *gln*A gene (structural gene for glutamine synthetase) some 8.5 kb upstream of an *ntr*C homolog. In enteric bacteria (as well as a number of nonenteric bacteria), *ntr*B and *ntr*C are linked to *gln*A in the form of a complex operon; *ntr*A is unlinked to this region. In *M. capsulatus* (Bath), *ntr*C is not as closely linked to *gln*A, a situation also found in *Rhizobium meliloti, Azorhizobium sesbaniae* ORS571, and *Rhodobacter capsulata*, organisms for which linkage of *gln*A to *ntr*C has not been demonstrated (Szeto *et al.*, 1987; Pawlowski *et al.*, 1987; Haselkorn, 1986). Attempts to complement *ntr* mutants of *E. coli* and *K. pneumoniae* using the insert of pCOS1 failed. However, the reason for this may have been that not all the *ntr*C gene of *Methylococcus* was present on this DNA fragment, and attempts are now being made to isolate the complete *ntr*C gene by chromosome walking techniques (Murrell *et al.*, unpublished).

The structural gene for glutamine synthetase (GS) isolated from *M. capsulatus* (Bath) was analyzed in detail (Cardy and Murrell, 1990). DNA sequencing of this *gln*A gene revealed an open reading frame (ORF) of 1407 bp preceded by a Shine–Dalgarno sequence (AGGAGGA) 10 bp upstream from the presumptive start codon (ATG). This ORF encoded a polypeptide of 468 amino acids with a predicted Mr of 51, 717, the *M. capsulatus* (Bath) GS enzyme subunit (Murrell and Dalton, 1983b). This was the first gene from a methanotroph to be fully sequenced. The DNA had a %mol G+C of 59. The codon usage was strongly biased toward the use of codons in which G and C predominate in the third position of all codons. (This was subsequently found for other *Methylococcus* and *Methylosinus* genes; see Section 4.2.2.)

In the Enterobacteriaceae, transcription of the *glnA* gene can occur from one of two functional promoters (Dixon, 1984). The upstream promoter (*glnA* p1) resembles the canonical −35 and −10 promoter, and the downstream promoter (*glnA* p2) resembles the NtrA- dependent promoters with conserved bases at −24 and −12 with respect to the transcription start site. Both types of promoter element were detected in the *M. capsulatus glnA* leader region (Fig. 3). Two putative *glnA* p1-type promoters (P_1 and P_3) were located on either side of a *glnA* p2 promoter element (P_2). This region of DNA also contained a single region of dyad symmetry with an 11 out of 15 bp match to the consensus sequence for the binding of the *ntrC* gene product TGCACTA-N_3-TGGTGCAA proposed by Dixon (1984).

Further sequence analysis of *Methylococcus glnA* and comparison with several other nucleotide-derived GS amino acid sequences revealed regions of conservation in the GS monomer, including the site of GS adenylylation at tyrosine residue 393 originally proposed to be a feature of the regulation of *Methylococcus* GS (Murrell and Dalton, 1983b). This comparison also revealed that the five regions of homology in the GS polypeptide that are highly conserved in prokaryotic and eukaryotic GS enzymes are also present in this organism. A conserved active-site histidine residue was also noted in the methanotroph enzyme.

Polypeptides encoded by pDC1, the plasmid carrying *glnA*, were studied in a Zubay-type *in vitro* transcription/translation system. pDC1 directed the *in vitro* synthesis of glutamine synthetase as determined by polyacrylamide gel electrophoresis, thus showing that this type of expression system can have widespread uses in the analysis of cloned methanotroph genes (see Section 4.2.2). More important, it was shown that the *glnA* gene of *Methylococcus* could be used to complement *E. coli glnA* mutants and that expression in *E. coli* was regulated by nitrogen levels in an Ntr⁺ but not an Ntr⁻ background, thus strengthening the proposal that a similar or anal-

---

Figure 3. Nucleotide sequence of the 2463-base *Sac*I–*Cla*I fragment containing the *glnA* gene of *M. capsulatus* (Bath). The sequence of the antisense strand is shown in the 5′ to 3′ direction along with the predicted amino acid sequence. Underlined sequences represent two sets of putative −10 and −35 hexamers (P_1 and P_3). Double-underlined sequences highlight a putative NtrA-dependent promoter. Arrows over sequences at nucleotide positions 361–378 indicate a region of dyad symmetry similar to the proposed NtrC binding sequence.

```
                                                                          80
GAGCTCGTGCCGCCGGCGTCATCTCACCAGCGCGCGCTTGAACAGCCTGGATTCAAGGCCGCTGGACGACCGGTTCTGGG
                                                                          160
TCGCTGCTTCGTGTCACGATGTGAGCGAACTGGAAAAGGCGGAGTCTCTGCAACTGGATTTCGCCGTTTTCGGTCCCGTG
                                                                          240
TTGCCTACCCGGTCCCATCCGGAGTCGGCTCCACTGGGCTGGGAGCATGTCTCCCAATGTCTCCAGTCCGTCAATCTCCC
                                                                          320
AGTCTATGCATTGGGTGGGATGCGGCAGAACATCTCGCCAGCGCCAGATCGGCCGGTGCCTGGGGGATAGCAGGCATTC
                 P1      GCTCCA                TATGAT              P2    400
GGGGATTTCTGTGATCGCAATGCTCCATAGAAGTGCCGTCGGCACTATGATAGTGCGCCATTCGAAACGAATGGCACTTA
                 P3                                                       
ACTCGCGGGTAAAATCAAGAAGATATTTGCCCGGTACCCTTGGTACAATTCCTGTGTAACTGTTGAGTTACCCAATCCTA
                                                                          560
GCTGGAGGAGGACGACTTTACCATGACACCGAAAGACGTATTGCAGATCATCAAGGAAAAGGAAGTCCGCTACGTGGACC
                         M  T  P  K  D  V  L  Q  I  I  K  E  K  E  V  R  Y  V  D
                                                                          640
TGCGTTTCGCCGACACCCGCGGCAAAGAACAGCACGTGACGGTCCCCGCCTCGACCATCGACGAGGCCGCCTTCGAAGAG
 L  R  F  A  D  T  R  G  K  E  Q  H  V  T  V  P  A  S  T  I  D  E  A  A  F  E  E
                                                                          720
GGCAAGATGTTCGACGGTTCCTCGATCGCCGGATGGAAAGGCATCAACGAATCTGACATGATTCTGATGCCGGATGCATC
 G  K  M  F  D  G  S  S  I  A  G  W  K  G  I  N  E  S  D  M  I  L  M  P  D  A  S
                                                                          800
CACGGCCGTCATGGATCCGTTCTTCGACGATCCGACGCTGATCCTCCGTTGTGACATCGTCGAGCCGGCCACCATGCAGG
 T  A  V  M  D  P  F  F  D  D  P  T  L  I  L  R  C  D  I  V  E  P  A  T  M  Q
                                                                          880
GTTATGAGCGGGATCCTCGCTCCATCGCCAAGCGGGCCGAAGCTTACATGAAGTCCACCGGCATTGCCGATACCGCGTTG
 G  Y  E  R  D  P  R  S  I  A  K  R  A  E  A  Y  M  K  S  T  G  I  A  D  T  A  L
                                                                          960
TTCGGGCCGGAAAACGAGTTCTTCATCTTCGACGACGTCCGCTGGGGCGCCAACATGTCCGGCTCGTTCTACAAGGTCGA
 F  G  P  E  N  E  F  F  I  F  D  D  V  R  W  G  A  N  M  S  G  S  F  Y  K  V  D
                                                                          1040
TTCCGAGGAGGCGGGCTGGAACTCGGAAAAGGTCTACGAAGACGGCAACATCGGCCATCGTCCGGGTGTGAAGGGCGGCT
 S  E  E  A  G  W  N  S  E  K  V  Y  E  D  G  N  I  G  H  R  P  G  V  K  G  G
                                                                          1120
ACTTCCCGGTGCCGCCGGTCGACTCCTTCCAGGATCTGCGCTCTGCCATGTGCAACACCCTGGAGGACATGGGCATGGTC
 Y  F  P  V  P  P  V  D  S  F  Q  D  L  R  S  A  M  C  N  T  L  E  D  M  G  M  V
                                                                          1200
GTGGAAGTCCATCACCACGAAGTAGCCACCGGCAGTGTGAAATCGGCGTCCGCTGCAATACCCTGGTGAAAAAGGC
 V  E  V  H  H  H  E  V  A  T  A  G  Q  C  E  I  G  V  R  C  N  T  L  V  K  K  A
                                                                          1280
CGACGAGGTACTGTTGCTCAAATACGCCGTTCAGAACGTGGCGCATGCCTATGGCAAGACCGCCACCTTCATGCCGAAAC
 D  E  V  L  L  L  K  Y  A  V  Q  N  V  A  H  A  Y  G  K  T  A  T  F  M  P  K
                                                                          1360
CCCTGGTCGGAGCAAACGGCAATGGCATGCATGTTCACCAGTCCGTGGCGAAAGACGGCAAGAATCTCTTCAGCGGTGAC
 P  L  V  G  A  N  G  N  G  M  H  V  H  Q  S  V  A  K  D  G  K  N  L  F  S  G  D
                                                                          1440
CTTTACGGCGGGTTGTCCGAAACCGCATTGCACTATATCGGCGGATACATCAGCATGCCCAAGGCGCTGAACGCGTTCTG
 L  Y  G  G  L  S  E  T  A  L  H  Y  I  G  G  Y  I  S  M  P  K  A  L  N  A  F  C
                                                                          1520
CAATGCCTCCACCAACAGCTACAAGCGGCTGGTGCCGGGCTTCGAAGCGCCGGTCATGCTGGCTTATTCCGCCCGTAACC
 N  A  S  T  N  S  Y  K  R  L  V  P  G  F  E  A  P  V  M  L  A  Y  S  A  R  N
                                                                          1600
GCTCGGCTTCGATCGGTATCCCCTACGTCATGAACCCGAAGGCCCGTCGCATCGAGGTCCGCTTCCCGGATTCCACGGCG
 R  S  A  S  I  G  I  P  Y  V  M  N  P  K  A  R  R  I  E  V  R  F  P  D  S  T  A
                                                                          1680
AATCCATACCTCGCTTTCGCGGCGATGCTGATGGCCGGTCTCGACGGCATCCAGAACAAGATCCATCCGGGCGATGCGAT
 N  P  Y  L  A  F  A  A  M  L  M  A  G  L  D  G  I  Q  N  K  I  H  P  G  D  A  M
                                                                          1760
GGACAAGGACCTGTACGACCTGCCGCCGGAAGAGGAAAAGGCCATTCCTCAGGTATGCTATTCCTTCGATCAGGCCCTGG
 D  K  D  L  Y  D  L  P  P  E  E  E  K  A  I  P  Q  V  C  Y  S  F  D  Q  A  L
                                                                          1840
AAGCGCTGGACAAGGACCGCGAGTTCCTCACGAGGGGCGGTGTTTTCACCGATGACATGATCGATGCGTACCTCGATCTC
 E  A  L  D  K  D  R  E  F  L  T  R  G  G  V  F  T  D  D  M  I  D  A  Y  L  D  L
                                                                          1920
AAAGGTCAGGAAGTGACCGCCTGCGATGAGCACTCATCCGTCGAGTTCGATATGTATTACAGCCTGTAATACAGGTCGA
 K  G  Q  E  V  T  A  C  D  E  H  S  S  V  E  F  D  M  Y  Y  S  L  *
                                                                          2000
GGTCGTGAGGACCGCTCGGTCTCGGACAGGCTGTTTCTCCTAAGTGACCGCTCGTAGCGACCTCGATCTCAAAGGAAGTG
                                                                          2080
ACCGCTGCGATGAGCACTCATCCCTCGAGTTCGATATGTATTACAGCTGTAATACAGGGTTTGAGGTCGTGGAGGACGCT
                                                                          2160
CGGTCTCGGACCAGGCGGGTTTTCTCCCATCGGCCTGATTTTCTGGAGAAAAGGACGCGAGGCCGACGGCAGCGAGAGCC
                                                                          2240
ATTCCCGCGCCGATCTGATCCGGGTGGAAGCGAACCTGGCTCGGTAAAACCGAGTCGGGAGCGATTAGTCTCGATCGTGC
                                                                          2320
GGGGAAAGATGGAGGCTGAGGTCGGAATCGAACCGGCATAGACGGATTTGCAATCCGCAGCATAACCACTCTGCCACTCA
                                                                          2400
GCCTTTGAAGGCAAAGCAAAAAGCCGCGACGTGAACGCGGCTTTTCCCAAAAAGCGTTTGGAGCGGGAAACGAGACTCGA
                                                                          2460
AACTCGCGACCCCAACCTTGGCAAGGTTGTGCTCTACCAACTGAGCTATTCCCGGCAATCGAT
```

ogous Ntr system may be present in methanotrophs. This complementation of *gln*A mutants in *E. coli* by the *Methylococcus gln*A gene was the first evidence for functional expression in *E. coli* of a gene from any methanotroph, and although *gln*A appears to be a very highly conserved gene throughout the prokaryotes, nevertheless the ability to express methanotroph genes in this way will be important for many future studies. It appears, therefore, from molecular studies on nitrogen metabolism genes in *Methylococcus*, that there are no major problems in the cloning, sequencing, and expression of methanotroph genes and that rapid progress should now be made with molecular studies on other sets of genes such as those involved in C_1 metabolism.

4.2. C_1-Specific Genes

4.2.1. Methanol Dehydrogenase Genes

Nunn and Lidstrom's study on the isolation and complementation analysis of methanol oxidation mutants of the facultative methanol utilizer *Methylobacterium* sp. strain AM1 allowed them to identify at least 10 genes involved in the metabolism of methanol by this organism. One of these genes, *mox*F, encodes the large subunit of the enzyme methanol dehydrogenase (Nunn and Lidstrom, 1986a,b; see also Chapter 6). An internal fragment of this *mox*F gene was found to hybridize to putative *mox*F genes from a variety of methanotrophs, including type I, type II, and type X [*M. capsulatus* (Bath)] methanotrophs. The probe showed little or no homology to nonmethanotrophs (Stephens *et al.*, 1988).

Hybridization conditions using stringencies allowing for approximately 30% base-pair mismatch enabled the cloning of putative *mox*F genes from *M. capsulatus* (Bath) and *M. albus* BG8 in the cosmid vectors pHC79 and pRS8600. Reprobing of restriction digests of these *mox*F clones with MDH structural gene probes from *Methylobacterium* sp. strain AM1 allowed the localization and, in the case of the *Methylococcus* clone, the orientation of the two *mox*F genes. Subclones of methanotroph *mox*F genes were isolated for expression studies in *E. coli* with the T7 expression vector pTZ18R (Tabor and Richardson, 1985). Expression of these constructs was heat-induced, and SDS-gel electrophoresis of cell-free extracts after protein expression was carried out. Subsequently, polypeptides were blotted to nitrocellulose and then challenged with antibody to *M. albus* B8G methanol dehydrogenase. Cross-reacting bands of 60 kDa, corresponding to the large subunit of MDH, were observed for both *M. albus* and *M. capsulatus mox*F gene products. The orientation of the MDH gene from *Methyl*-

ococcus was confirmed by these expression experiments, as was the orientation of the MDH gene from *Methylomonas*.

It was not possible to detect the presence of active methanol dehydrogenase in *E. coli* extracts obtained after expression studies. However, the availability of *mox*F mutants of *Methylobacterium* AM1 was exploited in complementation experiments. Recombinants containing *Methylococcus mox*F and *Methylomonas mox*F were mobilized into mutant UV26 using plasmid pRK2013. Methanol-positive colonies that oxidized methanol and contained *in vitro* methanol dehydrogenase activity (Stephens *et al.*, 1988) were obtained at high frequency. Although activities were lower than for the wild-type organism, it was a very important observation that these MDH structural genes were expressed in a heterologous host such as *Methylobacterium*, and this type of approach should prove useful for studying other methanol oxidation functions of methanotrophs. Indeed, Lidstrom and co-workers have identified a number of putative *mox*A genes in methanotrophs using a variety of *Methylobacterium mox*A gene probes (*mox*A genes being those involved in association of methanol dehydrogenase and its cofactor pyrroloquinoline quinone) (Lidstrom *et al.*, poster presentation, Society for General Microbiology, April 1988, University of Warwick).

Another strategy adopted for the identification of *mox*F structural genes from methanotrophs is to utilize degenerate oligonucleotide probes based on a highly conserved region of DNA at the 5' end of the *mox*F gene of methanotrophs (corresponding to the N-terminal region of the mature *Mox*F protein). In this way, the problems of varying %mol G+C, codon bias, and variation in amino acid sequence between different methanol dehydrogenases in different methanotrophs are obviated using a "universal" *mox*F oligonucleotide (Waechter-Brulla and Lidstrom, 1990). Using this strategy, the *mox*F gene from *Methylomonas* A4 has been identified and cloned. This *mox*F gene from a marine methanotroph also contained the highly conserved region near the 5' end, and the DNA upstream of the A4 *mox*F gene shares considerable sequence similarities with promoters of other *mox*F genes (Fig. 4). The transcription start site of this gene is currently being mapped by Northern blot and primer extension analysis (Waechter-Brulla and Lidstrom, 1990).

In a separate study, the methanol dehydrogenase structural gene was isolated from the type II methanotroph *M. trichosporium* OB3b (Al-Taho *et al.*, 1990). A genomic library of *M. trichosporium* was constructed in the expression vector λgt11 and then screened with antibody raised against purified methanol dehydrogenase from this organism. Lysates from cross-reacting plaques were prepared after growth of bacteriophages in the presence of IPTG and were subjected to Western blotting. A large-molec-

		-35		-10	+1	
Ec		TTGACA	XXXXXXXXXXXXXXXXX	TATAAT	XXXXXX A	
AM1	CT	AAAGACA	TCGCGTCCAATCAAAGC	TAGAAA	ATATA G	--- 170bp --- first Met
XX	GT	AAAGACA	TCTCCTTCAATCAACGCC	TAGAAA	CGAT A	--- 171bp --- first Met
DM4	AG	CTTGACA	GAGATGCATAGCCTTGTA	TAGAAC	TAGC C	--- 72bp --- first Met
Pd				GGGAAA	AACCC	--- 157bp --- first Met
BG8	AA	AAAGGAA	CTTTTCCCGACTCACA	CGGAAA	AGCCAT A	--- 212bp --- first Met
A4	AA	AAATCCA	AGGCATACGTTTTGATC	AAGAAA	AGCTGAT	--- 85bp --- first Met

Figure 4. DNA upstream of the *Methylomonas A4* moxF gene shares sequence similarities with promoters of the other *moxF* genes. Bases in italics at (or near) +1 represent verified transcriptional start sites. The −35 and −10 regions of the *E. coli* promoter are underlined; similar regions in other sequences are also underlined where they are identical to AM1 and XX. The role of the extended (72–212 bp) region between the transcriptional and translational start sites is not known. Ec, *Escherichia coli*; AM1, *Methylobacterium extorquens* AM1; XX, *Methylobacterium organophilum* XX; DM4, *Methylobacterium* sp. DM4 (dichloromethane dehydrogenase gene); Pd, *Paracoccus denitrificans*; BG8, *Methylomonas albus* BG8; A4, *Methylomonas A4*. (From Waechter-Brulla and Lidstrom, 1990.)

ular-weight protein, presumably a fusion protein, that cross-reacted with the antibody was produced. Subsequent subcloning and expression experiments resulted in the isolation of a 2.1-kb DNA fragment that contained a protein with a molecular weight that cross-reacted with MDH antibody when Western-blotted. It was presumed that expression of this protein was directed from an unidentified methanotroph promoter because the inducer IPTG was not required to obtain expression of MDH from recombinants in pUC18.

In the identification and isolation of methanol dehydrogenase, three strategies have now been used successfully; (1) probing using a heterologous gene probe; (2) probing using an oligonucleotide specific for a conserved region of the gene of interest; (3) probing using antibody raised against purified "target" protein. All these strategies should have wide applicability to methanotroph molecular biology if enough is known about the protein or gene of interest.

4.2.2. Methane Monooxygenase Genes

One of the major drawbacks in the genetic analysis of MMO has been the lack of suitable gene transfer systems and mutagenesis techniques for methanotrophs. Also, the obligate nature of these organisms often precludes selection of mutants defective in MMO. Probably the most well-characterized MMO is the soluble enzyme complex (sMMO) from *M. capsulatus* (Bath) (see Chapter 3), which grows very poorly on methanol. Therefore, the most successful strategy for the identification of sMMO genes has been to purify MMO polypeptides, obtain N-terminal amino acid sequence information, and then probe using degenerate oligonucleotides prepared from the derived DNA sequence. The following sections will summarize progress made to date on the cloning, sequencing, and expression of sMMO genes from *Methylococcus* and *M. trichosporium* OB3b.

4.2.2a. *Methylococcus*-soluble MMO Genes. The first report on the cloning of MMO genes was from Mullens and Dalton (1987), who had obtained a partial N-terminal amino acid sequence of the γ-subunit polypeptide of protein A of MMO from *M. capsulatus* (Bath). Using two non-overlapping regions of the sequence of the γ subunit, two oligonucleotides of low degeneracy were synthesized and used to probe a genomic clone bank of *M. capsulatus* (Bath) constructed in the vector pRD720. One plasmid, containing a 4.5-kb *Sau* 3A DNA fragment, pIM001, was isolated, restriction-mapped, and found to contain a 0.64-kb *Sal*1–*Pst*1 DNA fragment with good homology to the γ-specific oligonucleotides. Subsequent DNA sequence analysis of this fragment revealed that this fragment contained part of the gene encoding the γ subunit of sMMO. The first 57 bases of the gene, encoding the N-terminal 19 amino acids, were determined and exactly matched the known amino acid sequence.

Using a similar strategy to that described above, N-terminal amino acid sequence data for the β subunit of protein A from *M. capsulatus* (Bath) was obtained. A mixed oligonucleotide probe complementary to the DNA encoding this short amino acid region was then synthesized. Although the β-specific oligonucleotide was very short (14 nucleotides), the amino acid region chosen was one of very low degeneracy in the genetic code and the mixed oligonucleotide probe contained only eight different sequences. The theoretical melting temperature for the probe–target heteroduplex was estimated to be 37°C, and therefore low stringency/hybridization temperatures were used throughout probing experiments. Using this probe, a recombinant plasmid pCH4, containing a 11.9-kb *Eco*R1 DNA fragment, was isolated. Upon restriction mapping and reprobing, it was found to contain sequences homologous to the β-gene probe, and also when compared with pIM001, it was found to possess sequences homologous to the γ gene. The β and γ genes were therefore linked in the *Methylococcus* chromosome (Stainthorpe *et al.*, 1989).

DNA sequence analysis of the entire sMMO gene cluster of *Methylococcus* contained in pCH4 has shown that the genes encoding the α, β, and γ subunits of protein A, *mmo*X, Y, and Z, protein B (*mmo*B), and protein C (*mmo*C) are all linked (Stainthorpe *et al.*, 1989, 1990a). The organization of these genes is shown in Figure 5.

*mmo*X, the gene encoding the α subunit of protein A, consists of 1581 bases translating to a single polypeptide of 527 amino acids with a predicted Mr of 60, 633. This is in close agreement with the published size of the purified α polypeptide obtained by SDS-PAGE (Woodland and Dalton, 1984). Sequence homology searches did not reveal any significant homologies with other proteins in current databases, although two regions of the α polypeptide showed some similarity to conserved region of the ribonucleotide reductase B_2 proteins of *E. coli* and other organisms. This is intriguing because it is thought that the active site of protein A resides on the α subunit(s) and that a μ-hydroxo-bridge dinuclear iron center is closely associated with this polypeptide. This type of iron center occurs in proteins of diverse function, including hemerythrins, purple acid phosphatase, and ribonucleotide reductases (see Chapter 3 for discussion of this), and it has been shown that in the ribonucleotide reductase B_2 protein of *E. coli*, Fe 1 is ligated to His 118 and Asp 84 and Fe 2 is ligated to His 241 and Glu 204, with one glutamic acid residue, glu 115, being ligated to both iron atoms, forming a bridge between them (Nordlund *et al.*, 1990). It may therefore be more than a coincidence that two regions of the α subunit of protein A

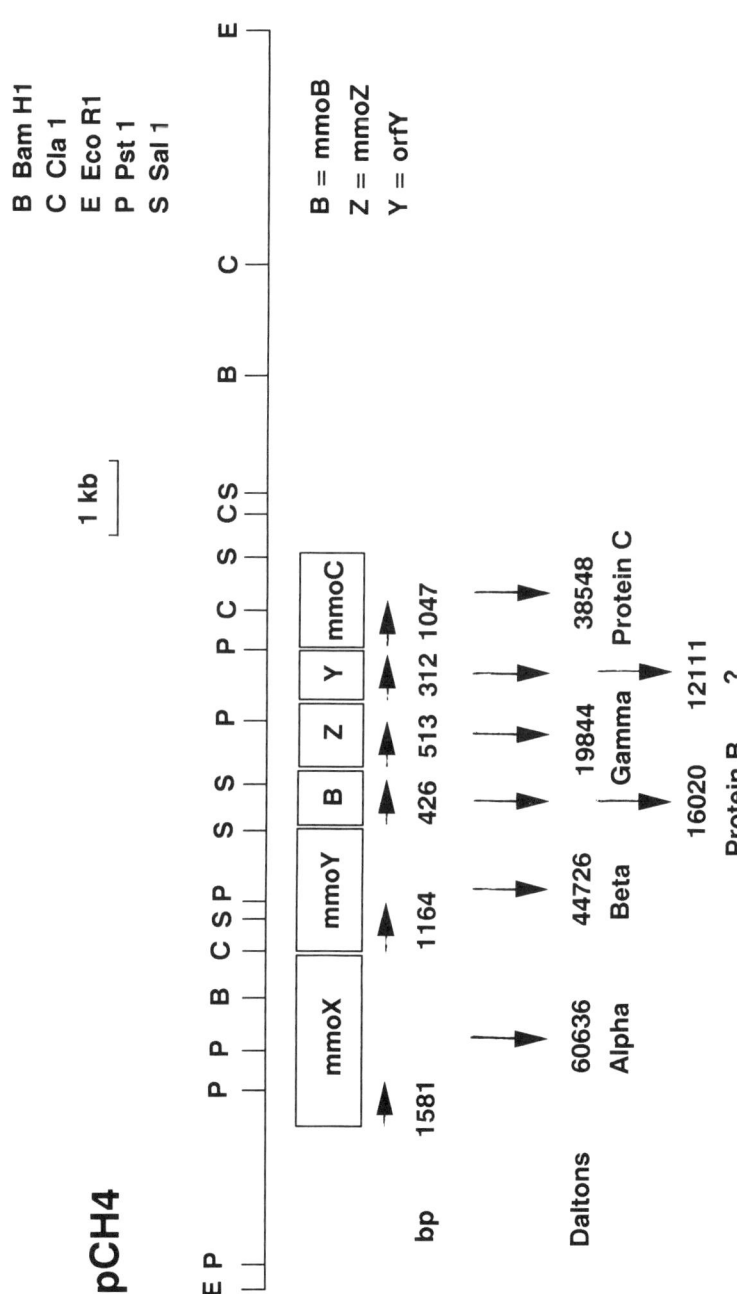

Figure 5. Plasmid pCH4 containing the MMO gene cluster of *M. capsulatus* (Bath).

from *Methylococcus* (and also *M. trichosporium* OB3b), centered around amino acid residues 144–151 and 242–246, share significant homology with Fe-binding sites in the B_2 protein of a variety of ribonucleotide reductases (Fig. 6). By the use of site-directed mutagenesis, it should eventually be possible to determine whether these residues do indeed ligate iron and whether they are essential for MMO activity.

Of the genes in the MMO cluster, only *mmo*X is preceded by 5' sequences homologous to *E. coli* −35 and −10 consensus sequences. Sequences homologous to the −12 and −24 conserved regions of *Pseudomonas* and to *nif*− and *ntr*− like consensus promoter regions were also noted in the A+T–rich sequence 5' to the *mmo*X gene (Stainthorpe *et al.*, 1990a). The promoter sequences that operate transcription of the highly expressed, copper- and methane-regulated proteins in *Methylococcus* may be entirely novel, and further investigations using S1 nuclease mapping and primer extension experiments are underway to resolve the structure of transcription control elements of MMO.

*mmo*Y, of 1164 bp encodes the β subunit of protein A, Mr 44, 726. It has no significant homology with any proteins in current databases. This gene is separated from *mmo*Z, a 513-bp ORF encoding the 19, 844-Da αsubunit of protein A by the gene encoding protein B, formerly *orf*X, (Stainthorpe *et al.*, 1989) now designated *mmo*B (Pilkington *et al.*, 1990).

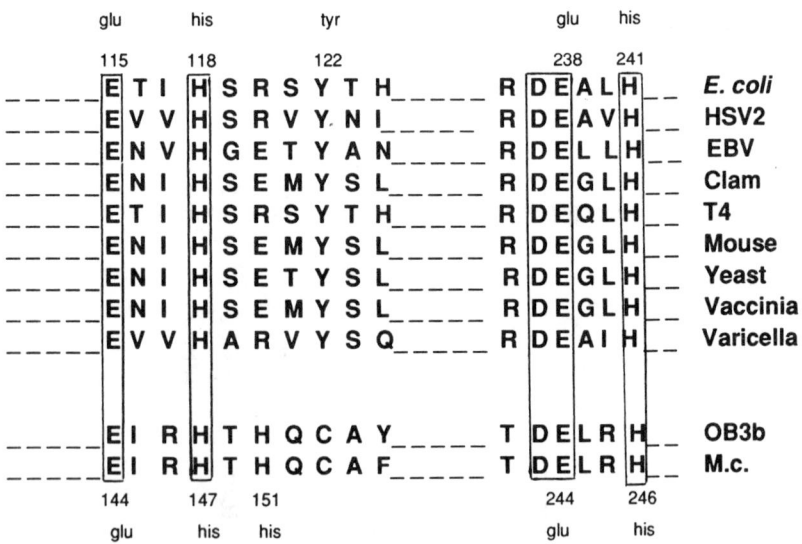

Figure 6. Sequence homology between putative Fe-binding sites in the B_2 protein of ribonucleotide reductase and the α subunit of protein A of MMO. (Ribonucleotide reductase data from Nordlund *et al.*, 1990.)

Since the α, β, and γ subunits of protein A are produced in stoichiometric amounts in *M. capsulatus*, it is intriguing that, although as might be expected *mmo*X, Y, and Z genes are linked, *mmo*Y and *mmo*Z are separated by *mmo*B, coding for the regulatory protein B. This is currently being investigated. Polypeptides β, γ, and B appear to be unique, and there are no significant homologies with any proteins in current databases.

3' of *mmo*Z and separated from it by *orf*Y (a gene of 312-bp encoding a 12-kDa protein of no known function) is *mmo*C, the protein C gene (Stainthorpe *et al.*, 1990a). The *mmo*C gene product of 38.5-kDa is the reductase component of sMMO and is responsible for donation of electrons from NADH to protein A. It is an iron-sulfur–containing protein and, as might be predicted, possesses considerable homologies to ferredoxins of plant and bacterial origin. The two strongly conserved domains center around amino acid residues 40–51 and 74–86 within the N-terminal end of the protein (Fig. 7). Three of the four Cys residues of the 2Fe–2S iron-sulfur center of ferredoxins occur within the first domain; the fourth cys (protein C amino acid 81) lies in the second conserved domain. The sequencing of *mmo*C has confirmed that protein C has a plant-like ferredoxin 2Fe–2S center as suggested by its absorbance and EPR spectra (Lund *et al.*, 1985). Data on the entire sMMO gene cluster are summarized in Table IV.

No major problems were found in DNA sequencing the entire 7.3-kb sMMO gene cluster of *M. capsulatus* (Bath) using the dideoxy chain termination method, despite the relatively high %mol G+C content of *Methyl-*

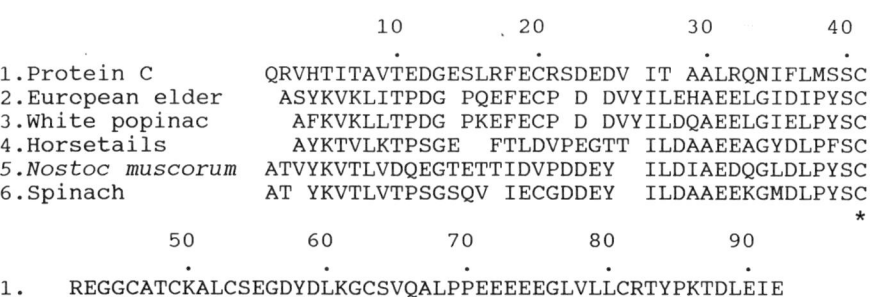

```
                        10           20            30            40
                         .            .             .             .
1.Protein C         QRVHTITAVTEDGESLRFECRSDEDV  IT AALRQNIFLMSSC
2.European elder    ASYKVKLITPDG PQEFECP D DVYILEHAEELGIDIPYSC
3.White popinac     AFKVKLLTPDG PKEFECP D DVYILDQAEELGIELPYSC
4.Horsetails        AYKTVLKTPSGE    FTLDVPEGTT ILDAAEEAGYDLPFSC
5.Nostoc muscorum   ATVYKVTLVDQEGTETTIDVPDDEY   ILDIAEDQGLDLPYSC
6.Spinach           AT YKVTLVTPSGSQV IECGDDEY   ILDAAEEKGMDLPYSC
                                                                *
              50          60          70           80           90
               .           .           .            .            .
1.  REGGCATCKALCSEGDYDLKGCSVQALPPEEEEEGLVLLCRTYPKTDLEIE
2.  RAGSCSSCAGKLVAGSVDQSDQS   FLDDEQIEEGWVLTCVAYPKSDVTIE
3.  RAGSCSSCAGKLVEGDLDQSDQS   FLDDEQIEEGWVLTCAAYPRSDVVIE
4.  RAGACSSCLGKVVSGSVDQSEGSF  LDDGQMEEGFVLTCIAIPESDLVIE
5.  RAGACSTCAGKIVSGTVDQSDQSF  LDDDQIEKGYVLTCVAYPTSDLKIE
6.  RAGACSSCAGKVTSGSVDQSDQSF  LEDGQMEEGWVLTCIAYPTGDVTIE
       * *                             *
```

Figure 7. Alignment of the N-terminal 92 amino acids of protien C with 2Fe2S ferredoxins from a variety of eukaryotic and prokaryotic sources. Strongly conserved domains (around the Cys residues of the iron-sulfur center) are boxed, identities are depicted in boldface, and conserved substitutions are underlined. (From Stainthorpe *et al.*, 1990a.)

TABLE IV. Summary of Data from Sequencing of Soluble MMO Gene Cluster from *M. capsulatus* (Bath)

Methylococcus sMMO Genes	bp	Da
*mmo*X(α)	1581	60,636
*mmo*Y(β)	1164	44,726
*mmo*Z(γ)	513	19,844
*mmo*B	426	16,020
*orf*Y	312	11,943
*mmo*C	1047	38,581

ococcus DNA. This higher percentage G+C is reflected in the nucleotide usage in codons of the genes in the sMMO gene cluster. Table V shows the nucleotide usage in the three positions of all codons in the *mmo* gene cluster of *Methylococcus*. There is a marked preference for either C or G in the third position of the codon (percentage G+C average of 80%), which might be expected for percentage G+C–rich DNA. The average %mol G+C for the six genes is approximately 59%, which is in fairly good agreement with previously published values. The high preference for C or G in the third position of the codon should be borne in mind when designing oligonucleotides for the use as probes from N-terminal amino acid sequence information (see Stainthorpe *et al.*, 1990a).

The availability of DNA sequence information on methanotrophs may also have use in the design of oligonucleotides or promoters and may be beneficial in methanotroph DNA sequence analysis in general. Table VI outlines the codon usage for all six ORFs of the *Methylococcus* sMMO gene

TABLE V. Nucleotide Usage in *mmo* Genes and *orf*Y Codons

	% G+C content[a]						
	*mmo*X	*mmo*Y	*mmo*B	*mmo* Z	*orf*Y	*mmo*C	Average of all genes
1st position	57	58	53	61	63	62	59
2nd position	40	46	28	33	45	44	40
3rd position	83	80	81	86	67	82	80
Average	60	61	54	60	58	63	59
Number of codons	528	388	142	171	106	349	1684

[a]Calculated using the Beckman Microgenie program.

TABLE VI. Codon Usage in the Six Genes of the MMO Gene Cluster of *Methylococcus*

TTT Phe	12	TCT Ser	0	TAT Tyr	21	TGT Cys	4
TTC Phe	75	TCC Ser	24	TAC Tyr	43	TGC Cys	19
TTA Leu	3	TCA Ser	2	TAA End	1	TGA End	5
TTG Leu	18	TCG Ser	29	TAG End	0	TGG Trp	41
CTT Leu	5	CCT Pro	4	CAT His	12	CGT Arg	14
CTC Leu	24	CCC Pro	16	CAC His	23	CGC Arg	57
CTA Leu	0	CCA Pro	5	CAA Gln	5	CGA Arg	2
CTG Leu	87	CCG Pro	48	CAG Gln	57	CGG Arg	26
ATT Ile	5	ACT Thr	4	AAT Asn	11	AGT Ser	5
ATC Ile	57	ACC Thr	62	AAC Asn	67	AGC Ser	24
ATA Ile	1	ACA Thr	3	AAA Lys	13	AGA Arg	0
ATG Met	40	ACG Thr	21	AAG Lys	70	AGG Arg	5
GTT Val	12	GCT Ala	16	GAT Asp	26	GGT Gly	29
GTC Val	44	GCC Ala	71	GAC Asp	76	GGC Gly	66
GTA Val	8	GCA Ala	14	GAA Glu	64	GGA Gly	8
GTG Val	43	GCG Ala	41	GAG Glu	69	GGG Gly	14

cluster. A marked preference for codons that are G–C rich, particularly at the third position, can be clearly seen.

The expression of polypeptides encoded by pCH4 has also been achieved using a Zubay-type *in vitro* transcription/translation system. An example of the radiolabeled products of such a reaction is shown in Figure 8, in which the products of *mmo*X, Y, and Z can be clearly seen as major bands. Minor bands of around 12 and 20 kDa could possibly be the products of *mmo*B and *orf*Y, but this awaits further verification after expression of selected subclones of pCH4.

4.2.2b. Soluble MMO Genes in Other Methanotrophs. The central 5.8-kb *Bam*H1 fragments of pCH4 containing the major portion of the MMO gene cluster of *Methylococcus* was used as a probe for the detection of soluble MMO genes in a number of representative strains of obligate methanotrophs in the University of Warwick culture collection, including soluble MMO *M. sporium* 5 and *M. sporium* 12, with major hybridizing *Eco*R1 restriction fragments of approximately 5.1, 4.0, and 1.4 kb. Other type II methanotrophs possessing specific DNA fragments with good homology to the *M. capsulatus*–soluble gene probe were a variety of *M. trichosporium* species including OB3b (Stainthorpe *et al.*, 1990b). No sMMO gene homologs were detected in *Methylocystis parvus* OBBP (type II) or any

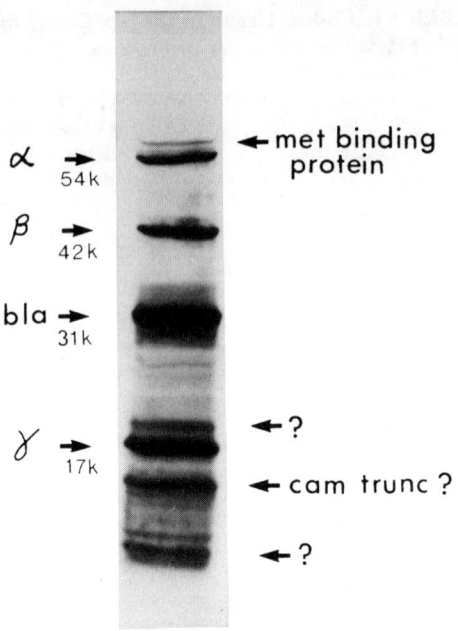

Figure 8. Fluorograph of ^{35}S-methionine labeled *in vitro* translation products of pCH4. Major radiolabeled polypeptides and their molecular weights are indicated. (Zubay-type *in vitro* transcription/translation kit as supplied by Amersham International.)

type I methanotrophs, including representatives of *M. albus*, *M. methanica*, *Methylomonas agile*, and *Methylobacter capsulatus*. These observations are consistent with physiological and biochemical observations that only *Methylococcus* and *Methylosinus* species in this culture collection possess a soluble MMO enzyme system; the others all oxidize methane using a particulate of membrane-bound system (see Chapter 3). Homology between *Methylococcus* MMO genes and *M. trichosporium* MMO homologs was high, and stringency of hybridization and nitrocellulose filter washing could be adjusted for conditions allowing approximately 20% base pair mismatch. This gene probe should therefore have wide applicability in the screening of methanotrophs for soluble MMO systems.

4.2.2c. Soluble MMO Genes of *M. trichosporium* OB3b. Using the *Methylococcus* sMMO gene probe described above, a recombinant clone harboring a 7.6-kb *Hind*III *M. trichosporium* OB3b genomic fragment exhibiting strong homology with the probe was isolated. This genomic fragment was mapped with respect to known endonuclease sites within the vector pBR329, and subsequent probing of Southern-blotted restriction fragments with specific *M. capsulatus* α and β gene probes revealed the presence and relative positions of α and β structural gene homologs within this plasmid pDVC200 (Cardy and Murrell, 1991a). Similarly, the genes encoding protein B and γ polypeptide were located on an adjacent *Eco*RI fragment. *mmo*C lies downstream (3') of these genes on an adjoining 3.5-kb *Hind*III DNA fragment. The gene organization in *M. trichosporium* OB3b is shown in the restriction map of the MMO gene cluster (Fig. 9). The ORFs indicated as boxes have been completely DNA-sequenced. The remaining MMO gene encoding protein C has recently been sequenced (Cardy *et al.*, 1991b).

Protein homology searches of the *M. trichosporium* OB3b *mmo*X, *mmo*Y, *mmo*B, and *mmo*Z deduced gene products using the Beckman Microgenie and DAP search program (Queen and Korn, 1986; Collins and Coulson, 1987) failed to detect any significant homologies with any proteins in current databases. However, the two conserved domains of *Methylococcus* *mmo*X that exhibited homology with the conserved regions in the B_2 subunit of *E. coli* ribonucleotide reductase (See Section 4.2.2a) are also conserved in the *mmo*X gene of *M. trichosporium* OB3b (Fig. 7). This again strengthens the argument that the binuclear iron center of sMMO resides within the α polypeptide(s) and that this region is a good candidate for site-directed mutagenesis studies on the active site of MMO.

Further sequence analysis of *M. trichosporium* OB3b MMO genes revealed that sequences homologous to the *E. coli* −35 and −10 consensus sequence and sequences highly homologous to the −12 and −24 consensus region of *Ntr*A-dependent promoters were present 5' of the *mmo*X gene. This promoter region is currently being characterized by S1 nuclease mapping and primer extension analysis.

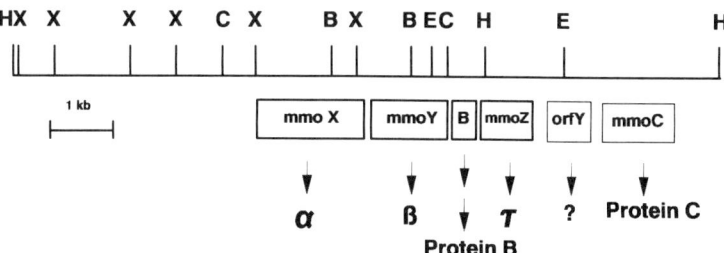

Figure 9. The soluble MMO gene cluster of *M. trichosporium* OB3b. The complete DNA sequence of *mmo*X through to *mmo*Z iscomplete. *mmo*C is currently being sequenced (Cardy *et al.* Murrell, 1991b). H, *Hind*III; X, *Xho*I; C, *Cla*I; B, *Bam*HI; E, *Eco*RI.

Analysis of the base usage in *mmo*X, *mmo*Y, *mmo*B, and *mmo*Z codons indicates a marked preference for G or C residues in the degenerate positions (Table VII). The average percentage G+C of the four complete ORFs is 61.4%, which is in good agreement with previous estimates derived from DNA melting experiments. Finally, a comparison was made between the MMO gene sequences of *M. capsulatus* (Bath) and *M. trichosporium* OB3b using the CLUSTAL programme of Higgins and Sharp (1988, 1989). Sequence alignments for the *mmo*X gene products (Fig. 10) indicate a 94% similarity in the two proteins, which indicates a high degree of conservation of this MMO gene between the species. Similar comparisons between the gene products of these two organisms showed an 83.5%, an 85%, and an 89.4% similarity between the *mmo*Y, *mmo*Z, and *mmo*B gene products, respectively (Cardy *et al.*, 1991a). A comparative analysis of the deduced polypeptide sequences of *M. trichosporium* OB3b sMMO components with the published *M. capsulatus* (Bath) sequences is summarized in Table VIII.

5. FUTURE DEVELOPMENTS

The technique of electroporation will undoubtedly become useful for the development of a gene transfer system for methanotrophs, and allied

TABLE VII. Codon Usage in *mmo*X, *mmo*Y, *mmo*Z, and *mmo*B genes of *M. trichosporium* OB3b

TTT Phe	1	TCT Ser	2	TAT Tyr	33	TGT Cys	0
TTC Phe	54	TCC Ser	11	TAC Tyr	19	TGC Cys	8
TTA Leu	0	TCA Ser	2	TAA End	0	TGA End	4
TTG Leu	3	TCG Ser	33	TAG End	0	TGG Trp	37
CTT Leu	5	CCT Pro	2	CAT His	15	CGT Arg	14
CTC Leu	48	CCC Pro	17	CAC His	19	CGC Arg	46
CTA Leu	1	CCA Pro	0	CAA Gln	3	CGA Arg	2
CTG Leu	35	CCG Pro	29	CAG Gln	42	CGG Arg	5
ATT Ile	5	ACT Thr	2	AAT Asn	17	AGT Ser	1
ATC Ile	61	ACC Thr	38	AAC Asn	27	AGC Ser	13
ATA Ile	0	ACA Thr	0	AAA Lys	11	AGA Arg	3
ATG Met	18	ACG Thr	33	AAG Lys	66	AGG Arg	3
GTT Val	7	GCT Ala	16	GAT Asp	28	GGT Gly	6
GTC Val	41	GCC Ala	58	GAC Asp	53	GGC Gly	69
GTA Val	0	GCA Ala	2	GAA Glu	17	GGA Gly	3
GTG Val	27	GCG Ala	53	GAG Glu	63	GGG Gly	1

```
  1   MAISLATKAATDALKVNRAPVGVEPQEVHKWLQSFNRDFKENRTKYPTKYHMANETKEQF
      MALSTATKAATDALAANRAPTSVNAQEVHRWLQSFNWDFKNNRTKYATKYKMANETKEQF
      **.* ********* .****..*..****.******.***.*****.***.*********

 61   KVIAKEYARMEAAKDERQFGTLLDGLTRLGAGNKVHPRWGETMKVISNFLEVGEYNAIAA
      KLIAKEYARMEAVKDERQFGSLQVALTRLNAGVRVHPKWNETMKVVSNFLEVGEYNAIAA
      *.**********.*******.*  .****.**.***.*.*****.***************

121   SAMLWDSATAAEQKNGYLAQVLDEIRHTHQCAFINHYYSKHYHDPAGHNDARRTRAIGPL
      TGMLWDSAQAAEQKNGYLAQVLDEIRHTHQCAYVNYYFAKNGQDPAGHNDARRTRTIGPL
      ..***** ***********************..*.*..*. .*************.****

181   WKGMKRVFADGFISRDAVECSVNLQLVGDTCFTNPLIVAVTEWAIGNGDEITPTVFLSVE
      WKGMKRVFSDGFISGDAVECSLNLQLVGEACFTNPLIVAVTEWAAANGDEITPTVFLSIE
      ********.***** ******.******..*************** .************.*

241   TDELRHMANGYQTVVSIANDPASAKFLNTDLNNAFWTQQKYFTPVLGYLFEYGSKFKVEP
      TDELRHMANGYQTVVSIANDPASAKYLNTDLNNAFWTQQKYFTPVLGMLFEYGSKFKVEP
      *************************.****************** .************* ************

301   WVKTWNRWVSEDWGGIWIGRLGKYGVES-RVLRDAKRDAYWAHHDLALAAYAMWPLAFAR
      WVKTWNRWVYEDWGGIWIGRLGKYGVESPRSLKDAKQDAYWAHHDLYLLAYALWPTGFFR
      ********* ****************** *.***.********** * ***.** .* *

361   LALPDEEDQAWFEANYPGWADHYGKIFNEWKKLGYEDPKSGFIPYQWLLANGHDVYIDRV
      LALPDQEEMEWFEANYPGWYDHYGKIYEEWRARGCEDPSSGFIPLMWFIENNHPIYIDRV
      *****.*..********* ******..* *.***.*****  *....*.*  *****

421   SQVPFIPSLAKGTGSLRVHEFNGKKHSLTDDWGERQWLIEPERYECHNVFEQYEGRELSE
      SQVPFCPSLAKGASTLRVHEYNGQMHTFSDQWGERMWLAEPERYECQNIFEQYEGRELSE
      ***** ******...******..*....*.**** ** *******.*.***********

481   VIAEGHGVRSDGKTLIAQPHTRGDNLWTLEDIKRAGCVFPDPLAKF-
      VIAELHGLRSDGKTLIAQPHVRGDKLWTLDDIKRLNCVFKNPVKAFN
      **** **.************.***.****.****..***..*. *
```

Figure 10. Amino acid sequence homology comparison of *M. trichosporium* OB3b sMMO α subunit with *M. capsulatus* (Bath) sMMO α subunit. Asterisks below the aligned sequences denote conserved residues, dots denote conservative substitutions. Overlines denote potential iron ligands.

TABLE VIII. Comparative Analysis of the Deduced Polypeptide Sequences of *M. trichosporium* OB36 and *M. capsulatus* sMMO Components

	M. capsulatus (Bath) sMMO			*M. trichosporium* OB3b sMMO		
Component	No. of aa's	Mwt (kDa)	Component	No. of aa's	Mwt (kDa)	Percentage similarity
Protein A			Protein A			
α	527	60.6	α	525	60	94
β	387	44.7	β	394	45	83.5
γ	170	19.8	γ	169	19.3	85
Protein B	141	16	Protein B	138	14.9	89.4
Protein C	348	38.5	Protein C	340	38	78

with the continued construction of broad-host-range plasmids, this should allow more success with the genetics of methanotrophs. Mutant isolation will probably remain difficult, mainly due to the obligate nature of these organisms. The main advances are likely to be achieved using the reverse genetics approach where marker-exchange mutagenesis results in the interruption of a cloned gene of interest with a transposon or interposon.

At the molecular level, our knowledge of the regulation of methane oxidation should be greatly increased now that the genes encoding MMO have been cloned and sequenced. We can begin to ask how oxygen, copper, methane, and its oxidation products regulate this gene cluster. The expression of these genes in a heterologous host should also allow site-directed mutagenesis studies on the active site of MMO and will strengthen our understanding of the mechanism of enzyme action. Once more is known about gene organization and expression in methanotrophs, then common-consensus promoter sequences can be identified and compared with their counterparts in methanol utilizers. At present, most of these types of studies are directed at soluble MMO, but it should be only a relatively short time before the genes encoding particulate MMO are cloned and characterized. The molecular events underlying the fascinating switch between these two modes of methanotrophy may then be analyzed.

REFERENCES

Allen, L. N., and Hanson, R. S., 1985, Construction of broad-host-range cosmid vectors: identification of genes necessary for growth of *Methylobacterium organophilum* on methanol, *J. Bacteriol.* **161**:955–962.

Al-Taho, N. M., and Warner, P. J., 1987, Restoration of phenotype in *Escherichia coli* auxotrophs by pULB113-mediated mobilisation from methylotrophic bacteria, *FEMS Microbiol. Lett.* **43**:235–239.

Al-Taho, N. M., Cornish, A., and Warner, P. J., 1990, Molecular cloning of the methanol dehydrogenase structural gene from *Methylosinus trichosporium* OB3b, *Curr. Microbiol.* **20**:153–157.

Anthony, C., 1986, Bacterial oxidation of methane and methanol, *Adv. Micro. Physiol.* **27**:113–210.

Bagdasarian, M., Lurz, R., Ruckert, B., Franklin, F. C. H., Bagdasarian, M. M., Frey, J., and Timmis, K. N., 1981, Specific-purpose plasmid cloning vectors II. Broad host range, high copy number, RSF1010-derived vectors and a host vector system for gene cloning in *Pseudomonas, Gene* **16**:237–247.

Bainbridge, B. W., 1983, The potential of methylotrophic bacteria: genetic approaches to improving bacteria of industrial interest, in: *Advances in Fermentation Conference Proceedings* (supplement to *Process Biochemistry*), Wheatland Journals, Rickmansworth, England, pp. 97–107.

Best, D. J., and Higgins, I. J., 1981, Methanol-oxidizing activity and membrane morphology in a methanol-grown obligate methanotroph, *Methylosinus trichosporium* OB3b, *J. Gen. Microbiol.* **125:**73–84.

Boulnois, G. J., Varley, J. M., Sharpe, G. S., and Franklin, F. C. H., 1985, Transposon donor plasmids, based on Co1Ib-P9 for use in *Pseudomonas putida* and a variety of other gram negative bacteria, *Mol. Gen. Genet.* **200:**65–67.

Brusseau, G. A., Tsien, H-G., Hanson, R. S., and Wackett, L. P., 1990, Optimization of trichloroethylene oxidation by methanotrophs and the use of a colorimetric assay to detect soluble methane monooxygenase activity, *Biodegradation* **1:**19–29.

Burrows, K. J., Cornish, A., Scott, D., and Higgins, I. J., 1984, Substrate specificities of the soluble and particulate methane monooxygenase of *Methylosinus trichosporium* OB3b, *J. Gen. Microbiol.* **130:**3327–3333.

Cardy, D. L. N., 1989, The molecular biology of ammonia assimilation in the obligate methane-oxidizing bacterium *Methylococcus capsulatus* (Bath), Ph.D. thesis, University of Warwick, Coventry, U.K.

Cardy, D. L. N., and Murrell, J. C., 1990, Cloning, sequencing and expression of the glutamine synthetase structural gene (*gln*A) from the obligate methanotroph *Methylococcus capsulatus* (Bath), *J. Gen. Microbiol.* **136:**343–352.

Cardy, D. L. N., Laidler, V., Salmond, G. P. C., and Murrell, J. C., 1991a, Molecular analysis of the methane monooxygenase (MMO) gene cluster of *Methylosinus trichosporium* OB3b, *Mol. Microbiol.* **5:**335–342.

Cardy, D. L. N., Laidler, V., Salmond, G. P. C., and Murrell, J. C., 1991b, The methane monooxygenase gene cluster of *Methylosinus trichosporium* OB3b; cloning and sequencing of the *mmo* C gene, *Arch. Microbiol.*, in press.

Chassey, B. M., Mercenier, A., and Flickinger, J., 1988, Transformation of bacteria by electroporation, *Trends Biotechnol.* **6:**303–309.

Collins, J. F., and Coulson, A. F. W., 1987, Molecular sequence comparison and sequence alignment, in: *Nucleic Acid and Protein Sequence Analysis: A Practical Approach* (M. J. Bishop and C. J. Rawkins, eds.), IRL Press, Oxford, pp. 323–358.

Dixon, R., 1984, The genetic complexity of nitrogen fixation, *J. Gen. Microbiol.* **130:**2745–2755.

Eccleston, M., and Kelly, D. P., 1972a, Assimilation and toxicity of exogenous amino acids in the methane-oxidizing bacterium *Methylococcus capsulatus*, *J. Gen. Microbiol.* **71:**521–554.

Eccleston, M., and Kelly, D. P., 1972b, Competition among amino acids for incorporation into *Methylococcus capsulatus*, *J. Gen. Microbiol.* **73:**303–314.

Fiedler, S., and Wirth, R., 1988, Transformation of bacteria with plasmid DNA by electroporation, *Anal. Biochem.* **170:**38–44.

Figurski, D., and Helinski, D. R., 1979, Replication of an origin-containing derivative of the plasmid RK2 dependent on a plasmid function provided *in trans.*, *Proc. Natl. Acad. Sci. USA* **76:**1648–1652.

Friedman, M., Long, S. R., Brown, S. E., Buikema, W. J., and Ausubel, F. M., 1982, Construction of a broad host range cosmid cloning vector and its use in the genetic analysis of *Rhizobium* mutants, *Gene* **18:**289–292.

Haber, C. L, Allen, L. N., and Hanson, R. S., 1983, Methylotrophic bacteria: diversity and genetics, *Science* **221:**1147–1152.

Hanson, R. S., 1980, Ecology and diversity of methylotrophic organisms, *Adv. Appl. Microbiol.* **26:**3–39.

Harwood, J. H., Williams, E., and Bainbridge, B. W., 1972, Mutation of the methane oxidizing bacterium, *Methylococcus capsulatus*, *J. Appl. Bact.* **35**(1):99–108.

Haselkorn, R., 1986, Organization of the genes for nitrogen fixation in photosynthetic bacteria and cyanobacteria, *Annu. Rev. Microbiol.* **40:**525–547.

Higgins, D. G., and Sharp, P. M., 1988, CLUSTAL: a package for performing multiple sequence alignments on a microcomputer, *Gene* **73:**237–244.

Higgins, D. G., and Sharp, P. M., 1989, Fast and sensitive multiple sequence alignments on a microcomputer, *CABIOS* **5:**151–153.

Holloway, B. W., 1981, The application of pseudomonad-based genetics to methanotrophs, in: *Microbial Growth on C_1 Compounds* (H. Dalton, ed.), Heyden, London, pp. 317–324.

Holloway, B. W., Kearney, P. P., and Lyon, B. R., 1987, The molecular genetics of C_1-utilizing micro-organisms—an overview, in: *Proceedings of the 5th International Symposium, Microbial Growth on C_1 Compounds* (H. W. Verseveld and J. A. Duine, eds.), Martinus Nijhoff, Dordrecht, pp. 223–229.

Hou, C. T., Laskin, A. I., and Patel, R., 1978, Growth and polysaccharide production by *Methylocystis parvus* OBBP on methanol, *Appl. Environ. Microbiol.* **37:**800–804.

Keen, N. T., Tamaki, S., Kobayashi, D., and Trollinger, D., 1988, Improved broad-host-range plasmids for DNA cloning in gram-negative bacteria, *Gene* **70:**191–197.

Knauf, V. C., and Nester, E. W., 1982, Wide host range cloning vectors: a cosmid clone bank of *Agrobacterium* Ti plasmids, *Plasmid* **8:**45–54.

Leak, D. J., and Dalton, H., 1983, *In vitro* studies of primary alcohols, aldehydes and carboxylic acids as electron donors for the methane monooxygenase in a variety of methanotrophs, *J. Gen. Microbiol.* **129:**3487–3497.

Lidstrom, M. E., and Wopat, A. E., 1984, Plasmids in methanotrophic bacteria: isolation, characterization and DNA hybridization analysis, *Arch. Microbiol.* **140:**27–33.

Lidstrom, M. E., Wopat, A. E., Nunn, D. N., and Toukdarian, A. E., 1984, Manipulation of methanotrophs, in: *Genetic Control of Environmental Pollutants* (G. S. Omenn and A. Hollaender, eds.), Plenum Press, New York, pp. 273–279.

Lidstrom-O'Connor, M. E., Fulton, G. F., and Wopat, A. E., 1983, *Methylobacterium ethanolicum:* a syntrophic association of two methylotrophic bacteria, *J. Gen. Microbiol.* **129:**3139–3148.

Lynch, M. J., Wopat, A. E., and O'Connor, M. E., Characterization of two new facultative methanotrophs, *Appl. Environ. Microbiol.* **40:**400–407.

Lund, J., Woodland, W. P., and Dalton, H., 1985, Electron transfer reactions in the soluble methane monooxygenase of *Methylococcus capsulatus* (Bath), *Eur. J. Biochem.* **147:**297–305.

McPheat, W. L., Mann, N. H., and Dalton, H., 1987a, Isolation of mutants of the obligate methanotroph *Methylomonas albus* defective in growth on methane, *Arch. Microbiol.* **148:**40–43.

McPheat, W. L., Mann, N. H., and Dalton, H., 1987b, Transfer of broad host range plasmids to the type I obligate methanotroph *Methylomonas albus, FEMS Micro. Lett.* **41:**185–188.

Merrick, M. J., 1988, Regulation of nitrogen assimilation by bacteria, in: *SGM Symposium 42, The Nitrogen and Sulphur Cycle* (J. A. Cole and S. J. Ferguson, eds.), Cambridge University Press, Cambridge, pp. 331–361.

Mullens, I. A., and Dalton, H., 1987, Cloning of the gamma-subunit methane monooxygenase from *Methylococcus capsulatus, Biotechnology* **5:**490–493.

Murrell, J. C., 1981, Nitrogen metabolism in the obligate methane-oxidizing bacteria, Ph.D. thesis, University of Warwick, Coventry, England.

Murrell, J. C., and Dalton, H., 1983a, Nitrogen fixation in obligate methanotrophs, *J. Gen. Microbiol.* **129:**3481–3486.

Murrell, J. C., and Dalton, H., 1983b, Purification and properties of glutamine synthetase from *Methylococcus capsulatus* (Bath), *J. Gen. Microbiol.* **129:**1187–1196.

Nicolaidis, A. A., and Sargent, A. W., 1987, Isolation of methane monooxygenase-deficient mutants from *Methylosinus trichosporium* OB3b using dichloromethane, *FEMS Microbiol. Lett.* **41:**47–52.

Nordlund, P., Sjöberg, B-M., and Eklund, H., 1990, Three-dimensional structure of the free radical protein of ribonucleotide reductase, *Nature* **345:**593–598.

Nunn, D. N., and Lidstrom, M. E., 1986a, Isolation and complementation analysis of 10 methanol oxidation mutant classes and identification of the methanol dehydrogenase structural gene of *Methylobacterium* sp. strain AM1, *J. Bacteriol.* **166:**581–590.

Nunn, D. N., and Lidstrom, M. E., 1986b, Phenotypic characterization of 10 methanol oxidation mutant classes in *Methylobacterium* sp. strain AM1, *J. Bacteriol.* **166:**591–597.

Oakley, C. J., and Murrell, J. C., 1988, *nif*H genes in the obligate methane-oxidizing bacteria, *FEMS Microbiol. Lett.* **49:**53–57.

O'Connor, M. L., 1981, Regulation and genetics in facultative methylotrophic bacteria, in: *Microbial Growth on C_1 Compounds* (H. Dalton, ed.), Heyden, London, pp. 294–300.

Pawlowski, K., Ratet, P., Sohell, J., and de Bruijn, F. J., 1987, Cloning and characterization of *nif*A and *ntr*C genes of the stem nodulating bacterium ORS571, the nitrogen fixing symbiont of *Sesbania rostrata*, *Mol. Gen. Genet.* **206:**207–216.

Pilkington, S. J., Salmond, G. P. C., Murrell, J. C., and Dalton, H., 1990, Identification of the gene encoding the regulatory protein B of soluble methane monooxygenase, *FEMS Microbiol. Lett.* **77:**345–349.

Queen, L., and Korn, L. J., 1986, A comprehensive sequence analysis program for the IBM personal computer, *Nucl. Acid. Res.* **12:**581–599.

Sharpe, G. S., 1984, Broad-host-range cloning vectors for gram-negative bacteria, *Gene*, **29:**93–102.

Simon, R., Priefer, U., and Pühler, A., 1983, Vector plasmids for *in vivo* and *in vitro* manipulation of gram-negative bacteria, in: *Molecular Genetics of the Bacteria–Plant Interaction* (A. Pühler, ed.), Springer-Verlag, Berlin, Heidelberg, pp. 98–106.

Stainthorpe, A. C., Murrell, J. C., Salmond, G. P. C., Dalton, H., and Lees, V., 1989, Molecular analysis of methane monooxygenase from *Methylococcus capsulatus* (Bath), *Arch. Microbiol.* **152:**154–159.

Stainthorpe, A. C., Lees, V., Salmond, G. P. C., Dalton, H., and Murrell, J. C., 1990a, The methane monooxygenase gene cluster of *Methylococcus capsulatus* (Bath), *Gene* **91:**27–34.

Stainthorpe, A. C., Salmond, G. P. C., Dalton, H., and Murrell, J. C., 1990b, Screening of obligate methanotrophs for soluble methane monooxygenase genes, *FEMS Microbiol. Lett.* **70:**211–216.

Stephens, R. L., Haygood, M. G., and Lidstrom, M. E., 1988, Identification of putative methanol dehydrogenase (*mox*F) structural genes in methylotrophs and cloning of *mox*F genes from *Methylococcus capsulatus* (Bath) and *Methylomonas albus* BG8, *J. Bacteriol.* **170:**2063–2069.

Szeto, W. W., Nixon, B. T., Ronson, C. W., and Ausubel, F. M., 1987, Identification and characterization of the *Rhizobium meliloti ntr*C gene: *R. meliloti* has separate regulatory pathways for activation of nitrogen fixation genes in free-living and symbiotic cells, *J. Bacteriol.* **169:**1423–1432.

Tabor, S., and Richardson, C. C., 1985, A bacteriophage T7 RNA polymerase/promoter system for controlled exclusive expression of specific genes, *Proc. Natl. Acad. Sci. USA* **82:**1074–1078.

Tatra, P. K., and Goodwin, P. M., 1984, R-factor mediated chromosome mobilization in the facultative methylotroph *Pseudomonas* sp. strain AM1, in: *Microbial Growth on C_1 Com-*

pounds (R. L. Crawford and R. S. Hanson, eds.), American Society for Microbiology, Washington, DC, pp. 224-227.

Tikhonenko, A. S., Bespalova, I. A., Tyutikov, F. M., Martynkina, L. P., Gal'chenko, V. F., and Kriviskii, A. S., 1982, Lysogeny in methanotrophic bacteria, *Mikrobiologiya* **51**:482-486.

Toukdarian, A. E., and Lidstrom, M. E., 1984a, DNA hybridization analysis of the *nif* region of two methylotrophs and molecular cloning of *nif*-specific DNA, *J. Bacteriol.* **153**:925-930.

Toukdarian, A. E., and Lidstrom, M. E., 1984b, Molecular construction and characterization of *nif* mutants of the obligate methanotroph *Methylosinus* sp. strain 6, *J. Bacteriol.* **157**:979-983.

Tyutikov, F. M., Belyaevf, N. N., Smirnova, L. C., Tikhonenko, A. S., and Krivisky, A. S., 1976, Isolation of bacteriophages of methane-oxidizing bacteria, *Mikrobiologiya* **6**:1056-1062.

Tyutikov, F. M., Bespalova, I. A., Rebentish, B. A., Aleksandrushkina, N. N., and Krivisky, A. S., 1980, Bacteriophages of methanotrophic bacteria, *J. Bacteriol.* **144**:375-381.

Tyutikov, F. M., Yesipova, V. V., Rebentish, B. A., Bepalova, I. A., Alexandrushkina, N. I., Galchenko, V. V., and Tikhonenko, A. S., 1983, Bacteriophages of methanotrophs isolated from fish, *Appl. Environ. Microbiol.* **46**:917-924.

Waechter-Brulla, D., and Lidstrom, M. E., 1990, Localization of the putative promoter for the *mox*F gene of *Methylomonas* sp. A4, in: *Abstracts of the Annual Meeting, American Society for Microbiology*, Anaheim, CA, p. 234.

Warner, P. J., Higgins, I. J., and Drozd, J. W., 1977, Examination of obligate and facultative methanotrophs for plasmid DNA, *FEMS Microbiol. Lett.* **1**:339-342.

Warner, P. J., Higgins, I. J., and Drozd, J. W., 1980, Conjugative transfer of antibiotic resistance to methylotrophic bacteria, *FEMS Microbiol. Lett.* **7**:181-185.

Whittenbury, R., and Dalton, H., 1981, The methylotrophic bacteria, in: *The Prokaryotes. A Handbook of Habitats, Isolation, and Identification of Bacteria* (M. P. Starr, H. Stolp, H. G. Trüper, A. Balows, and H. G. Schlegel, eds.), Springer-Verlag, Berlin, pp. 894-902.

Whittenbury, R., Phillips, K. C., and Wilkinson, J. F., 1970, Enrichment, isolation and some properties of methane-utilizing bacteria, *J. Gen. Microbiol.* **61**:205-218.

Williams, E., and Bainbridge, B. W., 1971, Genetic transformation in *Methylococcus capsulatus*, *J. Appl. Bacteriol.* **34**:683-687.

Williams, E., and Bainbridge, B. W., 1976, Mutation repair mechanisms and transformation in the methane-utilizing bacterium *Methylococcus capsulatus*, in: *Genetics of Industrial Microorganisms* (K. D. MacDonald, ed.), 2nd International Symposium, Academic Press, London, pp. 313-321.

Williams, E., and Shimmin, M. A., 1978, Radiation-induced filamentation in obligate methylotrophs, *FEMS Microbiol. Lett.* **4**:137-141.

Williams, E., Shimmin, M. A., and Bainbridge, B. W., 1977, Mutation in the obligate methylotrophs *Methylococcus capsulatus* and *Methylomonas albus*, *FEMS Microbiol. Lett.* **2**:293-296.

Wirth, R., Friessenegger, A., and Fiedler, S., 1989, Transformation of various species of gram-negative bacteria belonging to 11 different genera by electroporation, *Mol. Gen. Genet.* **216**:175-177.

Woodland, M. P., and Dalton, H., 1984, Purification and properties of component A of the methane monooxygenase from *Methylococcus capsulatus* (Bath), *J. Biol. Chem.* **259**:53-59.

Wünsche, L, Werner, B., Hauche, H., Kadelmann, W. J., Hilger, U., Krivisky, A. S., Jessipowa, W. W., Tikhonenko, A. S., and Zimmermann, I., 1977, Nachweis und erste Characterisierung von Bakteriophagen obligat methanassimilierender Bakterien, *Z. Allg. Mikrobiol.* **17**:321-323.

The Physiology and Biochemistry of Aerobic Methanol-Utilizing Gram-Negative and Gram-Positive Bacteria

5

L. DIJKHUIZEN, P. R. LEVERING, and G. E. DE VRIES

1. INTRODUCTION

One-carbon compounds (C_1) at all oxidation levels between methane and carbon dioxide occur abundantly throughout nature. Methane is present in fossil deposits and is formed by methanogenic bacteria. Methanol arises by the hydrolysis of methyl esters and ethers such as pectin and lignin, which are present in plants. Methylated amines occur in plants and animals and are produced by microbial degradation of choline derivatives present in plant membrane material and animal tissue. Formate is a major end-product of mixed-acid fermentation and carbon dioxide is present in the atmosphere and, as carbonates, in natural waters and soil. It is not surprising, therefore, that microorganisms are found in nature which are capable of growth on such compounds as carbon and energy sources.

According to the definition of Colby and Zatman (1972), C_1-utilizing microbes can arbitrarily be divided into two groups, methylotrophs and autotrophs, depending on the oxidation level of the C_1 compound assimilated. In their terminology, methylotrophs are recognized by their ability

---

L. DIJKHUIZEN and G. E. DE VRIES • Department of Microbiology, University of Groningen, 9751 NN Haren, The Netherlands. P. R. LEVERING • Microbiological R&D Laboratories, Organon International BV, 5340 BH Oss, The Netherlands.

Methane and Methanol Utilizers, edited by J. Colin Murrell and Howard Dalton. Plenum Press, New York, 1992.

to utilize carbon compounds containing one or more carbon atoms (but no carbon–carbon bonds) as sole carbon and energy source for growth, assimilating carbon as formaldehyde or a mixture of formaldehyde and carbon dioxide. The term "autotroph," on the other hand, is exclusively associated with those bacteria which can assimilate carbon dioxide for their growth. Although this description is generally accepted, an exact definition of a methylotroph is still a matter of discussion. Anthony (1982) defines methylotrophs as those microbes able to grow at the expense of C_1 compounds, irrespective of the way C_1 units are assimilated. This classification includes methanol-utilizing bacteria such as *Xanthobacter autotrophicus* (Weaver and Lidstrom, 1985) and *Paracoccus denitrificans* (Cox and Quayle, 1975), which assimilate carbon dioxide via the Calvin cycle. For the latter type of bacteria, alternative designations have been suggested, namely, C_1-utilizing autotrophs (Colby *et al.*, 1979) or pseudomethylotrophs (Zatman, 1981). In this chapter the original definition put forward by Colby and Zatman (1972) is followed. Not all methylotrophs are able to utilize methane as sole carbon and energy sources; only those which can are termed methanotrophs (Higgins *et al.*, 1981). Finally, there are certain organisms, bacteria as well as yeasts, which are able to grow on methylated amines as sole source of nitrogen, but fail to utilize such compounds as sole source of carbon and energy. Zatman (1981) has proposed calling these organisms methazotrophs.

A feature common to all microorganisms able to utilize C_1 compounds as sole source of carbon and energy is their unique capability to synthesize a metabolite containing a C_3 skeleton from C_1 units. Once such a compound has been produced, the synthesis of all cell constituents (polysaccharides, proteins, nucleic acids, and lipids) further proceeds via the general pathways of intermediary metabolism, as found in other organisms. Detailed studies by many research groups over the last 30 years have generated a great deal of knowledge on the enzymology of C_1 metabolism. To date, this had led to the discovery and biochemical description of four different pathways that effect the synthesis of compounds containing carbon–carbon bonds from C_1 units (Quayle, 1980a,b; Anthony, 1982). These pathways are the Calvin cycle or ribulose bisphosphate (RuBP) pathway of carbon dioxide fixation, the ribulose monosphate (RuMP) cycle of formaldehyde fixation, the xylulose monosphate (XuMP) cycle of formaldehyde fixation, and the serine pathway, in which both formaldehyde and carbon dioxide are fixed (Fig. 1). One of these pathways, the XuMP cycle, is present only in yeasts (van Dijken *et al.*, 1978).

The physiology and biochemistry of aerobic methylotrophic bacteria and yeasts have been dealt with in various reviews (Anthony, 1986; de Vries, 1986; de Vries *et al.*, 1990) and monographs (Anthony, 1982; Large,

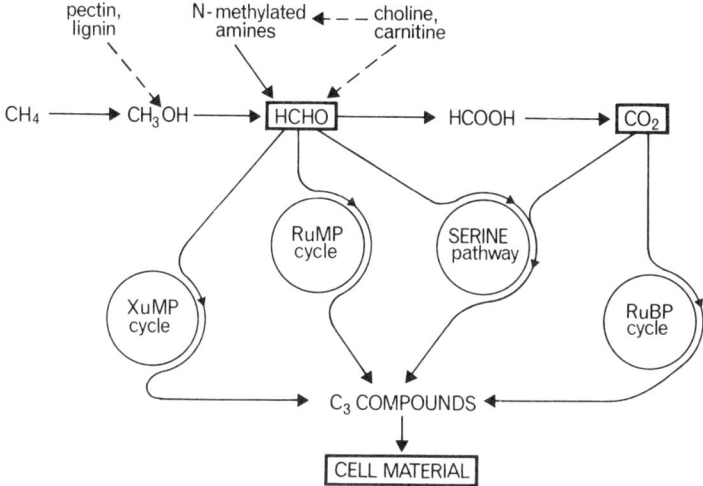

Figure 1. Dissimilatory and assimilatory pathways in C_1-utilizing microorganisms.

1983; Large and Bamforth, 1988). Various aspects of the regulation of C_1-specific enzymes in microorganisms employing the Calvin cycle (Dijkhuizen and Harder, 1984, 1985; Tabita, 1988), the serine pathway (Goodwin, 1990), and the XuMP cycle (de Koning and Harder, Chapter 7, this volume) have also been reviewed in recent years. Limited attention thus far has been focused on methanol utilization in bacteria employing the RuMP cycle. In this chapter we will review the physiology and biochemistry of methanol utilization by aerobic bacteria, with special emphasis on the enzymology and regulation of the RuMP cycle of formaldehyde fixation.

2. THE RIBULOSE MONOPHOSPHATE CYCLE OF FORMALDEHYDE FIXATION

2.1. Discovery and General Properties

The first observations on the ribulose monophosphate (RuMP) cycle of formaldehyde fixation were made by Quayle and his colleagues (Johnson and Quayle, 1965; Kemp and Quayle, 1966, 1967). They showed that short-term incubations of the methane-oxidizing bacterium *Methylomonas methanica* with ^{14}C-labeled methanol or formaldehyde resulted in the initial labeling of hexose phosphates. Subsequent studies with extracts of *M.*

methanica and of another methane utilizer, *Methylococcus capsulatus*, revealed the presence of an enzyme system that was able to catalyze the condensation of formaldehyde with pentose phosphate, resulting in a mixture of hexose phosphates consisting predominantly of fructose phosphate. The key reactions involved turned out to be those catalyzed by 3-hexulose-6-phosphate synthase (hexulose phosphate synthase; HPS) and phospho-3-hexuloisomerase (hexulose phosphate isomerase, HPI) (Lawrence *et al.*, 1970; Ferenci *et al.*, 1974; Kemp, 1972, 1974). Finally, by demonstrating the presence of essential cleavage and rearrangements enzymes in *M. methanica* and *M. capsulatus*, Strøm *et al.* (1974) were able to construct the net assimilation sequence shown in Figure 2. In the overall reaction of the RuMP cycle, one molecule of a C_3 compound, either dihydroxyacetone phosphate (which is readily isomerized to glyceraldehyde-3-phosphate) or pyruvate, is synthesized from three molecules of formaldehyde.

The RuMP cycle can be divided into three different stages (Fig. 2). Stage 1 ("fixation") is the aldol condensation of three molecules of formaldehyde with three molecules of ribulose-5-phosphate by way of HPS to yield three molecules of hexulose-6-phosphate, which are then isomerized by HPI to fructose-6-phosphate. This fixation sequence is common to all methylotrophic bacteria using the pathway. In stage 2 ("cleavage"), two C_3 compounds are produced from one of the molecules of fructose-6-phosphate. This may be achieved via two possible routes. Fructose-6-phosphate is either converted into fructose-1,6-bisphosphate (FBP), by phosphofructokinase, and subsequently cleaved by FBP aldolase (FBPA variant), or oxidized to 2-keto-3-deoxy-6-phosphogluconate (KDPG) by the enzymes of the Entner–Doudoroff pathway and ultimately cleaved in a reaction catalyzed by KDPG aldolase (KDPGA variant). This results in the formation of glyceraldehyde-3-phosphate plus the net product of the cycle, which is either dihydroxyacetone phosphate (from FBP) or pyruvate (from KDPG). In stage 3 ("rearrangement") the regeneration of three molecules of ribulose-5-phosphate is accomplished from the two remaining molecules of fructose-6-phosphate (from stage 1) and the one molecule of glyceraldehyde-3-phosphate produced in stage 2. Two variants of these sugar phosphate interconversions occur, both of which involve the enzymes transketolase, ribose-5-phosphate isomerase, and ribulose-5-phosphate, 3-epimerase. These rearrangement sequences differ in that they involve either transaldolase (TA variant) or sedoheptulose-1,7-bisphosphatase (SBPase variant). The combination of two cleavage and two rearrangement sequences thus leads to a total of four potential variants of the RuMP cycle which may be encountered in methylotrophic bacteria (Table I). Conversion of the end-product of the FBPA cleavage sequence,

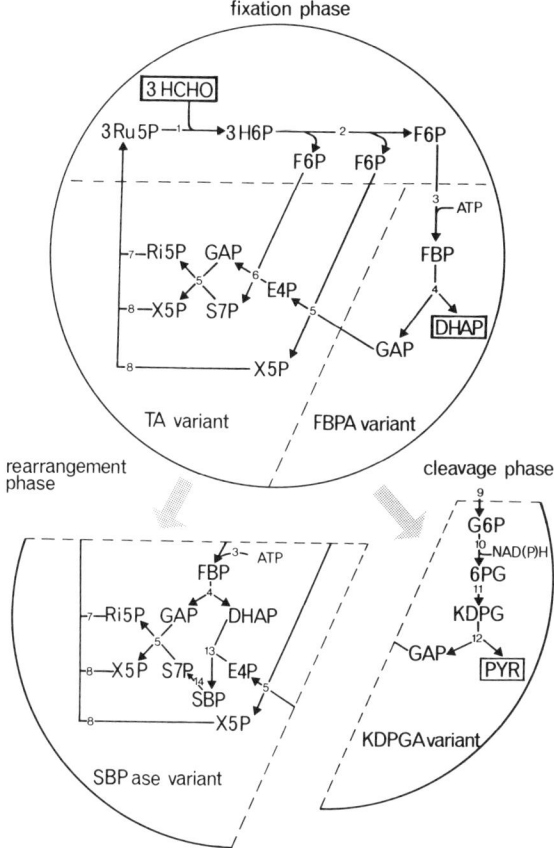

Figure 2. The RuMP cycle of formaldehyde fixation and its variants. The various reactions, indicated by numbers, are catalyzed by the following enzymes: 1. hexulose phosphate synthase; 2. hexulose phosphate isomerase; 3. phosphofructokinase; 4. fructors-1,6-bisphosphate (FBP) aldolase; 5. transketolase; 6. transaldolase (TA); 7. ribose-5-phosphat4e isomerase; 8. ribulose-5-phosphate 3-spimerase; 9. phosphoglucose isomerase; 10. glucose-6-phosphate dehydrogenase; 11. 6-phosphogluconate dehydratase; 12. 2-keto-3-deoxy-6-phosphogluconate (KDPG) aldolase; 13. sedoheptulose-1,7-bisphosphate aldolase; 14. sedoheptulose-1,7-bisphosphatase (SBPase). Abbreviations: DHAP, dihydroxyacetone phosphate; E4P, erythrose-4-phosphate; FBP, fructorse-1,6-bisphosphogluconate; 6PG, 6-phosphogluconate; PYR, pyruvate; Ri5P, ribose-5-phosphate; Ru5P, ribulose-5-phosphate; SBP, sedoheptulose-1,7-bisphosphate; S7P, sedoheptulose-7-phosphate, X5P, xylulose-5-phosphate.

TABLE I. Energetics of the RuMP Cycle of Formaldehyde Fixation[a]

Variant	Energy change (3HCHO → 1 pyruvate)	
	NAD(P)H$_2$	ATP
1. KDPGA/TA	+1	0
2. FBPA/SBPase	+1	0
3. KDPGA/SBPase	+1	−1
4. FBPA/TA	+1	+1

[a]Pyruvate is considered to be the end-product of each variant; see text. For abbreviations, see text.

dihydroxyacetone phosphate, into pyruvate by conventional glycolytic steps according to the overall reaction:

Dihydroxyacetone phosphate + NAD$^+$ + 2ADP → pyruvate + NADH + H$^+$ + 2ATP

allows a comparison of the respective yields, or requirements, for ATP and reducing power of these four variants. Only the combination of the FBP aldolase mode of sugar phosphate cleavage and the transaldolase mode of ribulose-5-phosphate formation is actually exergonic (provided pyruvate is the end-product), whereas the KDPGA/SBPase variant is energetically the most unfavorable. However, carbon assimilation via this latter variant of the RuMP cycle is, in terms of energy requirement, equally costly as the XuMP cycle, operating in yeasts (Anthony, 1982), and much more efficient than the serine pathway and the Calvin cycle (Quayle and Ferenci, 1978; Quayle, 1980a).

2.2. Occurrence of the RuMP Pathway of Formaldehyde Fixation

Although the presence of the key enzymes of the RuMP cycle was first demonstrated in obligate methane-utilizing bacteria, nonmethane utilizers soon appeared on the list of microorganisms using this pathway (Table II). This eliminated the possibility that the RuMP cycle might be uniquely associated with methanotrophy, as was initially thought (Colby and Zatman, 1972; Quayle, 1972). Over the years the number of methylotrophs reported to employ the RuMP cycle has increased considerably, with new representatives steadily being isolated. It must be emphasized, however, that the enzymology of many of these strains is rather poorly documented.

TABLE II. Distribution of Cleavage and Rearrangement Variants of the RuMP Cycle

Organism	Mode of cleavage	Mode of arrangement	Reference[b]
Obligate methanotrophs			
Methylococcus capsulatus (Texas)	KDPGA	TA	1
Methylomonas methanica	KDPGA	TA	1
Obligate methylotrophs			
Pseudomonas W6	KDPGA	TA	2
Pseudomonas C	KDPGA	TA	3
Pseudomonas oleovorans	KDPGA	TA	4
Organism 4B6	KDPGA	TA	5,6
Organism C2A1	KDPGA	TA	5,6
Organism W3A1	KDPGA	TA	5,6
Organism W6A	KDPGA	TA	5,6
Methylophilus methylotrophus	KDPGA	TA	7
Methylophilus methanolovorus	KDPGA	TA	8
Methylobacillus flagellatum KT	KDPGA	TA	9
Facultative methylotrophs			
Bacillus PM6	FBPA	SBPase	6
Bacillus S2A1	FBPA	SBPase	6
Bacillus C1	FBPA	TA	10
Acetobacter methanolicus MB58	FBPA	—[a]	11
Brevibacterium fuscum 24	FBPA	—	12
Mycobacterium vaccae 10	FBPA	—	12
Arthrobacter globiformis B-175	FBPA	—	12,13
Arthrobacter P1	FBPA	TA	14
Nocardia sp. 239	FBPA	TA	15
Mycobacterium gastri	—	—	16

[a] —, no data available. For abbreviations, see text.
[b] References: (1) Strøm *et al.* (1974); (2) Babel and Miethe (1974); (3) Chalfan and Mateles (1972), Stieglitz and Mateles (1973), Ben-Bassat and Goldberg (1977), Samuelov and Goldberg (1982a,b), Ben-Bassat *et al.* (1980); (4) Trotsenko (1983), Loginova and Trotsenko (1977a,b), Sokolov and Trotsenko (1978); (5) Colby and Zatman (1972, 1973); (6) Colby and Zatman (1975a,b); (7) Beardsmore *et al.* (1982); (8) Loginova and Trotsenko (1981), Loginova *et al.* (1981); (9) Govorukhina *et al.* (1987); (10) Dijkhuizen *et al.* (1988), Arfman *et al.* (1989); (11) Babel and Mothes (1978), Steudel *et al.* (1980), Uhlig *et al.* (1986); (12) Loginova and Trotsenko (1977b, 1979); (13) Loginova and Trotsenko (1976a,b); (14) Levering *et al.* (1981, 1982); (15) Kato *et al.* (1974, 1975, 1977b), Hazeu *et al.* (1983), de Boer *et al.* (1990a); (16) Kato *et al.* (1988).

In some of these organisms, for instance, the only evidence for the presence of the RuMP pathway has been a demonstration of rather low specific activities of HPS. Mutant evidence for the indispensable role of HPS in the RuMP pathway thus far has been obtained for only a single organism, *Arthrobacter* P1 (Levering *et al.*, 1987).

With respect to their physiology, the nonmethane utilizers using the RuMP cycle are conveniently divided here in two groups, the obligate and facultative methylotrophs. Organisms that in addition to their "methylotrophic" character were shown to grow (usually poorly) on glucose or fructose only will be considered here as obligate methylotrophs. This includes the "more restricted" facultative methylotrophs described by Colby and Zatman (1975a,b) (Table II). The same applies to *Pseudomonas* C and *Pseudomonas oleovorans,* organisms originally described as facultative methylotrophs employing the RuMP cycle (Chalfan and Mateles, 1972; Loginova and Trotsenko, 1977a; Müller and Sokolov, 1979), but later reported to be either obligate methylotrophs (Goldberg and Mateles, 1975; Samuelov and Goldberg, 1982b) or organisms that failed to grow on non-C_1 substrates in our laboratory, respectively.

Methylotrophic bacteria capable of growth on a variety of "heterotrophic" substrates are classified here as versatile methylotrophs. It is becoming increasingly clear that these facultative methylotrophs are found almost exclusively among gram-positive bacteria (Table II). Representatives are various bacilli, coryneform bacteria, and actinomycete species. The "less restricted" facultative methylotrophs *Bacillus* spp. PM6 and S2A1 (Colby and Zatman, 1975a,b) are placed in this category. Recently, we also reported the isolation of thermotolerant methylotrophic *Bacillus* spp. (i.e., strain C_1; Table II) employing the RuMP cycle (Dijkhuizen *et al.*, 1988). These *Bacillus* strains are able to grow in methanol mineral medium over a temperature range of 35–60°C. At the optimum growth temperatures (50–55°C), they display doubling times between 40 and 80 min. The metabolism of the strains studied is strictly respiratory. Further interesting features are their high molar growth yield with methanol and their marked resistance to high methanol concentrations. With *Bacillus* sp. C1, for instance, only a 50% reduction in growth rate is observed at a methanol concentration of 1.5 M. These versatile organisms are of interest from both a fundamental and an applied point of view. Gram-positive nonmethylotrophic bacteria have found wide application in the production of fine chemicals such as amino acids (de Boer and Dijkhuizen, 1990). Studies in this direction may be especially rewarding also with the methanol-utilizing coryneform, actinomycete, and *Bacillus* species now available. Facultative methylotrophs are more amenable to the extensive physiological and genetic manipulation required for strain development than the obligate gram-negative RuMP cycle methylotrophs (Levering and Dijkhuizen, 1986; Levering *et al.*, 1987; de Boer *et al.*, 1988, 1990a,b,c).

2.3. Distribution of Variants of the RuMP Pathway

In the original description of the RuMP cycle (Kemp and Quayle, 1966, 1967), involvement of FBP aldolase in the cleavage phase was proposed for the obligate methanotroph *M. methanica.* However, enzyme assays subsequently provided strong evidence for the operation of the alternative C_6 cleavage variant, which involves KDPG aldolase, both in this organism and in *M. capsulatus* (Strøm *et al.*, 1974). Although FBP aldolase is present also in these two organisms, the level of phosphofructokinase, the second key enzyme of this cleavage sequence, is so low that this route is probably only of minor physiological importance (Zatman, 1981). Transaldolase was found to be the key enzyme in the rearrangement reactions (Strøm *et al.*, 1974) in these methanotrophs. It is now evident that the KDPG aldolase/transaldolase variant of the RuMP cycle is the major C_1 assimilation route in all obligate methylotrophs studied (Table II). Variants involving FBP aldolase as the cleavage reaction are exclusively found among facultative methylotrophs. This includes the energetically most efficient combination, the FBP aldolase/transaldolase variant, which is encountered in *Arthrobacter* species, *Nocardia* sp. 239, and *Bacillus* sp. C1. Only three out of four possible combinations of cleavage and rearrangement phases of the RuMP cycle thus far have been reported to operate in these methylotrophs. The variant involving KDPG aldolase and SBPase is energetically the most unfavorable (Table I) and therefore perhaps the least likely to be found.

3. BIOCHEMISTRY OF ENZYMES INVOLVED IN METHANOL OXIDATION

3.1. Methanol Dehydrogenase in Gram-Negative Bacteria

Biochemical studies on the utilization of methanol in bacteria so far mainly have focused on gram-negative organisms. Although many microbes employ a convention NAD-linked alcohol dehydrogenase (EC 1.1.1.1) to support ethanol utilization, this enzyme generally has a very poor affinity for methanol, it at all. The gram-negative methylotrophs studies all appear to have opted for an entirely different enzyme system, involving pyrroloquinoline-quinone (PQQ; Duine and Frank, 1980; Jongejan and Duine, 1989) as a cofactor to drive electron transfer. The enzyme, methanol dehydrogenase (MDH, EC 1.1.98.8; Anthony and Zatman, 1967), is generally a dimeric protein consisting of two identical subunits with molecular weights ranging from 60,000 to 67,000. This enzyme is located in the periplasmic space (Alefounder *et al.*, 1981; Jones *et al.*, 1982), embedded in a pool of cytochrome c proteins (Quilter and Jones, 1984), which accept electrons from PQQ and donate these reducing equivalents directly to a terminal oxidase. PQQ-dependent methanol dehydrogenase oxidizes alkan-1-ols in general, although the affinity for the substrate decreases with increasing chain length. Formaldehyde may also serve as substrate presumably because of structural similarity with methanol when formaldehyde is in the hydrated form. Some enzymes studied have also been found to oxidize secondary alcohols (Sperl et al., 1974; Patel and Felix, 1976; Patel et al., 1979; Schär et al., 1985; Miyazaki et al., 1987). Because of this broad substrate specificity, the enzyme would be more accurately described as a quinoprotein alcohol dehydrogenase (Duine, 1988). The quinoprotein present in methylotrophs is nevertheless referred to as methanol dehydrogenase (MDH) because of its increased affinity for methanol (Km usually below 20 µM), which illustrates its apparent specialized function. The MDH proteins from different methylotrophs may, in addition to substrate range, differ considerably in isoelectric point (Yamanaka, 1981) or immunological cross-reactivity (Wolf and Hanson, 1978; Harms et al., 1987).

Methanol dehydrogenase activity *in vivo* most likely is further influenced by other factors. Only recently, a low-molecular-weight (12,000 Da) protein has been identified in *Methylobacterium extorquens* AM1, which is associated and copurifies with native MDH (Anderson and Lidstrom, 1988). Because of its low molecular weight, the protein may have escaped detection in previous studies [except in *Methylomonas* J (Ohta et al., 1981), *Xanthobacter autotrophicus* GJ10 (Janssen et al., 1987), and *Acetobacter methanolicus* (Elliot and Anthony, 1988)]. A heat-stable, oxygen-labile, low-molecular-weight factor X isolated from *Hyphomicrobium* X was found to stimulate electron transport from MDH to cytochrome c when tested in an assay system under physiological conditions (Dijkstra et al., 1988). Moreover, a modifier protein isolated from *M. extorquens* AM1 and *Methylophilus methylotrophus* was reported to act on the MDH protein, effectively decreasing the affinity of the enzyme for its product, formaldehyde (Ford et al., 1985; Page and Anthony, 1986).

For further information on the reaction mechanism of MDH, its interactions with special cytochromes in the electron transport chain, and the molecular biology of methanol dehydrogenase and PQQ synthesis, the reader is referred to the chapter by Lidstrom (Chapter 6, this book), the proceedings of a recent symposium on PQQ and Quinoproteins (Jongejan and Duine, 1989), and reviews by Anthony (1986), Goodwin (1990), de Vries (1986), and de Vries et al. (1990).

3.2. Methanol Dehydrogenase in Gram-Positive Bacteria

Only a limited number of gram-positive methylotrophic bacteria have been isolated in pure culture thus far (Table II). As a consequence, little is known about the enzymology of C_1 metabolism in gram-positive organisms. The available evidence indicates that these organisms are facultative methylotrophs, employing the RuMP cycle of formaldehyde fixation (with the exception of a *Corynebacterium* species, a serine pathway bacterium; Bastide *et al.*, 1989). Evidence is emerging now that especially the biochemistry of methanol oxidation in these strains is different from the enzyme systems normally encountered in gram-negative bacteria. Gram-positive bacteria do not possess a periplasmic space, thus raising the question of the location of the enzyme catalyzing the initial step of methanol oxidation. A single report has appeared on the cytoplasmic location of methanol-oxidizing activity in the actinomycete *Nocardia* sp. strain 239, presumably present in a multienzyme complex, with activities for NADH, formaldehyde, as well as methanol dehydrogenase (Duine *et al.*, 1984). Although addition of NAD was required for activity, accumulation of free NADH could not be detected with this system. Presumably, the cofactor was not in the usual free form, but remained enzyme-associated (Duine *et al.*, 1984). Cyclopropanol, a well-known inhibitor of the PQQ-dependent methanol dehydrogenase (EC 1.1.99.8) of gram-negative bacteria (Groeneveld *et al.*, 1984; Dijkstra *et al.*, 1984), did not inhibit this particular enzyme, although in denatured extracts PQQ was found to be present (Duine *et al.*, 1984). Further biochemical investigations of this enzyme system are certainly required in order to obtain a full understanding of the nature of this pathway of methanol oxidation.

All thermotolerant methylotrophic *Bacillus* strains (Dijkhuizen *et al.*, 1988; Al-Awadhi *et al.*, 1989; Brooke *et al.*, 1989) investigated were found to have an enzyme resembling classical (EC 1.1.1.1) NAD-dependent alcohol dehydrogenase (Arfman *et al.*, 1989). This NAD-dependent enzyme, however, can be distinguished from known alcohol dehydrogenases in its relatively high affinity for methanol. When challenged with a limited supply of methanol, the methylotrophic *Bacillus* strains may synthesize up to 30% of soluble protein in the form of this methanol dehydrogenase. The products formed apparently are efficiently removed by extremely high activities of HPS and NADH oxidase. Characterization of the purified enzyme revealed that it is composed of 10 identical subunits of 43,000 Da. Methanol dehydrogenase (but not formaldehyde reductase) activity requires the presence of Mg^{2+} and a second protein composed of two identical subunits of 25,000 Da.

The conversion of methanol and ethanol into their respective aldehydes does not greatly differ thermodynamically. The question therefore has to be asked why none of the gram-negative methylotrophs investigated have implemented an NAD-linked dehydrogenase for this reaction. Potentially this would be a more favorable mechanism with respect to energy generation. Methylotrophs are faced with two physiological problems. In order to grow relatively fast, a considerably higher flux of methanol conversion must be reached than would be required in heterotrophic organisms utilizing ethanol. The enzyme systems involved in methanol oxidation therefore should possess optimal turnover values. A highly active methanol oxidation system, on the other hand, may also cause problems, since toxic levels of formaldehyde could be produced when fluctuating methanol concentrations occur in the environment. The features of the PQQ-dependent methanol oxidation system in gram-negative bacteria seem to counteract these complications at least partly. First, the short and (energetically) wasteful electron transport pathway from methanol to oxygen may increase the turnover rate of the methanol oxidation complex. Most methylotrophs utilizing methanol as sole source of energy and reducing power are considered to be ATP- rather than NADH-limited (Anthony, 1982). Therefore, it may be favorable to immediately utilize some of the reducing equivalents, harnessed in methanol, in the form of ATP. Second, the noncytoplasmic location of formaldehyde formation might to some degree alleviate the cell from immediate damage if temporary increased amounts of formaldehyde were produced. The fact that a similar enzyme complex is employed by some methylotrophic bacteria for the utilization of ethanol further illustrates its potential advantage for alcohol utilization in general (Brooke and Attwood, 1983; Groen et al., 1984; Dijkstra et al., 1985; Weaver and Lidstrom, 1985). In gram-positive methylotrophs, which are devoid of a periplasmic space, the use of an NAD-linked dehydrogenase nevertheless appears to be a viable option (Dijkhuizen et al., 1988; Arfman et al., 1989).

3.3. Oxidation of Formaldehyde and Formate

Many methylotrophs generate reducing equivalents by the oxidation of formaldehyde via formate to CO_2. In these organisms the TCA cycle appears to play a minor role in energy metabolism during growth on C_1 compounds (Zatman, 1981; Shishkina and Trotsenko, 1982). Although the route for formaldehyde oxidation via formate to carbon dioxide appears straightforward, a larger number of distinct enzyme systems for the oxidation of formaldehyde to formate has been described in methylotrophs than for the conversion of any other C_1 compound.

Aldehyde dehydrogenases that may be involved in the direct oxidation of formaldehyde in methylotrophs (Babel and Mothes, 1978; Stirling and Dalton, 1978; Marison and Attwood, 1980; Anthony, 1982; Attwood and Quayle, 1984; Poels and Duine, 1989) can be distinguished by the nature of their electron acceptors. The reported enzymes may be divided into two groups: The NAD(P)-dependent dehydrogenases, which may or may not require reduced glutathione (GSH) or other cofactors (see below), and the dye-linked enzymes (EC 1.2.99.3). The latter group is generally measured by the reduction of certain dyes, such as 2,6-dichlorophen-olindophenol. *In vivo*, these enzymes may directly donate electrons to a component of the electron chain, leading to "direct" synthesis of ATP. Their precise location in the cell remains to be determined. Although methanol dehydrogenase itself is capable of formaldehyde oxidation *in vitro*, this appears not to be of physiological significance. In addition to the low specific activity and affinity of the enzyme with formaldehyde, mutant studies showed that strains lacking MDH were not affected in their ability to convert formaldehyde to formate (Heptinstall and Quayle, 1970; Harms *et al.*, 1985). Dye-linked aldehyde dehydrogenase activities are generally not specific for formaldehyde oxidation and their activities are rarely induced to higher levels during growth with methanol (Marison and Attwood, 1982; Weaver and Lidstrom, 1985). This makes it unlikely that these activities fulfill an important role in methanol metabolism. The situation is even less clear in certain autotrophic bacteria. In *Xanthobacter* H4-14, the presence of a formaldehyde dehydrogenase (dye-linked) could only be inferred from activity staining of polyacrylamide gels (Weaver and Lidstrom, 1985), while in *Paracoccus denitrificans,* enzyme activity could not be detected but its presence was suspected from the finding of aldehyde-sensitive mutants (Harms *et al.*, 1985). The fact that the latter mutants were unable to grow on methanol indicates the essential role of the supposed aldehyde-converting activity in this organism.

The high activities of cytoplasmic NAD-linked formaldehyde dehydrogenases (EC 1.2.1.46) reported for several organisms leave no doubt that part of the formaldehyde is imported and intracellularly converted to formate. This class of enzymes is diverse and can be further subdivided according to the additional required of cofactors, e.g., reduced glutathione (EC 1.2.1.1, Johnson and Quayle, 1964), tetrahydrofolate (NADP-dependent methylene-THF dehydrogenase, EC 1.5.1.20; Marison and Attwood, 1982; Attwood and Quayle, 1984), a low-molecular-weight peptide (Stirling and Dalton, 1978), a heat-stable, low-molecular-weight factor in the nonmethylotroph *Rhodococcus erythropolis* (Eggeling and Sahm, 1984, 1985). To complicate matters even further, different growth conditions may induce different types of enzymes (Levering *et al.*, 1987; van Ophem and Duine, 1989). Clearly, the wide variety of possible reactions for conversion of formaldehyde into formate necessitates a closer investigation of the route(s) of formaldehyde oxidation in methylotrophic bacteria (Duine, 1988).

More uniformity is found with respect to the oxidation of formate in those C_1-utilizing microbes that possess the direct oxidation route for formaldehyde. In most organisms, formate oxidation is mediated by a soluble NAD-dependent formate dehydrogenase, and rarely is a membrane-bound enzyme involved (Zatman, 1981; Karzanov *et al.*, 1989). Both membrane-associated (dye-linked) and NAD-linked activities have been reported, for instance in the autotrophs *Pseudomonas oxalaticus* (Dijkhuizen *et al.*, 1979). *Alcaligenes eutrophus* (Friedrich *et al.*, 1979), and *Paracoccus denitrificans* (Harms *et al.*, 1985). The NAD-linked enzyme in *Ps. oxalaticus* seems unusual because of the large molecular weight of the flavoprotein, its sensitivity toward oxygen, and the fact that NAD, oxygen, ferricyanide as well as redox dyes can serve as electron acceptors (Müller *et al.*, 1978). Both formate dehydrogenase enzymes from *Pa. denitrificans* contain a molybdenum cofactor (Harms *et al.*, 1985). A mutant that lacked the NAD-linked formate dehydrogenase activity still grew on formate and methanol, albeit with reduced efficiencies. In this strain NAD(P)H may therefore be generated by reversed electron flow, as is the case during autotrophic growth with hydrogen as sole source of energy.

3.4. Oxidation of Formaldehyde in Methylotrophs Using the RuMP Cycle

Two different routes of formaldehyde oxidation can be found in strains that use the RuMP pathway of formaldehyde fixation (Fig. 3). The direct route, which has formate as an intermediate, involves the two enzymes, formaldehyde dehydrogenase and formate dehydrogenase, discussed above (Fig. 3A). An alternative cyclic pathway for the complete oxidation of formaldehyde to CO_2 not involving these dehydrogenases was first described by Strøm *et al.* (1974) and Colby and Zatman (1975b). In this so-called dissimilatory RuMP cycle (Fig. 3B), hexulose phosphate synthase and hexulose phosphate isomerase are key enzymes and reducing equivalents are formed by the action of glucose-6-phosphate dehydrogenase and 6-phosphogluconate dehydrogenase.

To establish an overview of these various mechanisms used for formaldehyde oxidation (direct or cyclic) in RuMP cycle organisms, the data available in the literature on the presence of the various enzymes involved in both oxidative sequences in these organisms are presented in Table III. It appears that most obligate methanotrophs possess both the linear and cyclic pathways of formaldehyde oxidation. However, for these organisms very low activities of glucose-6-phosphate and 6-phosphogluconate dehydrogenases have been reported, while in the majority high levels of formaldehyde and formate dehydrogenases have been found (see references cited in Table III; see also Zatman, 1981). No formaldehyde dehydrogenase was detected in the Texas strain of *Methylococcus capsulatus*, but extracts

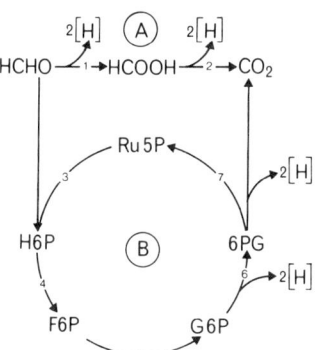

Figure 3. Mechanism of formaldehyde oxidation in RuMP-type methylotrophs. A. Linear oxidation pathway: B. Dissimilatory RuMP cycle. The numbers refer to the following enzymes: 1. formaldehyde dehydrogenase; 2. formate dehydrogenase; 3. hexulose phosphate synthase; 4. hexulose phosphate isomerase; 5. phosphoglucose isomerase; 6. glucose-6-phosphate dehydrogenase; 7. 6-phosphogluconate dehydrogenase; 7. 6-phosphogluconate dehydrogenase. For abbreviations see Fig. 2.

of this organism contained a high formate dehydrogenase activity. It therefore appears likely that *M. capsulatus* (Texas) mainly uses the linear route of formaldehyde oxidation via formate, as is the case in other obligate methanotrophs (Zatman, 1981).

The reverse situation exists in the (non-methane-utilizing) obligate methylotrophs. In these bacteria NAD(P)-linked glucose-6-phosphate and 6-phosphogluconate dehydrogenases are present at relatively high specific activities, and if enzymes responsible for the linear oxidation of formaldehyde are found, the levels of activity are low (Table III; Zatman, 1981). This suggests that the dissimilatory RuMP cycle is the major pathway for the generation of reducing equivalents in these organisms. Mutant studies with *Methylobacillus flagellatum* KT (Kletsova *et al.*, 1988) and isotopic labeling experiments with extracts from methanol-grown cells of *Pseudomonas* C (Ben-Bassat and Goldberg, 1977) and *Ps. oleovorans* (Sokolov and Trotsenko, 1978) have provided evidence that this is indeed the case. Subsequent studies with *Pseudomonas* C, involving addition of various ^{14}C-labeled compounds to methanol-limited chemostat cultures (Ben-Bassat *et al.*, 1980; Samuelov and Goldberg, 1982a) led to the conclusion that more methanol, and hence formaldehyde, is oxidized by way of formate than by the cyclic route. However, as outlined by Anthony (1982), objections can be made against part of the calculations that were performed to estimate the relative fluxes of carbon. Taking this into account, Anthony (1982) was able to estimate, from the data of Ben-Bassat *et al.* (1980), that more than 90% of the oxidation of formaldehyde may well have been accomplished via the dissimilatory RuMP cycle in *Pseudomonas* C.

The facultative RuMP-type methylotrophs constitute a rather heterogenous group with respect to their mode of formaldehyde oxidation (Table III). The data suggest that all *Bacillus* and *Arthrobacter* spp. investigated oxidize formaldehyde exclusively via the cyclic pathway (Colby

TABLE III. Oxidation of Formaldehyde in RuMP Cycle Methylotrophs[a]

Organism	Growth substrate	Formaldehyde dh		Formate dh		G6P dh		6PG dh		Reference[e]
		NAD	DCPIP	NAD	DCPIP	NAD	NADP	NAD	NADP	
Obligate methanotrophs										
Methylococcus capsulatus (Bath)	Methane	+[b]	−	+						1
M. capsulatus (Texas)	Methane	−	−	+					+	2,3
Methylomonas methanica	Methane	+[c]	+	+			+		+	2,3,4
Methylomonas GB3	Methane	+[c]		+					+	5
Methylomonas GB8	Methane	+[c]		+						5
Methylomonas albus	Methane	+		+		−		−		6
Methylococcus mobilis	Methane			+		−	+	−		7
Obligate methylotrophs										
Organism 4B6	TMA	+	−	+		+	+	+	+	8
Organism C2A1	TMA	−	+	+		+	+	+	+	8
Organism W3A1	TMA	+	+	+		+	+	+	+	8
Organism W6A	TMA	−		−	−	+	+	+	+	8
Methylophilus methylotrophus	Methanol	+[c]	−	+		+	+	+	+	9
Pseudomonas W6	Methanol	+		−		+	+	+	+	5,10
Methylomonas M15	Methanol			−		+	+	+		11
Pseudomonas C	Methanol	+[c]	+	+		+	+	+	+	12
Pseudomonas oleovorans	Methanol	+[c]	−	+		+	+	+	+	13
Methylomonas EP-1	Methanol	+		+						14
Methylomonas clara	Methanol	+		+		+		+		6

	Substrate						Ref
Methylobacillus glycogenes T-11	Methanol	+			+	+	15
Methylobacillus flagellatum KT	Methanol		+		+	+	16
Methylophilus methanolovorus	Methanol	+	−	+	+	+	17
Facultative methylotrophs							
Bacillus PM6	TMA	−	−	−	+	+	8
Bacillus S2A1	TMA	−	−	−	+	+	8
Bacillus C1	Methanol	−	−	−	−	+	18
Acetobacter methanolicus MB58	Methanol	$+^{b,c}$	+	−	+	+	5, 19
Brevibacterium fuscum 24	methylamine	$+^c$	−	+	+	+	20
Mycobacterium vaccae 10	Methanol	+	+	+	+	+	20
Arthrobacter globiformis B-175	Methylamine	−	+	−	−	+	21
Arthrobacter P1	Methylamine	−	−	−	−	+	22
Nocardia sp. 239	Methanol	$-^d$	(+)	+	+	+	23
Nocardia sp. 239	Methanol	−	−	+	+	+	24

[a] +, present; −, absent; open space, no data; TMA, trimethylamine; dh, dehydrogenase.
[b] Also NADP-dependent (GSH-independent) activity.
[c] GSH-dependent.
[d] Formate oxidized by whole cells.
[e] References: (1) Stirling and Dalton (1978); (2) Patel and Hoare (1971); (3) Strøm et al. (1974); (4) Johnson and Quayle (1964), Patel et al. (1979); (5) Babel and Mothes (1978); (6) Roitsch and Stolp (1985b); (7) Hazeu et al. (1980); (8) Colby and Zatman (1973, 1975b); (9) Large and Haywood (1981), Beardsmore et al. (1982); (10) Babel and Miethe (1974); (11) Sahm and Steinbach (1977); (12) Ben-Bassat and Goldberg (1977, 1980), Ben-Bassat et al. (1980); (13) Loginova and Trotsenko (1977a), Sokolov and Trotsenko (1978); (14) Papoutsakis et al. (1978); (15) Yordi and Weaver (1977); (16) Kiriuchin et al. (1988), Kletsova et al. (1988); (17) Trotsenko (1983), Loginova et al. (1981); (18) Arfman et al. (1989); (19) Gründig and Babel (1987a,b), Gründig and Doronina (1984), Uhlig et al. (1986); (20) Loginova and Trotsenko (1979); (21) Loginova and Trotsenko (1976b), Sokolov and Trotsenko (1985); (22) Levering et al. (1981); (23) Hazeu et al. (1983); (24) Kato et al. (1974, 1975, 1977b).

and Zatman, 1975b; Levering *et al.*, 1981; Sokolov and Trotsenko, 1985; Arfman *et al.*, 1989). Mutant studies with the acidophilic bacterium *Acetobacter methanolicus* MB58, on the other hand, showed that this organism employs the linear pathway for the oxidation of formaldehyde (Gründig and Babel, 1987a,b). Also, the actinomycete *Nocardia* sp. 239 presumably uses the direct route via formate. Methanol-grown cells of this organism are clearly able to oxidize formate, whereas only very low levels of glucose-6-phosphate dehydrogenase and 6-phosphogluconate dehydrogenase were detected in cell-extracts (Hazeu *et al.*, 1983). Although this suggests that the dissimilatory RuMP cycle is of minor importance only, the true identity of the enzymes involved in oxidation of the formaldehyde and formate produced from methanol remains to be established. Van Ophem and Duine (1989) recently reported the presence of multiple (form)aldehyde dehydrogenases in *Nocardia* sp. 239. The dye-linked formate dehydrogenase activity initially reported (Kato *et al.*, 1975) remained elusive in further work and no alternative formate-oxidizing enzyme system has yet been detected.

The data available on enzyme activities of the dissimilatory sequences of formaldehyde in *Brevibacterium fuscum* 24 and in *Mycobacterium vaccae* 10 do not allow a conclusion to be drawn as to which route is predominant (Table III). These organisms display activities of the enzymes of the cyclical as well as the linear pathway for formaldehyde oxidation. Evidence for a possible indispensable operation of either of these pathways may be obtained in mutant studies.

3.5. Hexulosephosphate Synthase and Hexulosephosphate Isomerase

The properties of hexulosephosphate synthase (HPS) and hexulosephosphate isomerase (HPI) purified from various sources have been discussed in detail by Anthony (1982). Some salient properties of HPS are given in Table IV. On the basis of a comparative study on biosynthetic exits for formaldehyde, Attwood and Quayle (1984) concluded that "the overall impression is one of a very highly efficient system for trapping free formaldehyde at low concentrations." HPS from *Methylococcus capsulatus* is a large, hexameric, particle-bound enzyme (Ferenci *et al.*, 1974), whereas the enzymes from all nonmethane-utilizing, obligately methylotrophic bacteria studied appear to be small, dimeric, soluble proteins. A similar enzyme is also present in the facultative bacterium *Mycobacterium gastri* MB19 (Kato *et al.*, 1988), whereas the enzyme purified from *Bacillus* sp. C1 most likely is a monomer (Arfman *et al.*, 1990). Also, the results of an immuno-

TABLE IV. Salient Properties of 3-Hexulose 6-Phosphate Synthases

Source	Specific activities		Km values (mM)		Mr (KDa)		Activator (M^{-1})	K equilibrium	Reference[e]
	a	b	RuMP	HCHO	c	d			
Methylococcus capsulatus	2.3	69	0.083	0.49	310	49	Mg^{2+}/Mn^{2+}	2.5×10^4	1
Methylomonas M15	4.5	66.5	1.6	1.1	40–43	22	Mg^{2+}/Mn^{2+}		2
Methylomonas aminofaciens	3.3	53	0.059	0.29	45–47	23	Mg^{2+}/Mn^{2+}		3
Pseudomonas oleovorans	—	196	—	1	45	—	Mg^{2+}/Mn^{2+}		4
Methylotrophus methylophilus	3.8	97.4	0.136	0.53	40	22.5	Mg^{2+}/Mn^{2+}		5
Arthrobacter 2B2	1.4	19	0.12	0.57	15.5	8.2	Mg^{2+}/Mn^{2+}		6
Bacillus C1	3.5	64	0.45	0.15	32	27	Co^{2+}/Mn^{2+} Mg^{2+}/Ni^{2+}		7
Mycobacterium gastri	3.2	74.2	??	1.4	43	24	Mg^{2+}/Mn^{2+}		8

[a,b]Specific activities [μmol. (min.mg protein)$^{-1}$] in crude extracts and most pure preparations, respectively.
[c,d]Molecular weight of holoenzymes and subunits, respectively.
[e]References: (1) Ferrenci et al. (1974); (2) Sahm et al. (1976); (3) Kato et al. (1977a, 1978); (4) Müller and Sokolov (1979); (5) Beardsmore et al. (1982); (6) Attwood and Quayle (1984); (7) Arfman et al. (1990); (8) Kato et al. (1988).

logical investigation indicate that the enzyme from the latter organism is unusual. Antibodies raised against purified HPS from *Methylophilus methylotrophus* cross-reacted with proteins in extracts from various other obligately methylotrophic bacteria and methanotrophic bacteria, but not with the purified enzyme from *Bacillus* sp. C1 (Burton *et al.*, 1989). HPS and HPI generally are highly specific for their substrates (Anthony, 1982). One exception appears to be the HPS enzyme from *M. gastri,* which was found to use glycolaldehyde and methylglyoxal as well as formaldehyde with ribulose-5-phosphate as the acceptor molecule. The product of the reaction with glycolaldehyde was tentatively identified as 4-heptulose 7-phosphate (Kato *et al.*, 1988).

4. METABOLIC REGULATION

In general, bacteria adjust the levels and expression of activity of various enzymes will be synthesized which are required to metabolize a particular carbon source. In the following sections the available information on the physiological regulation of methanol oxidation and the RuMP cycle is reviewed.

4.1. Regulation of Methanol Dehydrogenase Synthesis

Limited information is available on the control of enzymes involved in methanol oxidation. Studies on the regulation of methanol dehydrogenase (MDH) synthesis suggest that obligate and restricted RuMP cycle methylotrophs produce the enzyme constitutively. The level of its synthesis, however, is clearly subject to regulatory mechanisms. Increased MDH activity levels were observed, for instance, during growth of *Methanolomonas glucoseoxidans* in batch cultures on "heterotrophic" substrates that are metabolized slowly, and in chemostat cultures at low growth rates under methanol-limiting conditions (Roitsch and Stolp, 1985a). The latter phenomenon was also observed in studies with *Methylophilus methylotrophus* (Greenwood and Jones, 1986). The available data suggest that the concentration and activity of methanol dehydrogenase are inversely related to the standing concentration of methanol in the culture and/or the extent of energy limitation (Greenwood and Jones, 1986). Formaldehyde and formate dehydrogenase levels in these organisms may be controlled coordinately with MDH (Roitsch and Stolp, 1985a, 1986).

Constitutive synthesis of MDH is also apparent in serine pathway facultative methylotrophs. Roitsch and Stolp (1986) provided evidence that, in the absence of methanol, expression of methanol dehydrogenase activity is inversely related to the growth rates of the organisms studied. Formaldehyde dehydrogenase and formate dehydrogenase remained largely repressed during growth with multicarbon substrates. Addition of methanol to the medium may (Weaver and Lidstrom, 1985) or may not (Roitsch and Stolp, 1986) result in a strong further induction of the methanol dehydrogenase in *Methylobacterium extorquens* AM1.

Control of MDH synthesis in methanol-utilizing bacteria that assimilate carbon dioxide via the Calvin cycle appears to be completely different from the mechanisms observed in RuMP and serine pathway organisms. In these autotrophic bacteria, e.g., *Paracoccus denitrificans* and *Xanthobacter* strains, MDH activity generally cannot be detected during growth with multicarbon substrates (Harms *et al.*, 1985; Weaver and Lidstrom, 1985; de Vries *et al.*, 1988; Meijer *et al.*, 1990). Moreover, MDH activity levels in cells of *Xanthobacter* strain 25a were found to increase with increasing growth rates in methanol-limited chemostats (Croes *et al.*, 1990). MDH synthesis in these organisms appears to be controlled via a mechanism of induction (by methanol and/or formaldehyde), modified by catabolite repression. The available data also suggest that synthesis of MDH and formate dehydrogenase is regulated noncoordinately.

4.2. Regulation of the RuMP Cycle of Formaldehyde Fixation

In obligate methylotrophs (including methanotrophs) strict regulation of the RuMP cycle is not likely to occur at the level of enzyme synthesis, but rather directly on the activities of enzymes, especially those involved in balancing the flow of formaldehyde into assimilation and dissimilation. In methanotrophs, which oxidize formaldehyde via formate (see above; Table III), the key enzymes of the RuMP cycle HPS and HPI are thought to play mainly an anabolic role. Metabolic control exerted at these enzymes would therefore provide a mechanism for the regulation of carbon assimilation. However, no evidence for such a control mechanism was obtained in studies with HPS and HPI purified from methane-grown *Methylococcus capsulatus* (Texas) (Ferenci *et al.*, 1974). It was therefore speculated that the target for control may be glucose-6-phosphate dehydrogenase of KDPG aldolase, enzymes involved in the generation of glyceraldehyde-3-phosphate, required for the regeneration of the formaldehyde-acceptor molecule ribulose-5-phosphate.

Obligate methylotrophs employ the cyclical route for formaldehyde oxidation (Table III) and the KDPG aldolase cleavage variant of the RuMP cycle for carbon assimilation (Table II). It follows, then, that in these organisms the metabolic branch point is not at the level of formaldehyde, but at 6-phosphogluconate (Figs. 2, 3). In these bacteria HPS and HPI have an amphibolic role, and hence, it is unlikely that these enzymes are targets for regulation. Results of studies on properties of HPS purified from the obligate methanol utilizers *Methylophilus methylotrophus* (Beardsmore *et al.*, 1982) and *Methylomonas* M15 (Sahm *et al.*, 1976) support this view. The activities of these enzymes were not (or were hardly) affected by a number of sugar phosphates, ATP, ADP, AMP, NAD, NADP, or NADPH. The most likely site for regulation of the distribution of formaldehyde over the dissimilation and assimilation in these organisms appears to be glucose-6-phosphate dehydrogenase and 6-phosphogluconate dehydrogenase. These enzymes have been purified from various RuMP cycle methylotrophic bacteria, and all seem to have rather similar regulatory properties, inhibition by reduced pyridine nucleotides and nucleoside triphosphates being a common feature (Anthony, 1982; Kiriuchin *et al.*, 1988; Kletsova *et al.*, 1988). Probably the most extensively investigated organism among the obligate RuMP cycle methylotrophs, *Methylophilus methylotrophus*, may serve as an example to illustrate the regulation of carbon metabolism in these bacteria (Quayle, 1980a,b; Beardsmore *et al.*, 1982). The glucose-6-phosphate dehydrogenase of *M. methylotrophus* exhibits dual substrate specifity with respect to NADP and to NAD. The NADP-dependent activity of glucose-6-phosphate dehydrogenase is competitively inhibited by NADPH and ATP, and under conditions of high energy levels in the cells, this probably results in a decreased flow of formaldehyde carbon toward both assimilation and dissimilation. The specific activity of this enzyme is rather high, and therefore it appears likely that sufficient glucose-6-phosphate is still converted into 6-phosphogluconate to satisfy the need of the cells for carbon (via 6-phosphogluconate dehydratase and KDPG aldolase). *M. methylotrophus* has two 6-phosphogluconate dehydrogenases. One of these enzymes utilizes either NADP or NAD, while the other is active with NAD only. Dehydrogenation of 6-phosphogluconate under conditions of energy excess will be prevented, since both 6-phosphogluconate dehydrogenase enzyme activities are inhibited by NAD(P)H and ATP. Thus, increasing amounts of 6-phosphogluconate will be used for anabolism through cleavage via KDPG aldolase and less 6-phosphogluconate will be available for energy generation. Under conditions of energy limitation, the situation may be reversed. It should be emphasized here, however, that for a proper assessment of the *in vivo* regulation of this system, measurements of intracellular concentrations of substrates and inhibitors are still required.

Facultative methylotrophs are particularly suitable for studies on the regulation of the synthesis and activity of enzymes of the RuMP cycle. The regulatory phenomena that play a role in these organisms can be identified by a study of the metabolic changes that occur following a switch from heterotrophic to methylotrophic metabolism and during growth on mixed substrates. The available data on the specific activities of the key enzymes of the RuMP cycle in strains of facultative methylotrophic bacteria grown on various substrates indicate that the extent to which their synthesis is regulated varies (Table V). In *Bacillus* spp. PM6 and S2A1 the HPS/HPI levels were of the same order of magnitude after growth on trimethylamine, on citrate, or on glucose. This suggests that these enzymes are constitutive and not very sensitive to repression during growth of these organisms on "heterotrophic" substrates. Limited control of the synthesis of RuMP cycle enzymes is also evident in *Acetobacter methanolicus* MB58. Glucose as well as glycerol-grown cells of this organism exhibited approximately 40% HPS and HPI activities (assayed together) with respect to methanol-grown cells. The addition of excess HPI, purified from *Pseudomonas* W6, to the spectrophotometric assay allowed the separate determination of HPS activities. In this way it was shown that similar specific activities of this enzyme were present in methanol-, glucose-, and glycerol-grown cells of *A. methanolicus* MB58, indicating that the decreased combined activities of HPS and HPI during nonmethylotrophic growth are due to decreased HPI activities. Although very high HPS levels were detected in cells of *Bacillus* sp. C1 grown on methanol, the enzyme remained clearly present in cells grown on glucose as well.

TABLE V. Combined Activities of HPS and HPI in Facultative Methylotrophs Grown on Various Substrates in Batch Cultures

	Growth substrate			
Organism	Methanol	Trimethylamine	Glucose	Reference[f]
Bacillus PM6	—[a]	195	— (109[b])	1
Bacillus S2A1	—	95	71	1
Bacillus C1	17,300[c]		515[c]	2
Acetobacter methanolicus MB58	707	—	267 (285[d])	3
Brevibacterium fuscum 24	—	140 (135[e])	0	4
Mycobacterium vaccae 10	160	160	0	4
Arthrobacter globiformis B-175	—	— (55[e])	0	5
Arthrobacter P1	—	— (742[c,e])	0	6
Nocardia sp. 239	12,000[c]	—	0	7

[a] —, no data.
[b] Citrate-grown cells.
[c] HPS activity.
[d] Glycerol-grown cells.
[e] Methylamine-grown cells.
[f] References: (1) Colby and Zatman (1975b); (2) Arfman *et al.* (1989); (3) Gründig and Doronina (1984), Uhlig *et al.* (1986); (4) Loginova and Trotsenko (1979); (5) Bykovskaya and Voronkov (1977), Loginova and Trotsenko (1976b); (6) Levering *et al.* (1981); (7) de Boer *et al.* (1990a).

A common property of the four organisms discussed above is their limited ability to utilize heterotrophic substrates for growth (Colby and Zatman, 1975a; Steudel *et al.*, 1980; Dijkhuizen *et al.*, 1988). The data available on five other facultative methylotrophs (Table V) show that these organisms completely switch off the synthesis of HPS and/or HPI during growth on glucose. This indicates that much more effective control mechanisms operate in these bacteria. Whereas no detailed information is available on the ability of *Brevibacterium fuscum* 24 and *Mycobacterium vaccae* 10 to further utilize non-C_1 compounds, the other three organisms (*Arthrobacter* B-175, Loginova and Trotsenko, 1976b; *Arthrobacter* P1, Levering *et al.*, 1981; *Nocardia* sp. 239, Hazeu *et al.*, 1983; de Boer *et al.*, 1988, 1990a,b) are able to grow on numerous heterotrophic substrates and can therefore be regarded as versatile methylotrophs. These data strongly suggest that a correlation may exist between the nutritional versatility of facultative RuMP cycle organisms and the degree of regulation of the synthesis of certain RuMP cycle enzymes.

The regulation of the synthesis of the RuMP cycle enzymes has been studied in most detail in *Arthrobacter* P1 and *Nocardia* sp. 239 (Dijkhuizen and Levering, 1987; Dijkhuizen and Harder, 1988). During growth of *Arthrobacter* P1 on methylamine or formaldehyde, this organism does not possess a linear pathway for formaldehyde oxidation, but employs the dissimilatory RuMP cycle (see above; Table III). Synthesis of HPS and HPI in *Arthrobacter* P1 is solely controlled via induction by formaldehyde, and no repression was observed in the presence of glucose and acetate (Levering *et al.*, 1986). In *Nocardia* sp. 239, formaldehyde is an intermediate in both the dissimilatory and assimilatory pathways (see above; Table III). The results reported by de Boer *et al.* (1990a) show that in this organism the inducing effect of formaldehyde on the synthesis of the RuMP cycle enzymes is completely overruled by catabolite repression exerted by glucose under carbon-excess conditions. This reduces methanol to the status of an ancillary energy source, as is the case in the C_1 compound utilizing autotrophic bacteria (Dijkhuizen and Harder, 1984, 1985).

Regulation of formaldehyde assimilation via modulation of the activities of HPS and/or HPI has not yet been observed in facultative methylotrophs. These organisms assimilate formaldehyde via the FBP aldolase cleavage variant of the RuMP cycle (Table II). In those organisms employing the dissimilatory RuMP cycle of formaldehyde oxidation (e.g., the *Arthrobacter* and *Bacillus* species; Table III), fructose-6-phosphate is at the crossroads of assimilation and dissimilation. Allosteric regulation of the activities of phosphofructokinase and glucose-6-phosphate dehydrogenase would allow a balanced control of the flow of formaldehyde carbon over biosynthetic and energy-generating pathways (Figs. 2, 3). Control of glucose-6-phosphate dehydrogenase activity in *Arthrobacter globiformis* B-175 has been reported by Sokolov and Trotsenko (1985). However, no pertinent data about the involvement of such a control mechanism in other strains have become available yet.

Control of the RuMP cycle via regulation of the activities of the freely reversible "rearrangement" enzymes transaldolase, transketolase, ribulose-5-phosphate 3-epimerase, and ribulose-5-phosphate isomerase appears a less likely mechanism. The activity levels of these enzymes nevertheless generally are considerably increased during growth of RuMP cycle organisms on C_1 compounds (Levering *et al.*, 1982). Mutant studies with *Arthrobacter* P1 (Levering and Dijkhuizen, 1986) have provided evidence that this organism employs a C_1-inducible transaldolase isoenzyme in addition to a constitutive isoenzyme involved in growth on heterotrophic substrates. The presence of a complete set of C_1-inducible rearrangement enzymes in *Arthrobacter* P1 appears likely but requires further investigation.

In those bacteria which possess the irreversible enzyme SBPase, instead of transaldolase, this enzyme might be a target for control at the activity level, as it is in the RuBP cycle of CO_2 fixation in some autotrophs (Amachi and Bowien, 1979). However, no information on the properties of this enzyme in RuMP cycle methylotrophs has yet become available.

The data thus indicate that the regulation of the synthesis of RuMP cycle enzymes varies among different facultatively methylotrophic bacteria. The nature of the control mechanisms involved may also be influenced by the particular organization of C_1 catabolism in these organisms (linear versus cyclic pathway for formaldehyde oxidation).

REFERENCES

Al-Awadhi, N., Egli, T., Hamer, G., and Wehrli, E., 1989, Thermotolerant and thermophilic solvent-utilizing methylotrophic, aerobic bacteria, *System. Appl. Microbiol.* **11:**207–216.

Alefounder, P. R., McCarthy, J. E. G., and Ferguson, S. J., 1981, A periplasmic location for methanol dehydrogenase from *Paracoccus denitrificans*. Implications for proton pumping by cytochrome aa$_3$, *Biochem. Biophys. Res. Commun.* **98:**778–784.

Amachi, T., and Bowien, B., 1979, Characterization of two fructose bisphosphatase isoenzymes from the hydrogen bacterium *Nocardia opaca* 1b, *J. Gen. Microbiol.* **113:**347–356.

Anderson, D. J., and Lidstrom, M. E., 1988, The moxFG region encodes four polypeptides in the methanol-oxidizing bacterium *Methylobacterium* sp. strain AM1, *J. Bacteriol.* **170:**2254–2262.

Anthony C., 1982, *The Biochemistry of Methylotrophs*, Academic Press, London.

Anthony, C., 1986, Bacterial oxidation of methane and methanol, *Adv. Microb. Physiol.* **27:**113–210.

Anthony, C., and Zatman, L. J., 1967, The microbial oxidation of methanol. Purification and properties of the alcohol dehydrogenase of *Pseudomonas* sp. M27, *Biochem. J.* **104**:953–959.

Arfman, N., Watling, E. M., Clement, W., van Oosterwijk, R. J., de Vries, G. E., Harder, W., Attwood, M. M., and Dijkhuizen, L., 1989, Methanol metabolism in thermotolerant methylotrophic *Bacillus* strains involving a novel catabolic NAD-dependent methanol dehydrogenase as a key enzyme. *Arch. Microbiol.* **152**:280–288.

Arfman, N., Bystrykh, L., Govorukhina, N. I., and Dijkhuizen, L., 1990, 3-Hexulose-6-phosphate synthase from the thermotolerant methylotroph *Bacillus* sp. C1, in: *Methods in Enzymology, Vol. 188, Hydrocarbons and Methylotrophy* (M. E. Lidstrom, ed.), Academic Press, Orlando, FL, pp. 391–397.

Attwood, M. M., and Quayle, J. R., 1984, Formaldehyde as a central intermediary metabolite of methylotrophic metabolism, in: *Proceedings of the 4th International Symposium on Microbial Growth on C_1 Compounds* (R. L. Crawford and R. S. Hanson, eds.), American Society for Microbiology, Washington, DC, pp. 315–323.

Babel, W., and Miethe, D., 1974, Alternative zur Reaktionsfolge des Allulose-6-Phosphat-Weges bei einem methylotrophen Bakterium, *Z. Allg. Mikrobiol.* **14**:153–156.

Babel, W., and Mothes, G., 1978, Dissimilatorische Sequenzen in methylotrophen Bakterien, *Z. Allg. Mikrobiol.* **18**:17–26.

Bastide, A., Laget, M., Patte, J-C., and Duménil, G., 1989, Methanol metabolism in *Corynebacterium* sp. XG, a facultatively methylotrophic strain, *J. Gen. Microbiol.* **135**:2869–2874.

Beardsmore, A. F., Aperghis, P. N. G., and Quayle, J. R., 1982, Characterization of the assimilatory and dissimilatory pathways of carbon metabolism during growth of *Methylophilus methylotrophus* on methanol, *J. Gen Microbiol.* **128**:1423–1439.

Ben-Bassat, A., and Goldberg, I., 1977, Oxidation of C_1 compounds in *Pseudomonas* C, *Biochim. Biophys. Acta* **497**:586–597.

Ben-Bassat, A., and Goldberg, I., 1980, Purification and properties of glucose-6-phosphate dehydrogenase ($NADP^+/NAD^+$) and 6-phosphogluconate dehydrogenase ($NADP^+/NAD^+$) from methanol-grown *Pseudomonas* C, *Biochim. Biophys. Acta* **611**:1–10.

Ben-Bassat, A., Goldberg, I., and Mateles, R. I., 1980, Distribution of methanol carbon between assimilation and oxidation pathways in methanol-grown *Pseudomonas* C, *J. Gen. Microbiol.* **116**:213–223.

Boer, L. de, and Dijkhuizen, L., 1990, Microbial and enzymatic processes for L-phenylalanine production, *Adv. Biochem. Eng./Biotechnol.* **41**:1–27.

Boer, L. de, Harder, W., and Dijkhuizen, L., 1988, Phenylalanine and tyrosine metabolism in the facultative methylotroph *Nocardia* sp. 239, *Arch. Microbiol.* **149**:459–465.

Boer, L. de, Euverink, G. J., van der Vlag, J., and Dijkhuizen, L., 1990a, Regulation of methanol metabolism in the facultative methylotroph *Nocardia* sp. 239 during growth on mixed substrates in batch- and continuous cultures, *Arch. Microbiol.* **153**:337–343.

Boer, L. de, Dijkhuizen, L., Grobben, G., Goodfellow, M., Stackebrandt, E., Parlett, J. H., Whitehead, D., and Witt, D., 1990b, *Amycolatopsis methanolica* sp. nov., a facultatively methylotrophic Actinomycete, *Int. J. Syst. Bact.* **40**:194–204.

Boer, L. de, Grobben, G., Vrijbloed, J. W., and Dijkhuizen, L., 1990c, Biosynthesis of aromatic amino acids in *Nocardia* sp. 239: Effects of amino acid analogues on growth and regulatory enzymes, *Appl. Microbiol. Biotechnol.* **33**:183–189.

Brooke, A. G., and Attwood, M. M., 1983, Regulation of enzyme synthesis during the growth of *Hyphomicrobium* X on mixtures of methylamine and ethanol, *J. Gen. Microbiol.* **129**:2399–2404.

Brooke, A. G., Watling, E. M., Attwood, M. M., and Tempest, D. W., 1989, Environmental control of metabolic fluxes in thermotolerant methylotrophic *Bacillus* strains, *Arch. Microbiol.* **151:**268–273.
Burton, S. M., Carver, M. A., and Jones, C. W., 1989, Immunological investigations of methanol dehydrogenase and hexulose phosphate synthase in methylotrophs, in: *Abstracts 6th International Symposium on Microbial Growth on C_1 Compounds,* Göttingen, P209.
Bykovskaya, S. V., and Voronkov, V. V., 1977, Properties of hexulose phosphate synthetase of methylotrophic yeast and bacteria, *Microbiologiya* **46:**46–50.
Chalfan, Y., and Mateles, R. I., 1972, New pseudomonad utilizing methanol for growth, *Appl. Microbiol.* **23:**135–140.
Colby, J., and Zatman, L. J., 1972, Hexose phosphate synthase and tricarboxylic acid–cycle enzymes in bacterium 4B6, an obligate methylotroph, *Biochem. J.* **128:**1373–1376.
Colby, J., and Zatman, L. J., 1973, Trimethylamine metabolism in obligate and facultative methylotrophs, *Biochem. J.* **132:**101–112.
Colby, J., and Zatman, L. J., 1975a, Tricarboxylic acid–cycle and related enzymes in restricted facultative methylotrophs, *Biochem. J.* **148:**505–511.
Colby, J., and Zatman, L. J., 1975b, Enzymological aspects of the pathways for trimethylamine oxidation and C_1 assimilation in obligate methylotrophs and restricted facultative methylotrophs, *Biochem. J.* **148:**513–520.
Colby, J., Dalton, H., and Whittenbury, R., 1979, Biological and biochemical aspects of microbial growth on C_1 compounds, *Annu. Rev. Microbiol.* **33:**481–517.
Cox, R. B., and Quayle, J. R., 1975, The autotrophic growth of *Micrococcus denitrificans* on methanol, *Biochem. J.* **150:**569–571.
Croes, L. M., Meijer, W. G., and Dijkhuizen, L., 1991, Regulation of methanol oxidation and carbon dioxide fixation in *Xanthobacter* strain 25a grown in continuous culture. *Arch. Microbiol.* **155:**159–163.
Dijken, J. P. van, Harder, W., Beardsmore, A. J., and Quayle, J. R., 1978, Dihydroxyacetone: an intermediate in the assimilation of methanol by yeasts? *FEMS Microbiol. Lett.* **4:**97–102.
Dijkhuizen, L., and Harder, W., 1984, Current views on the regulation of autotrophic carbon dioxide fixation via the Calvin cycle in bacteria, *Antonie van Leeuwenhoek* **50:**473–487.
Dijkhuizen, L,. and Harder, W., 1985, Microbial metabolism of carbon dioxide, in: *Comprehensive Biotechnology,* Vol. 1 (M. Moo-Young, ed.), Pergamon Press, Oxford, pp. 409–423.
Dijkhuizen, L., and Harder, W., 1988, Regulation of C_1 metabolism in methylotrophic microorganisms, in: *Continuous Culture* (P. Kyslik, E. A. Dawes, V. Krumphanzl, and M. Novak, eds.), Academic Press, London, pp. 105–118.
Dijkhuizen, L., and Levering, P. R., 1987, Metabolic regulation in facultative methylotrophs, in: *Microbial Growth on C_1 Compounds* (H. W. van Verseveld and J. A. Duine, eds.), Martinus Nijhoff Publishers, Dordrecht, pp. 95–104.
Dijkhuizen, L., Timmerman, J. W. C., and Harder, W., 1979, A pyridine nucleotide–independent membrane-bound formate dehydrogenase in *Pseudomonas oxalaticus* OX1, *FEMS Microbiol. Lett.* **6:**53–56.
Dijkhuizen, L., Arfman, N., Attwood, M. M., Brooke, A. G., Harder, W., and Watling, E. M., 1988, Isolation and initial characterization of thermotolerant methylotrophic *Bacillus* strains, *FEMS Microbiol. Lett.* **52:**209–214.
Dijkstra, M., Frank, J., Jongejan, J. A., and Duine, J. A., 1984, Inactivation of quinoprotein alcohol dehydrogenases with cyclopropane-derived suicide substrates, *Eur. J. Biochem.* **140:**369–373.

Dijkstra, M., van den Tweel, W. J. J., de Bont, J. A. M., Frank, J., and Duine, J. A., 1985, Monomeric and dimeric quinoprotein alcohol dehydrogenase from alcohol-grown *Pseudomonas* BB1, *J. Gen. Microbiol.* **131**:3163–3169.

Dijkstra, M., Frank, J., and Duine, J. A., 1988, Methanol oxidation under physiological conditions using methanol dehydrogenase and a factor isolated from *Hyphomicrobium* X, *FEBS Lett.* **227**:198–202.

Duine, J. A., 1988, Unity and diversity in biological redox catalysis: comparative enzymology of some microbial oxidoreductases showing variation in cofactor identity, in: *The Roots of Modern Biochemistry* (Kleinkauf, H., von Döhren, K., and Jaenicke, R., eds.), Walter de Gruyter, Berlin, pp. 671–682.

Duine, J. A., and Frank, J., 1980, The prosthetic group of methanol dehydrogenase. Purification and some of its properties, *Biochem. J.* **187**:221–226.

Duine, J. A., Frank, J., and Berkhout, M. P. J., 1984, NAD-dependent, PQQ-containing methanol dehydrogenase: a bacterial dehydrogenase in a multienzyme complex, *FEBS Lett.* **168**:217–221.

Eggeling, L., and Sahm, H., 1984, An unusual formaldehyde oxidizing system in *Rhodococcus erythropolis* grown on compounds containing methyl groups, *FEMS Microbiol. Lett.* **25**:253–257.

Eggeling, L., and Sahm, H., 1985, The formaldehyde dehydrogenase of *Rhodococcus erythropolis*, a trimeric enzyme requiring a cofactor and active with alcohols, *Eur. J. Biochem.* **150**:129–134.

Elliot, E. J., and Anthony, C., 1988, The interaction between methanol dehydrogenase and cytochrome c in the acidophilic methylotroph *Acetobacter methanolicus*, *J. Gen. Microbiol.* **134**:369–377.

Ferenci, T., Strøm, T., and Quayle, J. R., 1974, Purification and properties of 3-hexulose phosphate sythase and phospho-3-hexuloisomerase from *Methylococcus capsulatus*, *Biochem. J.* **144**:477–486.

Ford, S., Page, M. D., and Anthony, C., 1985, The role of a methanol dehydrogenase modifier protein and aldehyde dehydrogenase in the growth of *Pseudomonas* AM1 on 1,2 propanediol, *J. Gen. Microbiol.* **131**:2173–2182.

Friedrich, C. G., Bowien, B., and Friedrich, B., 1979, Formate and oxalate metabolism in *Alcaligenes eutrophus*, *J. Gen. Microbiol.* **115**:185–192.

Goldberg, I., and Mateles, R. I., 1975, Growth of *Pseudomonas* C on C_1 compounds: a correction, *J. Bacteriol.* **124**:1028–1029.

Goodwin, P. M., 1990, The biochemistry and genetics of C_1 metabolism in the pink pigmented facultative methylotrophs, in: *Advances in Autotrophic Microbiology and One-Carbon Metabolism*, Vol. 1 (G. A. Codd, L. Dijkhuizen, and F. R. Tabita, eds.), Kluwer, Dordrecht, pp. 143–162.

Govorukhina, N. I., Kletsova, L. V., Tsygankov, Y. D., Trotsenko, Y. A., and Netrusov, A. I., 1987, Characteristics of a new obligate methylotrophic bacterium, *Mikrobiologiya* **56**:849–854.

Greenwood, J. A., and Jones, C. W., 1986, Environmental regulation of the methanol oxidase system of *Methylophilus methylotrophus*, *J. Gen Microbiol.* **132**:1247–1256.

Groen, B. Frank, J., and Duine, J. A., 1984, Quinoprotein alcohol dehydrogenase from ethanol-grown *Pseudomonas aeruginosa*. *Biochem. J.* **223**:921–924.

Groeneveld, A., Dijkstra, M., and Duine, J. A., 1984, Cyclopropanol in the exploration of bacterial alcohol oxidation, *FEMS Microbiol. Lett.* **25**:311–314.

Gründig, M. W., and Babel, W., 1987a, Routes of formaldehyde oxidation to CO_2 in *Acetobacter methanolicus* MB58, *J. Basic Microbiol.* **8**:457–459.

Gründig, M. W., and Babel, W., 1987b, The linear oxidation of formaldehyde to CO_2 as the proper energy generating sequence for the assimilation of methanol in *Acetobacter methanolicus* MB58, *Arch. Microbiol.* **149:**149–155.

Gründig, M. W., and Doronina, N. V., 1984, Dissimilation of methanol in *Acetobacter* sp. MB58, *Z. Allg. Microbiol.* **24:**77–84.

Harms, N., de Vries, G. E., Maurer, K., Veltkamp, E., and Stouthamer, A. H., 1985, Isolation and characterization of *Paracoccus denitrificans* mutants with defects in the metabolism of one-carbon compounds, *J. Bacteriol.* **164:**1064–1070.

Harms, N., de Vries, G. E., Maurer, K., Hoogendijk, J., and Stouthamer, A. H., 1987, Isolation and nucleotide sequence of the methanol dehydrogenase structural gene from *Paracoccus denitrificans, J. Bacteriol.* **169:**3969–3975.

Hazeu, W., Batenburg-van der Vegte, W. H., and de Bruyn, J. C., 1980, Some characteristics of *Methylococcus mobilis* sp. nov., *Arch. Microbiol.* **124:**211–220.

Hazeu, W., de Bruyn, J. C., and van Dijken, J. P., 1983, *Nocardia* sp. 239, a facultative methanol utilizer with the ribulose monophosphate pathway of formaldehyde fixation, *Arch. Microbiol.* **135:**205–210.

Heptinstall, J., and Quayle, J. R., 1970, Pathways leading to and from serine during growth of *Pseudomonas* AM1 on C_1 compounds or succinate, *Biochem. J.* **117:**563–572.

Higgins, I. J., Best, D. J., Hammond, R. C., and Scott, D., 1981, Methane-oxidizing microorganisms, *Microbiol. Rev.* **45:**556–590.

Janssen, D. B., Keuning, S., and Witholt, B., 1987, Involvement of a quinoprotein alcohol dehydrogenase and an NAD-dependent aldehyde dehydrogenase in 2-chloroethanol metabolism in *Xanthobacter autotrophicus* GJ10, *J. Gen. Microbiol.* **133:**85–92.

Johnson, P. A., and Quayle, J. R., 1964, Microbial growth on C_1 compounds. Oxidation of methanol, formaldehyde and formate by methanol-grown *Pseudomonas* AM1, *Biochem. J.* **93:**281–290.

Johnson, P. A., and Quayle, J. R., 1965, Microbial growth on C_1 compounds. Synthesis of cell constituents by methane- and methanol-grown *Pseudomonas methanica, Biochem. J.* **95:**859–867.

Jones, C. W., Kingsbury, S. A., and Dawson, M. J., 1982, The partial resolution and dye-mediated reconstitution of methanol oxidase activity in *Methylophilus methylotrophus, FEMS Microbiol. Lett.* **13:**195–200.

Jongejan, J. A., and Duine, J. A., eds., 1989, Special issue: PQQ and quinoproteins, *Antonie van Leeuwenhoek,* **56**(1).

Karzanov, V. V., Bogatsky, Yu. A., Tishkov, V. I., and Egorov, A. M., 1989, Evidence for the presence of a new NAD-dependent formate dehydrogenase in *Pseudomonas* sp. 101 cells grown on a molybdenum-containing medium, *FEMS Microbiol. Lett.* **60:**197–200.

Kato, N., Tsuji, K., Tani, Y., and Ogata, K., 1974, A methanol-utilizing actinomycete, *J. Ferment. Technol.* **52:**917–920.

Kato, N., Tsuji, K., Tani, Y., and Ogata, K., 1975, Utilization of methanol by an actinomycete, in: *Microbial Growth on C_1 Compounds* (The Organizing Committee, eds.), The Society of Fermentation Technology, Tokyo, pp. 91–98.

Kato, N., Ohashi, H., Hori, T., Tani, Y., and Ogata, K., 1977a, Properties of 3-hexulose phosphate synthase and phospho-3-hexuloisomerase of a methanol-utilizing *Bacterium,* 77a, *Agric. Biol. Chem.* **41:**1133–1140.

Kato, N., Tsuji, K., Ohashi, K., Tani, Y., and Ogata, K., 1977b, Two assimilation pathways of C_1-compounds in *Streptomyces* sp. no. 239 during growth on methanol. *Agric. Biol. Chem.* **41:**29–34.

Kato, N., Ohashi, H., Tani, Y., and Ogata, K., 1978, 3-Hexulose phosphate synthase from *Methylomonas aminofaciens* 77a. Purification properties and kinetics, *Biochim. Biophys. Acta* **523**:236–244.

Kato, N., Miyamoto, N., Shimao, M., and Sakazawa, C., 1988, 3-Hexulose phosphate synthase from a new facultative methylotroph, *Mycobacterium gastri* MB19, *Agric. Biol. Chem.* **52**:2659–2661.

Kemp, M. B., 1972, The hexose phosphate synthetase of *Methylococcus capsulatus*, *Biochem. J.* **127**:64p–65p.

Kemp, M. B., 1974, Hexose phosphate synthase from *Methylococcus capsulatus* makes D-arabino-3-hexulose phosphate, *Biochem. J.* **139**:129–134.

Kemp, M. B., and Quayle, J. R., 1966, Microbial growth on C_1 compounds. Incorporation of C_1 units into allulose phosphate by extracts of *Pseudomonas methanica*, *Biochem. J.* **99**:41–48.

Kemp, M. B., and Quayle, J. R., 1967, Microbial growth on C_1 compounds. Uptake of (^{14}C)formaldehyde and (^{14}C)formate by methane-grown *Pseudomonas methanica* and determination of the hexose labelling pattern after brief incubation with (^{14}C)methanol, *Biochem. J.* **102**:94–102.

Kiriuchin, M. Y., Kletsova, L. V., Chistoserdov, A. Y., and Tsygankov, Y. D., 1988, Properties of glucose 6-phosphate and 6-phosphogluconate dehydrogenases of the obligate methylotroph *Methylobacillus flagellatum* KT, *FEMS Microbiol. Lett.* **52**:199–204.

Kletsova, L. V., Chibisova, E. S., and Tsygankov, Y. D., 1988, Mutants of the obligate methylotroph *Methylobacillus flagellatum* KT defective in genes of the ribulose monophosphate cycle of formaldehyde fixation, *Arch. Microbiol.* **149**:441–446.

Large, P. J., 1983, *Methylotrophy and Methanogenesis*, Van Nostrand–Reinhold, Wokingham.

Large, P. J., and Bamforth, C. W., 1988, *Methylotrophy and Biotechnology*, Longman, Harlow.

Large, P. J., and Haywood, G. W., 1981, *Methylophilus methylotrophus* grows on methylated amines, *FEMS Microbiol. Lett.* **11**:207–209.

Lawrence, A. J., Kemp, M. B., and Quayle, J. R., 1970, Synthesis of cell constituents by methane-grown *Methylococcus capsulatus* and *Methanomonas methano-oxidans*, *Biochem. J.* **116**:631–639.

Levering, P. R., and Dijkhuizen, L., 1986, Regulation and function of transaldolase isoenzymes involved in sugar and one-carbon metabolism in the ribulose monophosphate cycle methylotroph *Arthrobacter* P1, *Arch. Microbiol.* **144**:116–123.

Levering, P. R., Dijken, J. P. van, Veenhuis, M., and Harder, W., 1981, *Arthrobacter* P1, a fast growing versatile methylotroph with amine oxidase as a key enzyme in the metabolism of methylated amines. *Arch. Microbiol.* **129**:72–80.

Levering, P. R., Dijkhuizen, L., and Harder, W., 1982, Enzymatic evidence for the operation of the FBP aldolase cleavage and TK/TA rearrangement variant of the RuMP cycle in *Arthrobacter* P1, *FEMS Microbiol. Lett.* **14**:257–261.

Levering, P. R., Croes, L. M., and Dijkhuizen, L., 1986, Regulation of methylamine and formaldehyde metabolism in *Arthrobacter* P1. Formaldehyde is the inducing signal for the synthesis of the RuMP cycle enzyme hexulose phosphate synthase, *Arch. Microbiol.* **144**:272–278.

Levering, P. R., Tiesma, L., Woldendorp, J. P., Steensma, M., and Dijkhuizen, L., 1987, Isolation and characterization of mutants of the facultative methylotroph *Arthrobacter* P1 blocked in one-carbon metabolism, *Arch. Microbiol.* **146**:346–352.

Loginova, N. V., and Trotsenko, Yu. A., 1976a, Facultative methylotroph belonging to the genus *Arthrobacter*, *Microbiologiya* **44**:892–896.

Loginova, N. V., and Trotsenko, Yu. A., 1976b, Pathways of oxidation and assimilation of methylated amines in *Arthrobacter globiformis*, *Microbiologiya* **45**:217–223.

Loginova, N. V., and Trotsenko, Yu. A., 1977a, Methanol metabolism in *Pseudomonas oleovorans*, *Microbiologiya* **46**:170–175.
Loginova, N. V., and Trotsenko, Yu. A., 1977b, The metabolic pathways of methylated amines in bacteria, in: *Abstracts of the 2nd International Symposium on Microbial Growth on C_1-Compounds*, Pushchino, USSR, pp. 37–39.
Loginova, N. V., and Trotsenko, Yu. A., 1979, Carbon metabolism is methylotrophic bacteria isolated from activated sludge, *Microbiologiya* **50**:13–18.
Loginova, N. V., Govorukhina, N. J., and Trotsenko, Yu. A., 1981, Metabolism of the obligate methylotroph *Methylophilus methanolovorus*, *Microbiologiya* **50**:217–220.
Marison, I. W., and Attwood, M. M., 1980, Partial purification and characterization of a dye-linked formaldehyde dehydrogenase from *Hyphomicrobium* X, *J. Gen. Microbiol.* **117**:305–313.
Marison, I. W., and Attwood, M. M., 1982, A possible alternative mechanism for the oxidation of formaldehyde to formate, *J. Gen. Microbiol.* **128**:1441–1446.
Meijer, W. G., Croes, L. M., Jenni, B., Lehmicke, L. G., Lidstrom, M. E., and Dijkhuizen, L., 1990, Characterization of *Xanthobacter* strains H4-14 and 25a and enzyme profiles after growth under autotrophic and heterotrophic conditions, *Arch. Microbiol.* **153**:360–367.
Miyazaki, S. S., Toki, S. I., Izumi, Y., and Yamada, H., 1987, Purification and characterization of methanol dehydrogenase of a serine producing methylotroph *Hyphomicrobium methylovorum*, *J. Ferment. Technol.* **65**:371–377.
Müller, R., and Sokolov, A. P., 1979, Kinetic properties of the purified 3-hexulosephosphate synthase from *Pseudomonas oleovorans*, *Z. Allg. Mikrobiol.* **19**:261–267.
Müller, W., Willnow, P., Ruschig, V., and Höpner, T., 1978, Formate dehydrogenase from *Pseudomonas oxalaticus*, *Eur. J. Biochem.* **83**:485–498.
Ohta, S., Fujita, T., and Tobari, J., 1981, Methanol dehydrogenase of *Methylomonas* J: purification, crystallization and some properties, *J. Biochem.* **90**:205–215.
Ophem, P. W. van, and Duine, J. A., 1989, Three different types of aldehyde dehydrogenases from *Nocardia* sp. 239, in: *Abstracts 6th International Symposium on Microbial Growth on C_1 Compounds*, Göttingen, P243.
Page, M. D., and Anthony, C., 1986, Regulation of formaldehyde oxidation by the methanol dehydrogenase modifier proteins of *Methylophilus methylotrophus* and *Pseudomonas* AM1, *J. Gen. Microbiol.* **132**:1553–1563.
Papoutsakis, E., Lim, H. C., and Tsao, G. T., 1978, Role of formaldehyde in the utilization of C_1 compounds via the ribulose monophosphate cycle, *Biotechnol. Bioeng.* **20**:421–442.
Patel, R. N., and Felix, A., 1976, Microbial oxidation of methane and methanol: crystallization and properties of methanol dehydrogenase from *Methylosinus sporium*, *J. Bacteriol.* **128**:413–424.
Patel, R. N., and Hoare, D. S., 1971, Physiological studies of methane and methanol–oxidizing bacteria: oxidation of C_1 compounds by *Methylococcus capsulatus*, *J. Bacteriol.* **107**:187–192.
Patel, R. N., Hoare, S. L., and Hoare, D. S., 1979, (1-^{14}C)-Acetate assimilation by obligate methanotrophs, *Pseudomonas methanica* and *Methylosinus trichosporium*, *Antonie van Leeuwenhoek* **45**:499–511.
Poels, P. A., and Duine, J. A., 1989, NAD-linked, GSH- and factor-independent aldehyde dehydrogenase of the methylotrophic bacterium, *Hyphomicrobium* X, *Arch. Biochem. Biophys.* **271**:240–245.
Quayle, J. R., 1972, The metabolism of one-carbon compounds by microorganisms, *Adv. Microbial Physiol.* **7**:119–203.
Quayle, J. R., 1980a, Microbial assimilation of C_1 compounds, *Biochem. Soc. Trans.* **8**:1–10.

Quayle, J. R., 1980b, Aspects of the regulation of methylotrophic metabolism, *FEBS Lett.* **117:**K16–K27.
Quayle, J. R., and Ferenci, T., 1978, Evolutionary aspects of autotrophy, *Microbiol. Rev.* **42:**251–273.
Quilter, J. A., and Jones, C. W., 1984, The organisation of methanol dehydrogenase and c-type cytochromes on the respiratory membrane of *Methylophilus methylotrophus*, *FEBS Lett.* **174:**167–172.
Roitsch, T., and Stolp, H., 1985a, Overproduction of methanol dehydrogenase in glucose grown cells of a restricted RuMP type methylotroph, *Arch. Microbiol.* **142:**34–39.
Roitsch, T., and Stolp, H., 1985b, Distribution of dissimilatory enzymes in methane and methanol oxidizing bacteria, *Arch. Microbiol.* **143:**233–236.
Roitsch, T., and Stolp, H., 1986, Synthesis of dissimilatory enzymes of serine type methylotrophs under different growth conditions, *Arch. Microbiol.* **144:**245–247.
Sahm, H., and Steinbach, R., 1977, Purification and regulation of glucose-6-phosphate dehydrogenase from an obligate methanol utilizing bacterium *Methylomonas* M15, in: *Abstracts of the 2nd International Symposium on Microbial Growth on C_1 Compounds*, Pushchino, USSR, pp. 50–51.
Sahm, H., Schütte, H., and Kula, M.-R., 1976, Purification and properties of 3-hexulosephosphate synthase from *Methylomonas* M15, *Eur. J. Biochem.* **66:**591–596.
Samuelov, N., and Goldberg, I., 1982a, Effect of growth conditions on the distribution of methanol carbon between assimilation and oxidation pathways in *Pseudomonas* C, *Biotechnol. Bioeng.* **24:**731–736.
Samuelov, N., and Goldberg, I., 1982b, Is *Pseudomonas* C an obligate or facultative methylotroph? *Biotechnol. Bioeng.* **24:**2605–2608.
Schär, H. P., Chemla, P., and Ghisalba, O., 1985, Methanol dehydrogenase from *Hyphomicrobium* MS 223, *FEMS Microbiol. Lett.* **26:**117–122.
Shishkina, V. N., and Trotsenko, Y. A., 1982, Multiple enzymic lesions in obligate methanotrophic bacteria, *FEMS Microbiol. Lett.* **13:**237–242.
Sokolov, A. P., and Trotsenko, Yu. A., 1978, Cyclic pathway of formaldehyde oxidation in *Pseudomonas oleovorans*, *Mikrobiologiya* **46:**1119–1121.
Sokolov, A. P., and Trotsenko, Yu. A., 1985, Purification and properties of glucose-6-phosphate dehydrogenase and 6-phosphogluconate dehydrogenase from the facultative methylotroph *Arthrobacter globiformis*, *Biokhimiya* **50:**1269–1277.
Sperl, G. T., Forrest, H. S., and Gibson, D. T., 1974, Substrate specificity of the purified primary alcohol dehydrogenases from methanol-oxidizing bacteria, *J. Bacteriol.* **118:**541–550.
Steudel, A., Miethe, D., and Babel, W., 1980, Bakterium MB58, ein methylotrophes "Essigsäurebakterium," *Z. Allg. Mikrobiol.* **20:**663–672.
Stieglitz, B., and Mateles, R. I., 1973, Methanol metabolism in *Pseudomonas* C *J. Bacteriol.* **114:**390–398.
Stirling, D. I., and Dalton, H., 1978, Purification and properties of an $NAD(P)^+$-linked formaldehyde dehydrogenase from *Methylococcus capsulatus* (Bath), *J. Gen. Microbiol.* **107:**19–29.
Strøm, T., Ferenci, T., and Quayle, J. R., 1974, The carbon assimilation pathways of *Methylococcus capsulatus*, *Pseudomonas methanica* and *Methylosinus trichosporium* (OB3B) during growth on methane, *Biochem. J.* **144:**465–476.
Tabita, F. R., 1988, Molecular and cellular regulation of autotrophic carbon dioxide fixation in microorganisms, *Microbiol. Rev.* **52:**155–189.
Trotsenko, Yu. A., 1983, Metabolic features of methane- and methanol-utilizing bacteria, *Acta Biotechnol.* **3:**269–277.

Uhlig, H., Karbaum, K., and Steudel, A., 1986, *Acetobacter methanolicus* sp. nov., an acidophilic facultatively methylotrophic bacterium, *Int. J. Syst. Bacteriol.* **36:**317–322.

Vries, G. E. de, 1986, Molecular biology of bacterial methanol oxidation, *FEMS Microbiol. Rev.* **39:**235–258.

Vries, G. E. de, Harms, N., Maurer, K., Papendrecht, A., and Stouthamer, A. H., 1988, Physiological regulation of *Paracoccus denitrificans* methanol dehydrogenase synthesis and activity, *J. Bacteriol.* **170:**3731–3737.

Vries, G. E. de, Kües, U., and Stahl, U., 1990, Physiology and genetics of methylotrophic bacteria, *FEMS Microbiol. Rev.* **75:**57–101.

Weaver, C. A., and Lidstrom, M. E., 1985, Methanol dissimilation in *Xanthobacter* H4-14: activities, induction and comparison to *Pseudomonas* AM1 and *Paracoccus denitrificans*, *J. Gen. Microbiol.* **131:**2183–2197.

Wolf, H. J., and Hanson, R. S., 1978, Alcohol dehydrogenase from *Methylobacterium organophilum*, *Appl. Environ. Microbiol.* **36:**105–114.

Yamanaka, K., 1981, Comparative aspects of methanol dehydrogenase, in: *Microbial Growth on C_1 Compounds* (H. Dalton, ed.), Heyden, London, pp. 21–30.

Yordi, J. R., and Weaver, T. L., 1977, *Methylobacillus:* a new genus of obligately methylotrophic bacteria, *Int. J. Syst. Bacteriol.* **27:**247–255.

Zatman, L. J., 1981, A search for patterns in methylotrophic pathways, in: *Microbial Growth on C_1 Compounds* (H. Dalton, ed.), Heyden, London, pp. 42–54.

The Genetics and Molecular Biology of Methanol-Utilizing Bacteria

6

MARY E. LIDSTROM

1. INTRODUCTION

In the past decade, the development of genetic capabilities in methylotrophic bacteria has greatly expanded the scope of research in these organisms. Work is now underway to define the molecular mechanisms involved in regulatory phenomena, to manipulate the physiology of the organisms, and to sort out the complex steps involved in C_1 metabolism. This chapter will review recent advances in genetic technology as applied to C_1 metabolism in the methanol-utilizing bacteria, with emphasis on the information gained concerning the physiology of these organisms.

Genetic techniques applicable to the methanol-utilizing bacteria have been developed only recently. The single most important breakthrough was the discovery that the broad-host-range vectors developed for other gram-negative bacteria were transferrable to methanol-utilizing bacteria and were stably maintained (Windass *et al.*, 1980; Holloway, 1984). Table I lists some of the vectors that have been used successfully in these bacteria, and specific examples of their applications will be presented in the next sections of this chapter. In most methanol-utilizing bacteria, these vectors are mobilizable by conjugation at frequencies sufficiently high to allow use of this transfer method for genetic manipulations. In some cases, such as *Paracoccus denitrificans*, an autotrophic methanol-utilizer, low transfer fre-

---

MARY E. LIDSTROM • Environmental Engineering Science, California Institute of Technology, Pasadena, California 91125.

Methane and Methanol Utilizers, edited by J. Colin Murrell and Howard Dalton. Plenum Press, New York, 1992.

quencies were observed, which were increased in restriction-negative mutants (DeVries, 1986). Mobilization is carried out on solid media using either triparental matings, involving *Escherichia coli* donor and mobilizer strains and the methylotrophic recipient, or biparental matings involving an *E. coli* donor such as S17-1 (Simon *et al.*, 1983) containing the mobilization genes integrated into the chromosome. With the identification of replicons and mobilization systems that could be used successfully in the methanol-utilizing bacteria, a variety of other vectors became available that were based on these systems. These vectors are useful for carrying out a variety of molecular genetic manipulations, including cosmid cloning, subcloning, promoter cloning, and generating translational fusions to reporter genes (see Table I).

Another key technique that has proven useful in the methanol utilizers is marker exchange (marker rescue). Marker exchange is a method for generating specific mutations in a targeted chromosomal gene (Ruvkun and Ausubel, 1981). In this procedure, a fragment containing the gene of interest is cloned into a mobilizable but nonreplicating plasmid vector, which is transferred into the host strain. Homologous recombination will occur between the cloned and chromosomal sequences, and a double crossover event will result in exchange or replacement of the sequences. If the plasmid-borne fragment contains an insertion or deletion, a mutation is constructed, and this can be identified by screening for specific phenotypes. In methanol-utilizing bacteria, the plasmid pBR322 is mobilizable but does not replicate. Therefore, fragments cloned into this vector are suitable for marker exchange experiments, and this approach has been used to generate specific mutants in *Methylobacterium extorquens* AM1 (a facultative serine pathway methanol utilizer; Nunn and Lidstrom, 1986a), *Methylophilus methylotrophus* [an obligate ribulose monophosphate (RuMP) cycle methanol-utilizer; Bohanon *et al.*, 1987], and *Methylobacillus flagellatum* (an obligate RuMP cycle methanol-utilizer; Y. Tsygankov, personal communication). This approach is especially valuable to confirm the identity of cloned genes and to search for other markers on a cloned fragment.

Another important method for analyzing cloned genes is to determine the gene products encoded. This is normally accomplished by expressing the cloned genes in an alternate host that lacks those genes (usually *E. coli*) and visualizing the expressed proteins on gels. DNA from *Methylophilus* strains is expressed in *E. coli* without apparent difficulty (Kearney and Holloway, 1987). However, expression in *E. coli* has not been successful with DNA from *Methylobacterium* strains (Anderson and Lidstrom, 1988; Machlin *et al.*, 1988) or *Paracoccus denitrificans* (Harms *et al.*, 1987). Work

TABLE I. Vectors Used in Methylotrophic Bacteria (All Are Mob$^+$)

Vector	Incompatibility group	Selectable markers[a]	Size (kb)	Other features	Useful cloning sites	Ref.[b]
Cloning vectors						
pVK100	P1	Km,Tc	23	Cosmid	*Hind*III,*Sal*I,*Xho*I	1
pLA2917	P1	Km,Tc	21	Cosmid	*Sau*3a,*Bgl*II,*Pst*I,*Hind*III,*Sal*I	2
pGSS8	Q	Sm,Tc	9.5		*Sal*I,*Sst*I,*Eco*RI,*Bam*HI,*Sst*II	3
pAYC30	Q	Hg,Sm,Su	17		*Sal*I,*Sst*I,*Sst*II	4
Subcloning vectors						
pRK404	P1	Tc	10.6	*lacZ*	*Bam*HI,*Hind*III,*Pst*I	5
pRK310	P1	Tc	20.4	*lacZ*	*Bam*HI,*Hind*III,*Pst*I	5
Promoter cloning vectors						
pGD500	P1	Tc	28.1	*lacZ* = reporter gene (transcriptional fusion)	*Bam*HI,*Bgl*II	5
pGD926	P1	Tc	28.1	*lacZ* = reporter gene (translational fusion)	*Hind*III/*Bam*HI	5
pAYC36	Q	Ap	9.4	Sm = reporter gene (transcriptional fusion)	*Bam*HI,*Eco*RI,*Hpa*I,*Kpn*I,*Sma*I	4
Mobilizing vectors						
pRK2013	ColE1	Km	48	RK2 transfer system		5
pRK2073	ColE1	Sm,Tp	61	RK2 transfer system		5
R751 = RP4 = RK2	P1	Tc,Km,Ap	56	RK2 transfer system		6
Transposon delivery vectors						
pM075 (R91-5::*Tn*5)	P10	Cb,Km	—	*Tn*5	—	7
pMD100 (pRK2013::*Tn*5)	ColE1	Km,Hg	—	*Tn*501	—	8

[a]Km, kanamycin; Tc, tetracycline; Sm, streptomycin; Hg, mercury; Su, sulfanilamide; Tp, trimethoprim; Ap, ampicillin; Cb, carbenicillin.
[b]1, Knauf and Nester, 1982; 2, Allen and Hanson, 1985; 3, Windass *et al.*, 1980; 4, Chistoserdov and Tsygankov, 1986; 5, Ditta *et al.*, 1985; 6, Hennam *et al.*, 1982; 7, Whitta *et al.*, 1985; 8, Ely, 1985.

in our laboratory showed that transcriptional termination occurred in *E. coli* within cloned *M. extorquens* AM1 insert DNA, as assessed using transcriptional test vectors (Nunn and Lidstrom, unpublished), and this was probably the reason that expression was not obtained. We were able to circumvent this problem using the dual T7 promoter/polymerase expression system developed by Tabor and Richardson (1985). With this system, the insert DNA is cloned downstream of a phage T7 promoter, and expression occurs via the phage T7 RNA polymerase, the gene for which is present on a second plasmid. Since the T7 RNA polymerase does not recognize terminators, good expression of cloned *M. extorquens* AM1 DNA is obtained (Anderson and Lidstrom, 1988).

Other important techniques developed in the methanol utilizers will be discussed in the next sections.

2. PLASMIDS IN METHANOL-UTILIZING BACTERIA

Indigenous plasmids are common in methanol-utilizing bacteria, but no functions have been identified for any of them. In most methanol utilizers, simple plasmid isolation methods result in poor yields, but variations have been reported that result in reproducibly high plasmid yields (Ueda *et al.*, 1987; Kim and Lidstrom, 1989). *M. extorquens* AM1 contains three plasmids, first identified as 23, 29, and 32.5 kb (Warner and Higgins, 1977), but later shown to be larger (33, 40, and 65 kb; Kim and Lidstrom, 1989). The middle-sized plasmid has been missing in some laboratory strains, but no phenotypic changes could be associated with the absence of this plasmid (Kim and Lidstrom, 1989).

Methylomonas methylovora (an obligate RuMP cycle strain) contains a conjugative plasmid of 133 kb (Monteiro *et al.*, 1982), and an unidentified RuMP cycle methanol utilizer isolated from soil contains a 9-kb plasmid (pUK1) which was restriction-mapped (Ueda *et al.*, 1987). The obligate RuMP cycle methanol utilizer *Methylomonas clara* contains two plasmids, of 46 and 16 kb, respectively, but the smaller plasmid (pBE-3) has been shown to be a deletion derivative of the larger (Marquardt *et al.*, 1984). pBE-3 has been restriction-mapped and has been shown to be transcriptionally active in *M. clara* (Metzler *et al.*, 1988). However, the gene product(s) have not been identified. A putative promoter sequence was identified for this transcript, which showed considerable homology to the *E. coli* consensus promoter. So far, the genes involved in methylotrophic metabolism that have been cloned do not appear to be plasmid-encoded, but future experiments are needed to address this question more fully.

3. MUTAGENESIS IN *METHYLOBACILLUS FLAGELLATUM*

Mutants have been difficult to obtain in methanol utilizers, especially the obligate strains (Holloway, 1984; DeVries, 1986), and this represents a continuing obstacle to genetic studies in these organisms. In most cases, the reason for poor mutant isolation frequencies is not known. However, in *M. flagellatum*, a systematic approach to mutagenesis has been carried out, with positive results. Conditions were developed to optimize mutagenesis procedures, and both auxotrophs and C_1 mutants were isolated (Tsygankov and Kazakova, 1987; Kletsova *et al.*, 1988). Initially, auxotrophs were isolated at low frequency in this organism. However, Tsygankov and co-workers showed that auxotrophic mutants become hypersensitive to the presence of other supplements, especially amino acids. Therefore, the normal selection procedures using master plates containing casamino acids severely limited the number and type of auxotrophs obtained (Tsygankov and Kazakova, 1987). Auxotrophs were isolated at much higher frequencies if the mutagenized cells were plated on minimal medium containing single supplements. In some cases, it appeared that poor-affinity uptake systems were also a problem, and this could be circumvented by increasing the supplement concentration.

In the case of the mutants defective in RuMP cycle enzymes, a different approach was taken. Since *M. flagellatum* is an obligate methylotroph, mutations in C_1 pathways are lethal and can only be isolated as conditional mutations. Tsygankov and co-workers isolated 4000 nitrosoguanidine mutants capable of growth at 30°C but not at 42°C. 500 of these grew on minimal medium at the permissive temperature with low reversion frequencies and were used for further screening. Each of these was assayed individually for the enzymes of the RuMP cycle. Nine mutants were detected that were defective in phosphoglucoisomerase and two mutants were defective in glucose 6-phosphate dehydrogenase. In all cases, the enzymes in question were inactivated at the higher growth temperature, suggesting that the mutations occurred in structural genes for the enzymes. The availability of mutants makes possible the cloning of the genes encoding those enzymes by complementation with broad-host range vectors.

4. TRANSPOSON MUTAGENESIS

Transposon mutagenesis is an attractive alternative approach to chemical mutagenesis for methanol utilizers, because it offers the possibility of obtaining a pool of preselected mutant strains. Nonreplicating vectors containing transposon inserts are transferred to the recipient, by either transformation or mobilization, and selection for the marker on the transposon identifies those cells in which transposition has occurred. A bank of 5000–7000 such insertion strains should theoretically serve as a mutagenesis pool, in which most genes in the chromosome have been mutagenized (DeVries, 1986). The members of this pool can then be screened individually for specific defects. However, transposon mutagenesis has not been successful in most methylotrophs (DeVries, 1986), and in some cases mutants that are isolated are unstable. Methanol mutants and auxotrophs of the autotrophic methylotroph *Xanthobacter* H4-14 generated by *Tn5* insertions were too unstable for genetic characterization (Anderson and Lidstrom, unpublished), and Bohanon *et al.* (1987) have shown that *Tn5* insertions in a *trp* gene of *M. methylotrophus* that were inserted into the chromosome by marker exchange were also unstable. These mutations could be stabilized by deletion of the IS5OR portion of the transposon suggesting that the problem was due to subsequent excisions.

Despite the problems noted above, two successful transposon mutagenesis protocols have been developed. A *Tn5* delivery system (pMO75; see Table I) based on a *Pseudomonas aeruginosa* plasmid has been used to obtain *Tn5* mutants in *M. extorquens* AM1 (Whitta *et al.*, 1985). Although the frequency of Km^r was low (10^{-7} per recipient), six methanol mutants and six auxotrophs were identified out of a total of 4070 Km^r strains screened. Another plasmid, pMD100, containing the Hg^r transposon *Tn501* inserted in pRK2013 (see Table I) has been used successfully to generate mutants in *M. flagellatum* (Y. Tsygankov, personal communication). The general applicability of these transposon delivery vehicles to other methanol-utilizing bacteria is not known at this time.

5. CHROMOSOMAL MAPPING

Because of the problems in isolating auxotrophs of methanol utilizers, chromosomal mapping has been difficult. In addition, most of the gene transfer systems that are useful involve direct cloning, which restricts mapping to fragments of 20–30 kb. However, it has been possible in some cases to take advantage of broad-host-range vectors that have chromosome mobilization capabilities to generate partial genetic maps. Three of these systems will be discussed below.

5.1. Auxotrophy Markers in *Methylophilus* Strains

Holloway and co-workers have carried out an elegant series of studies to map auxotrophy markers in *Methylophilus* strains using a combination of chromosome mobilization, cosmid cloning, and surrogate genetics. They have used the IncP1 plasmid R68.45, which has the property of integrating into host chromosomes. At low frequency, these integrated plasmids give rise to R-prime (R') plasmids, which consist of R68.45 carrying a portion of the chromosome, as much as 100 kb in some strains (Holloway, 1984; DeVries, 1986). R68.45-mediated chromosome mobilization has been used to transfer markers from donor *Methylophilus viscogenes* and *M. methylotrophus* strains to a variety of *E. coli* and *Pseudomonas aeruginosa* auxotrophs for complementation analysis (Moore *et al.*, 1983; Kearney and Holloway, 1987; Lyon *et al.*, 1988). In addition, a cosmid library has been used in similar complementation tests (Lyon *et al.*, 1988). By analyzing the data obtained from these different approaches, partial genetic maps have been constructed (Table II). In those cases in which overlap was observed, the gene order in the two *Methylophilus* strains was similar. Not all the tested auxotrophs were complemented, and it is not known whether this was due to a lack of functionally similar genes or to problems in expression.

5.2. R-Prime Mapping in *Methylobacterium extorquens* AM1

R68.45-mediated chromosomal transfer has also been used to map several markers in *M. extorquens* AM1, including auxotroph, drug resis-

TABLE II. Linkage of Auxotrophy Markers in *Methylophilus viscogenes* and *M. methylotrophus* AS1[a]

	M. viscogenes	*M. methylotrophus*
Group 1	trpF argG argF leu10 trpB trpA met9011	pur66 trpF argG argF leu10 pheA trpB trpA purF met9011
Group 2	pyrE trpE trpD trpC metA met28 phe3	pyrE trpE trpD trpC metA met28
Group 3	pyrB proC	pyrB proC pur66
Group 4	proB pur136	leu8 proB pur136
Group 5	pyrD hisV leu8	
Group 6	pyrF phe2	
Group 7	hisI hisIII hisIIB hisIIA	
Group 8		argA argB
Group 9		metE metC metB thr48

[a]From Lyon *et al.*, 1988.

tance, and C_1 genes (Tatra and Goodwin, 1983, 1985). Although R' plasmids were not isolated, their formation was assumed. The recombination frequencies observed were variable, so specific map distances could not be computed. However, the data suggested the gene order shown in Table III. These data have relevance to cloning studies involving genes encoding functions for methanol oxidation and the assimilation of C_1 units (see Sections 7 and 9).

5.3. Hfr-Like Donors in *M. flagellatum*

Broad-host-range plasmids can also be used to generate Hfr-like donors, which transfer chromosomal markers in a polar fashion, dependent upon the site of insertion and the direction of transfer (Holloway, 1984). The advantage of Hfr-like strains over R' plasmids is the potential for linking much larger sections of the chromosome. By generating a series of these Hfr-like donor strains, it is theoretically possible to map the entire chromosome.

Tsygankov and co-workers have used plasmid pAS8-121 to generate Hfr-like donor strains in *M. flagellatum* (Serebrijski *et al.*, 1989). This plasmid is a hybrid between pBR322 and RP4, in which the *trf*A gene (involved in replication control) has been inactivated with *Tn*7. Such plasmids integrate into the chromosome, yielding Hfr-like donors. Of the six Hfr strains isolated, only one showed a gradient of marker transfer, and data were presented to suggest that in the other cases, autonomously replicating pAS8-121 was present in the cytoplasm, in addition to a chromosomally integrated copy. However, the sixth Hfr strain did not contain autonomously replicating pAS8-121 and transferred markers with a frequency

TABLE III. Proposed Gene Order in *M. extorquens* AM1[a]:
str -*mmf*-1-*pho*-*cyc* -*mtd*-1-*mcl*-1-*mmf*-2

Antibiotic resistance
   str, streptomycin resistance
   pho, phosphonomycin resistance
   cyc, cycloserine resistance

One-carbon metabolism
   mmf-1, defective in the conversion of acetyl CoA to glyoxylate
   mtd-1, *mox*A mutant; defective in apomethanol dehydrogenase/PQQ assembly or modification
   mcl-1, defective in malyl CoA lyase
   mmf-2, defective in glycerate kinase

[a]From Tatra and Goodwin, 1985.

of from 10^{-4} to 10^{-6} per donor. This strain should now prove useful for mapping the auxotrophic mutants isolated in this strain (Tsygankov and Kazakova, 1987).

6. EXPRESSION OF FOREIGN GENES IN METHANOL-UTILIZING BACTERIA

A variety of foreign genes have been expressed in methanol utilizers, and in general, it appears that heterologous expression is not a problem given appropriate constructions and vector promoters. Most antibiotic resistance genes found on cloning vectors are expressed, although the ampicillin resistance gene found on pBR322 is an exception and is not expressed in *M. extorquens* AM1 (Anderson and Lidstrom, unpublished). The glutamate dehydrogenase gene of *E. coli* and two eukaryotic cDNAs encoding chicken ovalbumin and mouse dihydrofolate reductase were cloned and expressed in *M. methylotrophus* (Windass *et al.*, 1980; Hennam *et al.*, 1982). In all these cases, expression apparently originated from vector promoters and was lower than that observed for the same construction in *E. coli*. In *Hyphomicrobium* X, a hyphal budding serine cycle methylotroph, the pyruvate dehydrogenase complex from *E. coli* was expressed at low levels, but the promoters involved were not determined (Dijkhuizen *et al.*, 1984).

In some cases, foreign promoters have been shown to be functional in methanol utilizers. In *M. flagellatum,* the *tac* promoter was used to express cDNA encoding human interferon αF (Chistoserdov *et al.*, 1987). Significant expression was achieved, with levels two- to threefold higher than those obtained in *E. coli*. In addition, both *tac* and *trp* promoters have been shown to be functional in *M. methylotrophus* (Byrom, 1984). A streptomycin resistance gene was expressed from the *tac* promoter, and complementation of a Trp mutant was shown to be dependent on the *trp* promoter. However, no well-studied regulatable expression systems are available at this time for methanol-utilizing bacteria, and development of such systems is important for future manipulations of this group.

7. METHANOL OXIDATION GENES

The methanol oxidation (Mox) system has proven amenable to genetic analysis in methanol-utilizing bacteria, largely due to the availability of a positive selection method for isolating Mox⁻ mutants. Methanol dehydrogenase (MeDH) oxidizes a variety of primary alcohols, including allyl alcohol, to the corresponding aldehyde (Anthony, 1986). The product of allyl alcohol oxidation, allyl aldehyde (acrolein), is toxic and kills the cells. Therefore, when mutagenized cells are plated on an alternate substrate in the presence of allyl alcohol, cells that express the Mox system are killed and Mox⁻ mutants grow normally (Nunn and Lidstrom, 1986a; Machlin *et al.*, 1988). This selection technique depends on the availability of an alternate substrate that does not repress MeDH expression, and for most methanol utilizers, methylamine is useful for this purpose. However, in the case of the autotrophic methanol utilizers, which do grow on methylamine, allyl alcohol selection has not been particularly successful due to the repression of MeDH on all known alternate substrates (Harms *et al.*, 1985; Weaver and Lidstrom, unpublished). Two Mox systems have been studied in detail, in *M. extorquens* AM1 and *Methylobacterium organophilum* XX; in addition, the structural gene for one of the subunits of the methanol dehydrogenase has been studied in *Paracoccus denitrificans*.

7.1. Localization and Identification of Mox Genes in *Methylobacterium* Strains

The allyl alcohol selection method described above has been used to isolate a number of mutants in both *M. extorquens* AM1 (Nunn and Lidstrom, 1986a) and *M. organophilum* XX (Machlin *et al.*, 1988). These have been placed into complementation groups using genomic clone banks constructed in broad-host-range plasmids (Nunn and Lidstrom, 1986a; Machlin *et al.*, 1988), and more recently, the mutants from each strain have been complemented by the clones from the other strain and new mutants of *M. extorquens* AM1 have been complemented to extend the known complementation groups (Bastien *et al.*, 1989; Lee and Lidstrom, unpublished). The results of these studies are listed in Table IV. As may be seen from these data, a minimum of 20 Mox genes are present in *M. extorquens* AM1, and a minimum of 14 are present in *M. organophilum* XX. In some cases, mutants were isolated in only one of the strains, but many overlapping classes were identified.

The genes on the complementing clones have been mapped in both strains, as shown in Figure 1. For clarity, each of the proposed genes has been given a single letter designation, but the original designations for both sets of genes are noted in Table IV. The gene order is remarkably similar, the only exception being the inverse order of the genes in the Mox D region between the two strains. It is not yet known whether the genes designated *mox*K, H, J, R, S, or T in *M. extorquens* AM1 are present in *M. organophilum* XX.

Transcript sizes have been determined for several of the Mox genes in

TABLE IV. Apparent Mox Classes

Class	Mutants[a] AM1	Mutants[a] XX	Complementing AM1 clone (previous designation)	Complementing XX clone	XX transcript (kb)	Growth on methanol + PQQ
A	PG1	SM29	A (A1)	V B1	1.1	—
K	UV21	—	A (A2)	V B1		—
L	M15a	—	A (A3)	V A2		—
B	UV25	SM37	B	V A1		—
	UV4	SM13	B	V A1		—
C	AA18	—	C	—		+
	UV40		C	—		+
	UV41		C	—		+
	UV42		C	—		+
	UV44		C	—		+
	UV45		C	—		+
	UV50		C	—		+
P	—	PT34	C	VI D	1.35	+
		PT47	C	VI D		+
M	—	SM10	D	VI A	0.6	—
		PT45	D	VI A		—
N	—	SM2	D	VI B	1.5	—
D	UV9	SM48	D	VI C	1.5	—
	UV27	SM3	D	VI C		—
O	UV46	—	D (D)	VI D		+
Q	—	SM4	E	VII A	1.1	—
		SM18	E	VII A		—
E	AA31	RH33	E	VII B	0.58	—
	AA32	SM16	E	VII B		—
F	UV26	SM28	FJGI	V C	2.1	—
G	UV10	—	FJGI	V B3		—
	UV24	—	FJGI	V B3		—
	UV19	—	FJGI	V B3		—
I	—	SM35	FJGI	V B2	0.74	—
J	—	—	—	—		
H	UV48	—	H	—		+
R	—	—	(new)	—		—
S	—	—	(new)	—		—
T	—	—	D	—		+

[a]AM1, *M. extorquens* AM1; XX, *Methylobacterium organophilum* XX.

M. organophilum XX (Table IV; Bastien *et al.*, 1989). In all cases these appeared to be monocistronic, except for the *mox*N and D genes, which appeared to be cotranscribed. In *M. extorquens* AM1, the only transcript that has been studied is that of the *mox*FJGI region. In this case, preliminary data suggest a large transcript covering all four genes is induced

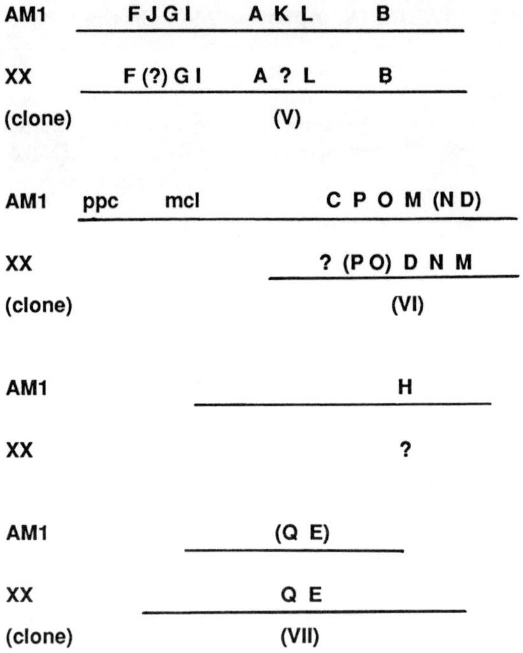

```
AM1              F J G I      A K L        B
                _____

XX               F (?) G I    A ? L        B
(clone)                         (V)
                _____


AM1       ppc      mcl              C P O M (N D)
                                    _____

XX                                  ? (P O) D N M
(clone)                             _____
                                         (VI)


AM1                                       H
                                    _____

XX                                        ?


AM1                              (Q E)
                                _____

XX                                Q E
(clone)                         _____
                                    (VII)
```

Figure 1. The Mox genes in two *Methylobacterium* strains. For simplicity, the *M. extorquens* AM1 gene designations have been used in both cases, but the corresponding *M. organophilum* XX clone numbers are also noted. AM1, *M. extorquens* AM1; XX, *M. organophilum* XX; ?, equivalent gene is not known to be present; ( ), gene order is not known.

in methanol-grown cells, but a smaller transcript is also made that appears to cover only the first two genes (Anderson and Lidstrom, unpublished). Further work will be necessary to confirm this and determine the 3' ends of these transcripts.

7.2. Phenotypic Characterization of Mox Mutants in *Methylobacterium* Strains

The discovery of a large number of Mox genes in *Methylobacterium* strains was surprising, since this system carries out a single biochemical step in these organisms. Although the function of many of these genes is still not clear, the phenotypes of the mutants have provided clues. Table V lists proposed functions for each of the Mox genes in *M. extorquens* AM1. In most cases, the phenotypes of the corresponding Mox mutants in *M. organophilum* XX are similar, but more work is necessary to determine their precise function, in both strains.

TABLE V. Proposed *mox* Functions in *M. extorquens* AM1

Gene	Proposed function	Encoded protein (kDa)
*mox*A	Assembly/modification	?
*mox*K	Assembly/modification	?
*mox*L	Assembly/modification	19
*mox*C	PQQ production	?
*mox*O	PQQ production	?
*mox*P	PQQ production	?
*mox*H	PQQ production	?
*mox*T	PQQ production	?
*mox*F[a]	MeDH large subunit	60
*mox*G[a]	Cytochrome cL	21
*mox*I[a]	Putative MeDH small subunit	12
*mox*J	?	30
*mox*B	Regulation	?
*mox*M	? (mutants not available)	?
*mox*N	? (mutants not available)	?
*mox*D	? (pleiotropic; contains novel cyt$c$)	?
*mox*Q	? (mutants not available)	?
*mox*E	? (pleiotropic)	?
*mox*R	?	?
*mox*S	?	?

[a]Function has been confirmed.

7.2.1. *mox*F, J, G, and I

In both *Methylobacterium* strains, the structural genes (*mox*F and I) for the large (alpha) and small (beta) subunits of the MeDH have been identified by a combination of Western blots and sequencing (Nunn and Lidstrom, 1986a; Machlin *et al.*, 1988; Machlin and Hanson, 1988; Anderson *et al.*, 1990; Nunn *et al.*, 1989). In *M. extorquens* AM1, the structural gene (*mox*G) for the cytochrome C_L, the immediate electron acceptor of the MeDH, has also been identified by sequencing (Nunn and Anthony, 1988). The fourth gene in the *mox*FJGI operon in *M. extorquens* AM1 (*mox*J) has been identified by expression in *E. coli* (Anderson and Lidstrom, 1988) and by sequencing (Anderson *et al.*, 1990), but no clues as to its function have arisen from these analyses. In *P. denitrificans*, sequencing has revealed an open reading frame downstream of *mox*F that would encode a protein of approximately 29 kDa (Harms *et al.*, 1989). This sequence shows considerable similarity to the *mox*J sequence of *M. extorquens* AM1 and probably is the *mox*J homolog (Anderson *et al.*, 1990).

7.2.2. *mox*A, K, L, C, P, O, H, and T

A number of the Mox genes in *M. extorquens* AM1 appear to involve the MeDH cofactor, pyrroloquinoline quinone (PQQ). Three closely linked genes, *mox*A, K, and L (formerly *mox*A1, A2, and A3), are required for MeDH activity, and the inactive protein synthesized in the mutants shows an altered PQQ spectrum (Nunn and Lidstrom, 1986b). None of the mutants in these classes grows on methanol containing PQQ in the medium. Other Mox mutants, however, do grow on methanol containing PQQ (Table IV; Lee and Lidstrom, unpublished). These include the *mox*C, P, O (formerly classed as a *mox*D mutant), H, and T mutants. These mutants may be defective in PQQ synthesis genes. Although no *M. organophilum* XX mutants are available in *mox*H, O, C, or T, mutants in *mox*P were able to grow on methanol plus PQQ (Lidstrom, unpublished), suggesting that a similar situation may exist in this strain.

7.2.3. Other Mox Genes

The functions of the other eight genes (*mox*B, D, E, M, N, Q, R, and S) are more obscure. Most of the mutants in these genes show pleiotropic phenotypes, and therefore they may be involved in regulation, stability, processing, or other unknown functions. Work is currently ongoing to characterize the biochemical defects in these strains.

7.3. Characterization of Mox Genes

The large subunit of the MeDH is similar in most gram-negative methylotrophs (Anthony, 1986), so it was not surprising to discover that the genes encoding this protein were also conserved. This gene, designated *mox*F, has been cloned from *M. extorquens* AM1 (Nunn and Lidstrom, 1986a), *P. denitrificans* (Harms *et al.*, 1987), *M. organophilum* XX (Machlin *et al.*, 1988), and *Methylobacterium* DSM 760 (Mazodier *et al.*, 1988). Sequence data are now available for three of these genes, from *P. denitrificans* (Harms *et al.*, 1987), *M. organophilum* (Machlin and Hanson, 1988), and *M. extorquens* AM1 (Anderson *et al.*, 1990). All three sequences show a predicted signal peptide of 27–32 amino acids. The two *Methylobacterium* signal sequences are identical except for one substitution, while the *P. denitrificans* sequence shows little similarity (Table V). All three signal sequences contain an Ala-Xaa-Ala tripeptide before the cleavage site. A comparison of the sequences of the mature proteins shows that they are remarkably conserved, with each consisting of 599 amino acids. The two *Methylobacterium* sequences show 96% similarity at the amino acid level, and each of these shows an overall similarity with the *P. denitrificans* gene of 75–80%. Some stretches within all three genes are over 90% conserved at both the nucleotide and amino acid sequence level. Sequence data are not available for other Mox F proteins, but surveys using either the *M. extorquens* AM1 or *M. organophilum* XX *mox*F genes as probes have shown that specific hybridization can be detected to DNA from a variety of methane- and methanol-utilizing bacteria (Stephens *et al.*, 1988; Tsuji *et al.*, 1990). These results suggest that at least portions of these genes are conserved in gram-negative methylotrophs.

Sequence data are also available for the *mox*G and I gene products of *M. extorquens* AM1, and Table VI shows a comparison of the predicted signal sequences for all five genes. Although some similarities exist, they mainly reflect the expected hydrophobic and hydrophilic stretches, along with the conservation of Ala-Xaa-Ala at the cleavage sites noted above. Table VII shows the codon usage for four of these genes. As is common for genes with a high % G+C ratio, strong codon bias exists for Gs and Cs in the third position, and some codons are used rarely, if at all.

The promoter sequence is not yet known for any of the Mox genes. However, the transcriptional start site has been determined for the two *Methylobacterium mox*F genes (Machlin and Hanson, 1988; Anderson, Ph.D. thesis, Univ. of Washington, 1988), and these are offset by only one nucleotide (Table VII). A scan of the regions upstream of the start sites shows

TABLE VI. Proposed Signal Sequences of Methylotrophic Genes

Methylobacterium organophilum **XX:**

*mox*F   Met Ser Arg Phe Val Thr Ser Val Ser Ala Leu Ala Met Leu Ala Leu Ala Pro Ala Ala Leu Ser Ser Val **Ala** Tyr **Ala**

Methylobacterium extorquens **AM1:**

*mox*F   Met Ser Arg Phe Val Thr Ser Val Ser Ala Leu Ala Met Leu Ala Leu Ala Pro Ala Ala Leu Ser Ser Gly **Ala** Tyr **Ala**

*mox*G   Met Met Asn Arg Val Lys Ile Gly Thr Ala Leu Leu Gly Leu Thr Leu Ala Gly Ile Ala Leu **ProAla** Leu **Ala**

*mox*I   met Lys Thr Thr Leu Ile Ala Ala Ala Ile Val Ala Leu Ser Gly Leu Ala Ala **ProAla** Leu **Ala**

Paracoccus denitrificans:

*mox*F   Met Asn Arg Asn Thr - Pro Lys Ala Arg Gly Ala Ser Ser Leu Ala Met Ala Val Ala Met Gly Leu Ala Val Leu Thr Thr Ala **Pro Ala** Thr **Ala**

TABLE VII. Codon Usage Analysis of *mox* Genes

	moxG AM1	moxF AM1	moxF Pd	moxF XX		moxG AM1	moxF AM1	moxF Pd	moxF XX		moxG AM1	moxF AM1	moxF Pd	moxF XX		moxG AM1	moxF AM1	moxF Pd	moxF XX
UUU (F)	0	1	2	1	UCU (S)	0	0	0	0	UAU (V)	1	4	14	6	UGU (C)	0	1	1	0
UUC (F)	5	20	20	20	UCC (S)	4	6	3	8	UAC (V)	5	28	11	26	UGC (C)	4	3	3	4
UUA (L)	0	0	0	0	UCA (S)	0	1	0	0										
UUG (L)	0	1	2	1	UCG (S)	2	14	15	12						UGG (W)	3	19	20	19
CUU (L)	2	4	3	0	CCU (P)	0	0	0	0	CAU (H)	0	1	7	2	CGU (R)	0	4	1	3
CUC (L)	10	25	2	28	CCU (P)	4	10	17	9	CAC (H)	4	9	6	9	CGC (R)	4	10	19	10
CUA (L)	1	0	0	0	CCA (P)	0	0	1	0	CAA (Q)	2	1	1	0	CGA (R)	0	0	0	0
CUG (L)	7	19	38	21	CCG (P)	8	23	20	23	CAG (Q)	7	15	18	18	CGG (R)	1	3	5	4
AUU (I)	2	0	1	0	ACU (T)	0	0	1	0	AAU (N)	3	1	2	4	AGU (S)	1	0	0	0
AUC (I)	5	21	21	18	ACC (T)	8	24	34	31	AAC (N)	8	32	26	29	AGC (S)	0	4	4	6
AUA (I)	0	0	0	0	ACA (T)	0	1	0	1	AAA (K)	0	1	3	0	AGA (R)	0	0	0	0
AUG (M)	5	18	28	17	ACG (T)	8	17	12	10	AAG (K)	13	44	28	44	AGG (R)	0	1	2	1
GUU (V)	0	0	0	0	GCU (A)	0	3	3	1	GAU (D)	1	14	4	6	GGU (G)	0	14	4	5
GUC (V)	3	21	23	29	GCC (A)	11	36	33	36	GAC (D)	10	37	43	45	GGC (G)	20	60	58	65
GUA (V)	1	0	0	0	GCA (A)	1	2	2	1	GAA (E)	2	3	9	2	GGA (G)	1	1	1	1
GUG (V)	5	18	19	15	GCG (A)	3	15	17	15	GAG (E)	9	15	20	17	GGG (G)	3	1	7	0

ALA = A GLU = E LEU = L SER = S
ARG = R GLN = Q LYS = K THR = T
ASN = N GLY = G MET = M TRP = W
ASP = D HIS = H PHE = F TYR = Y
CYS = C ILE = I PRO = P VAL = V

AM1 = *Methylobacterium extorquens* AM1
Pd = *Paracoccus denitrificans*
XX = *Methylobacterium organophilum* XX

areas around -10 and -35 with some similarity to the consensus *E. coli* promoter. However, the most strongly conserved bases of the consensus sequence are not well represented, and it is known that neither of these genes is expressed in *E. coli* from its own promoter. The -10 and -35 sequences do show strong similarity to each other, and a sequence with some similarity also exists upstream of the *P. denitrificans moxF* gene (Table VIII). It is possible that these conserved sequences are important in transcriptional initiation in these bacteria, but further studies will be necessary to confirm this. All three regions contain repeated sequences upstream of the translational start site, and all three contain a strong putative stem and loop structure just downstream of the *moxF* coding sequences. It is not known whether these putative structures are involved in regulation of expression of these genes.

8. PQQ SYNTHESIS GENES

The cofactor PQQ is found in a variety of enzymes known as quinoproteins (Duine *et al.*, 1986), but it was first discovered as the MeDH cofactor. The biosynthetic pathway for this cofactor is just beginning to be elucidated, but it appears that tyrosine and glutamate are the starting materials (Houck *et al.*, 1989). In *Acinetobacter calcoaceticus,* a nonmethylotroph that contains a PQQ-linked glucose dehydrogenase, it has been shown that only four genes are required for synthesis of PQQ in *E. coli* (Goosen *et al.*, 1989). These genes were linked on the chromosome and were all required for PQQ synthesis. Only three of the genes encoded proteins, while the fourth encoded a 24-residue peptide.

In *M. organophilum* DSM 760, PQQ mutants have been isolated that

TABLE VIII. Comparison of the Sequence 5' to the Start of *moxF* Genes from Three Different Methylotrophs

	-35	-10	+1
COLI	TTGACA XXX XX XXXX XX XXXX XX X TATAAT XXXXXG		
AM1	CT **AAAGACA** TCGCGTCCAATCAAAGCC **TAGAAA** AT AT A*G*—170bp—first Met		
XX	GT **AAAGACA** TCTCCTTCAATCAACGCC **TAGAAA** *CGATA*—171bp—first Met		
Pd	GATCGGACGGG**GAAA** AACCCC—178bp—first Met		

bold face = similarities
italics = mapped transcriptional start sites
moxF = methanol dehydrogenase structural gene

XX = *Methylobacterium organophilum* XX (serine pathway methanol-utilizer)
AM1 = *Methylobacterium extorquens* AM1 (serine pathway methanol-utilizer)
Pd = *Paracoccus denitrificans* (autotrophic methanol-utilizer)

have been placed into six complementation groups (Biville et al., 1989). Five of the genes are clustered (*pqq*ABCDE), and the sixth (*pqq*F) is located about 19 kb away. One of these genes (*pqq*A) has been shown to be located within 30 kb of *mox*F (Mazodier et al., 1988). The functions of these genes have not been determined, and it is not known whether they are equivalent to the *A. calcoaceticus* genes, or whether they are conserved among methylotrophs. However, it may be that they are equivalent to *mox*C, O, P, H, or T of *M. extorquens* AM1, described earlier.

9. GENES FOR ASSIMILATION OF C_1 UNITS

9.1. Serine Cycle Genes

The methanol utilizers that contain the serine cycle for assimilation of C_1 units include *Methylobacterium* and *Hyphomicrobium* strains (Anthony, 1982), but so far genetic studies have been carried out only in the *Methylobacterium* strains. Mutants in assimilation genes have been isolated in both *M. extorquens* AM1 and *M. organophilum* XX (DeVries, 1986). Early attempts at genetic mapping of these mutants in *M. organophilum* XX using a low-frequency transformation system placed them into three cotransformation groups (Table IX; O'Connor and Hanson, 1978). Since the mutants were pleiotropic, it was assumed that the genes were clustered in an operon, and a model for a linkage map was proposed (O'Connor and Hanson, 1978; O'Connor, 1981). However, it is now clear that pleiotropy is common in C_1 mutants for reasons that are not known (Nunn and Lidstrom, 1986b; Stone and Goodwin, 1989), and at least some of these genes are separated on the chromosome (Allen and Hanson, 1985). Confusion currently exists concerning these *M. organophilum* XX mutants, as some may not be true *M. organophilum* XX strains, although they appear to be *Methylobacterium* strains (Allen and Hanson, 1985). In addition, the reported phenotypes for two of these mutants (8A and 8Z) have been reversed (Allen and Hanson, 1985), and therefore it is possible that these two mutants have been inadvertently switched at some point (Stone and Goodwin, 1989). Unfortunately, many of these mutants are no longer in culture, so it has not been possible to retest them all. Therefore, the only conclusions that can be drawn from this earlier work are that a gene for glycerate kinase may be located near a Mox gene, but these are not closely linked to the cytochrome *c*, the acetyl CoA oxidation pathway, and the known Mox genes (Table IX). These conclusions have been supported by complementation studies with cosmid clone banks (Table IX), which have shown that genes for the malyl

CoA lyase, glycerate kinase, and the acetyl CoA oxidation pathway are not on common cosmid clones and are each separated by at least 30 kb on the chromosome (Allen and Hanson, 1985). None of these genes are on the cosmid clones containing cytochrome c and the Mox system (Machlin *et al.*, 1988).

In *M. extorquens* AM1, assimilation genes have been studied with both narrow- and broad-range mapping tools, and more is known concerning their genetic organization. These studies have shown that the known C_1 genes in *M. extorquens* AM1 are found in at least four gene clusters (Fig. 1). Two serine pathway genes (for malyl CoA lyase and PEP carboxylase) are located within 15 kb of the *mox*C–D gene cluster (Fulton *et al.*, 1984; Lidstrom *et al.*, 1987) and genes for glycerate kinase and the acetyl CoA

TABLE IX. Apparent Linkage Groups in Methylobacterium *organophilum* XX (from O'Connor and Hanson, 1978; Allen and Hanson, 1985; Machlin *et al.*, 1988 and Stone and Goodwin, 1989)

	By cosmid mapping		By transformation[c]	
	mutant	defect	mutant	defect
Group 1	PT1001[a]	(glycerate kinase)	8Z	(glycerate kinase)
	2121M[b]	(glycerate kinase)	17M	(unknown Mox gene)
			4M	(pleiotropic)
Group 2	2111B[b]	(uncharacterized)		
Group 3	PCT57[a]	(Malyl CoA lyase)		
	PG5[a]	(pleiotropic)		
Group 4	8A	(acetyl CoA recycle)	8A	(acetyl CoA recycle)
	PCT48[a]	(acetyl CoA recycle)		
Group 5	PG1[a]	(Mox A)	24C	(cytochrome c)
	M15A[a]	(Mox L)	22C	(extra cytochrome c)
	SM37	(Mox B)		
	2111I[b]	(Mox F; MeDH 60 K subunit)		
	UV10[a]	(Mox G; cytochrome c)		
	SM35	(Mox I; MeDH 10 K subunit)		
Group 6	SM10	(Mox M)		
	SM2	(Mox N)		
	SM48	(Mox D)		
	PT34	(Mox P)		
Group 7	SM4	(Mox Q)		
	RH33	(Mox E)		

Complementation groups as per Allen and Hanson (1985) and Machlin *et al.* (1988).
[a]*M. extorquens* AM1 mutant.
[b]*M. organophilum* DSM 760 mutant.
[c]Except for mutant 8A, none of the mutants in this column have been mapped with the cosmid clones, and so the associations shown are those suggested by the phenotypes.

oxidation pathway are located at least 10–15 kb from each other and from other C_1 genes (Stone and Goodwin, 1989). The R′ mapping described previously suggested that *mox*A was located between the acetyl CoA recycle and malyl CoA lyase genes. Further work will be necessary to determine whether this gene order is correct. Table IX shows it is possible that a similar gene organization exists in *M. organophilum* XX, although the malyl CoA lyase gene is apparently more than 15 kb from *mox*C. More studies will be necessary to determine the chromosomal map in this strain.

9.2. Ribulose Monophosphate Cycle Genes

Much less is known concerning genetics of the ribulose monophosphate cycle. In the obligate strains, this pathway is essential for growth, and therefore only conditional mutants can be isolated. As mentioned previously, two such mutants have been isolated in *M. flagellatum*, but these have not yet been mapped (Kletsova *et al.*, 1988). Mutants are also available in the facultative RuMP cycle organism *Arthrobacter* P1 (Levering *et al.*, 1987), but no genetic system is available for this strain, so these have also not been mapped.

10. SUMMARY AND FUTURE PROSPECTS

The application of recombinant DNA techniques to the methylotrophic bacteria has resulted in a quantum leap forward in our understanding of molecular genetics in these organisms. In several strains, genes have been cloned and mapped, promoter sequences are beginning to emerge, and chromosomal maps are being developed. However, much remains to be learned. Many key systems in methanol utilizers have not yet been studied, especially in the RuMP cycle strains, and even in the serine cycle strains few details are known at this time. Once more is known concerning gene organization and expression in these strains, the developing base of genetic information should provide interesting comparisons. For instance, will a common consensus promoter sequence be present, or will varying sequences be identified, perhaps relying on separate sigma factors? Will gene order and operon structure be conserved, or has divergence occurred between the different groups of methylotrophs? In the facultative methylotrophs, in which C_1 functions are apparently under transcriptional regulation, what will the regulatory elements be and how will they act? Will similar systems be present in the obligate methanol utilizers? What controls high level expression and protein turnover? What gene products are involved in the many steps required for both dissimilatory and assimilatory methylotrophic metabolism? The next decade promises to be an exciting and challenging period in the area of methylotrophic studies. It is expected that the advances made with genetic approaches will have strong applications to both commercial and environmental uses of these important organisms.

REFERENCES

Allen, L., and Hanson, R., 1985, Construction of broad-host-range cosmid cloning vectors: identification of genes necessary for growth of *Methylobacterium organophilum* on methanol. *J. Bacteriol.* **161**:955–962.

Anderson, D., and Lidstrom, M., 1988, The *mox*FG region encodes four polypeptides in the methanol-oxidizing bacterium *Methylobacterium* sp. strain AM1, *J. Bacteriol.* **170**:2254–2262.

Anderson, D., Morris, C., Nunn, D., Anthony, C., and Lidstrom, M., 1990, Nucleotide sequence of the *Methylobacterium extorquens* AM1 *mox*F and *mox*J genes involved in methanol oxidation, *Gene* **90**:173–176.

Anthony, C., 1982, *The Biochemistry of Methylotrophs*, Academic Press, London, 382 pp.

Anthony, C., 1986, Bacterial oxidation of methane and methanol, *Adv. Microb. Physiol.* **27**:113–210.

Bastien, C. A., Machlin, S. M., Zhang, Y., Donaldson, K., and Hanson, R. S., 1989. The organization of genes required for the oxidation of methanol to formaldehyde in three type II methylotrophs, *Appl. Environ. Microbiol.* **55**:3124–30.

Biville, F., Turlin, E., and Gasser, F., 1989, Cloning and genetic analysis of six pyrroloquinoline quinone biosynthesis genes in *Methylobacterium organophilum* DSM760, *J. Gen. Microbiol.* **135**:2917–29.

Bohanon, M., Bastien, C., Yoshida, R., and Hanson, R., 1987, Isolation of auxotrophic mutants of *Methylophilus methylotrophus* by modified-marker exchange, *Appl. Environ. Microbiol.* **54**:271–273.

Byrom, D., 1984, Host-vector systems for *Methylophilus methylotrophus*, in: *Microbial Growth on C1 Compounds* (R. L. Crawford and R. S. Hanson, eds.), ASM Press, Washington, DC, pp. 221–223.

Chistoserdov, A. Y., and Tsygankov, Y. D., 1986, Broad host range vectors derived from an RSF1010::Tn1 plasmid, *Plasmid* **16**:161–167.

Chistoserdov, A., Eremashvili, M., Mashko, S., Lapidus, A., Skvortsova, M., and Sterkin, V., 1987, Expression of human interferon F gene in obligate methylotroph *Methylobacillus flagellatum* KT and *Pseudomonas putida* (in Russian), *Mol. Genet. Microbiol. Virol.* **8**:36–42.

DeVries, G., 1986, Molecular biology of bacterial methanol oxidation, *FEMS Microbiol. Rev.* **39**:235–258.

Dijkhuizen, L., Harder, W., DeBoer, L., Van Boven, A., Clement, W., Bron, S., and Venema, G., 1984, Genetic manipulation of the restricted facultative methylotroph *Hyphomicrobium* X by the R-plasmid mediated introduction of the *Escherichia coli pdh* genes, *Arch. Microbiol.* **139**:311–318.

Ditta, G., Schmidhauser, T., Yakobson, E., Lu, P., Liang, X., Finlay, D., Guiney, D., and Helinski, D., 1985, Plasmids related to the broad host range vector, pRK290, useful for gene cloning and monitoring gene expression, *Plasmid* **13**:149–153.

Duine, J., Frank, J., and Jongejan, J., 1986, PQQ and quinoprotein enzymes in microbial oxidations, *FEMS Microbiol. Rev.* **32**:165–178.

Ely, B., 1985, Vectors for transposon mutagenesis of non-enteric bacteria, *Mol. Gen. Genet.* **200**:302–304.

Fulton, G., Nunn, D., and Lidstrom, M., 1984, Molecular cloning of a malyl coenzyme A lyase gene from *Pseudomonas* sp. strain AM1, a facultative methylotroph, *J. Bacteriol.* **160**:718–723.

Goosen, N., Horsman, H., Huinen, R., De Groot, A., and Van De Putte, P., 1989, Genes involved in the biosynthesis of PQQ from *Acinetobacter calcoaceticus*, *Ant. van. Leeuw.* **56**:85–89.

Harms, N., DeVries, G., Maurer, K., Veltkamp, E., and Stouthamer, A., 1985, Isolation and characterization of *Paracoccus denitrificans* mutants with defects in the metabolism of one-carbon compounds, *J. Bacteriol.* **164**:1064–1070.

Harms, N., DeVries, G., Maurer, K., Hoogendijk, J., and Stouthamer, A., 1987, Isolation and nucleotide sequence of the methanol dehydrogenase structural gene from *Paracoccus denitrificans*, *J. Bacteriol.* **169**:3969–3975.

Harms, N., Van Spanning, R. J. M., Oltmann, L. F., and Stouthamer, A. H., 1989, Regulation of methanol dehydrogenase synthesis in *Paracoccus denitrificans*, *Ant. van Leeuw.* **56**:47–50.

Hennam, J. F., Cunningham, A. E., Sharpe, G. S., and Atherton, K. T., 1982, Expression of eukaryotic coding sequences in *Methylophilus methylotrophus*, *Nature* **297**:80–82.

Holloway, B., 1984, Genetics of methylotrophs, in: *Methylotrophs: Microbiology, Biochemistry and Genetics* (C. T. Hou, ed.), CRC Press, Boca Raton, FL, pp. 87–106.

Houck, D., Hanners, J., Unkefer, C., Van Kleef, M., and Duine, J., 1989, PQQ; Biosynthetic studies in *Methylobacterium* AM1 and *Hyphomicrobium* X using specific ^{13}C labelling and NMR, *Ant. van Leeuw.* **56**:93–99.

Kearney, P., and Holloway, B., 1987, Expression in *Escherichia coli* and *Pseudomonas aeruginosa* of *Methylophilus methylotrophus* AS1 genes carried on prime plasmids, *FEMS Microbiol. Lett.* **44**:7–11.

Kim, Y., and Lidstrom, M., 1989, Plasmid analysis in pink facultative methylotrophic bacteria using a modified acetone-alkaline hydrolysis method, *FEMS Microbiol. Lett.* **60**:125–129.

Kletsova, L., Chibisova, E., and Tsygankov, Y., 1988, Mutants of the obligate methylotroph *Methylobacillus flagellatum* KT defective in genes of the ribulose monophosphate cycle of formaldehyde fixation, *Arch. Microbiol.* **149**:411–446.

Knauf, V. C., and Nester, E. W., 1982, Wide host range cloning vectors: a cosmid clone bank of *Agrobacterium* Ti plasmid, *Plasmid* **8**:45–54.

Levering, P. R., Tiesma, L., Woldendrop, J. P., Steensma, M., and Dijkhuizen, L., 1987, Isolation and characterization of mutants of the facultative methylotroph *Arthrobacter* P1 blocked in one-carbon metabolism, *Arch. Microbiol.* **146**:346–352.

Lidstrom, M., Nunn, D., Anderson, D., Stephens, R., and Haygood, M., 1987, Molecular biology of methanol oxidation, in : *Microbial Growth on C-1 Compounds* (H. Van Verseveld and J. Duine, eds.), Nijhoff, Amsterdam, pp. 246–254.

Lyon, B., Kearney, P., Sinclair, M., and Holloway, B., 1988, Comparative complementation mapping of *Methylophilus* spp. using cosmid gene libraries and prime plasmids, *J. Gen. Microbiol.* **134**:123–130.

Machlin, S., and Hanson, R., 1988, Nucleotide sequence and transcriptional start site of the *Methylobacterium organophilum* XX methanol dehydrogenase structural gene, *J. Bacteriol.* **170**:4739–47.

Machlin, S., Tam, P., Bastien, C., and Hanson, R., 1988, Genetic and physical analyses of *Methylobacterium organophilum* XX genes encoding methanol oxidation, *J. Bacteriol.* **170**:141–148.

Marquardt, R., Wohner, G., and Winnacker, E., 1984, Properties of plasmids from *Methylomonas clara*, *J. Biotechnol.* **1**:317–320.

Mazodier, P., Biville, F., Turlin, E., and Gasser, F., 1988, Localization of a pyrroloquinoline quinone biosynthesis gene near the methanol dehydrogenase structural gene in *Methylobacterium organophilum* DSM 760, *J. Gen. Microbiol.* **134**:2513–24.

Metzler, T., Marquardt, R., Prave, P., and Winnacker, E-L., 1988, Characterisation of a promoter from *Methylomonas clara*, *Mol. Gen. Genet.* **21**:210–214.

Monteiro, M., Typas, M., Moffett, B., and Bainbridge, B., 1982, Isolation and characterization of a high molecular weight plasmid from the obligate methanol-utilising bacterium *Methylomonas (Methanomonas) methylovora*, *FEMS Microbiol. Lett.* **15**:235–237.

Moore, A. T., Nayudu, M., and Holloway, B. W., 1983, Genetic mapping in *Methylophilus methylotrophus* AS1, *J. Gen. Microbiol.* **129**:785–799.

Nunn, D., and Anthony, C., 1988, The nucleotide sequence and deduced amino acid sequence of the cytochrome C_L gene of *Methylobacterium extorquens* AM1, a novel class of c-type cytochrome, *Biochem. J.* **256**:673–676.

Nunn, D., and Lidstrom, M., 1986a, Isolation and complementation analysis of 10 methanol oxidation mutant classes and identification of the methanol dehydrogenase structural gene of *Methylobacterium* sp. strain AM1, *J. Bacteriol.* **166**:581–590.

Nunn, D., and Lidstrom, M., 1986b, Phenotypic characterization of 10 methanol oxidation mutant classes in *Methylobacterium* sp. strain AM1, *J. Bacteriol.* **166**:591–597.

Nunn, D. N., Day, D., and Anthony, C., 1989, The second subunit of methanol dehydrogenase of *Methylobacterium extorquens* AM1, *Biochem. J.* **260**:857–862.

O'Connor, M., 1981, Extension of the model concerning linkage of genes coding for C-1 related functions in *Methylobacterium organophilum*, *Appl. Environ. Microbiol.* **41**:437–441.

O'Connor, M., and Hanson, R., 1978, Linkage relationships between mutants of *Methylobacterium organophilum* impaired in their ability to grow on one-carbon compounds, *J. Gen. Microbiol.* **104**:105–111.

Ruvkun, G., and Ausubel, F., 1981, A general method for site-directed mutagenesis in prokaryotes, *Nature* **298**:85–88.

Serebrijski, I., Kazakova, S., and Tsygankov, Y., 1989, Construction of Hfr-like donors of the obligate methanol-oxidizing bacterium *Methylobacillus flagellatum* KT, *FEMS Microbiol. Lett.* **59**:203–206.

Simon, R., Priefer, U., and Puehler, A., 1983, A broad host range mobilization system for *in vivo* genetic engineering: transposon mutagenesis in gram-negative bacteria, *Biotechnology* **1**:784–91.

Stephens, R., Haygood, M., and Lidstrom, M., 1988, Identification of putative methanol dehydrogenase (*mox*F) structural genes in methylotrophs and cloning of *mox*F genes from *Methylococcus capsulatus* Bath and *Methylomonas albus* BG8, *J. Bacteriol.* **170**:2063–69.

Stone, S., and Goodwin, P., 1989, Characterization and complementation of mutants of *Methylobacterium* AM1 which are defective in C-1 assimilation, *J. Gen. Microbiol.* **135**:227–234.

Tabor, S., and Richardson, C., 1985, A bacteriophage T7 RNA polymerase/promoter system for controlled exclusive expression of specific genes, *Proc. Natl. Acad. Sci. USA* **82**:1074–1078.

Tatra, P., and Goodwin, P., 1983, R-plasmid mediated chromosome mobilization in the facultative methylotroph *Pseudomonas* AM1, *J. Gen. Microbiol.* **129**:2629–32.

Tatra, P., and Goodwin, P., 1985, Mapping of some genes involved in C-1 metabolism in the facultative methylotroph *Methylobacterium* sp. strain AM1 (*Pseudomonas* AM1), *Arch. Microbiol.* **143**:169–177.

Tsuji, K., Tsien, H., Hanson, R., DePalma, S., Scholtz, R., and LaRoche, S., 1990, 16s ribosomal RNA sequence analysis for determination of phylogenetic relationship among methylotrophs, *J. Gen. Microbiol.* **136:**1–10.

Tsygankov, Y., and Kazakova, S., 1987, Development of gene transfer systems in *Methylobacillus flagellatum* KT: isolation of auxotrophic mutants, *Arch. Microbiol.* **149:**112–119.

Ueda, S., Kitamoto, N., Tamura, Y., Sakakibara, Y., and Shimizu, S., 1987, Isolation and characterization of a plasmid in a methylotrophic bacterium, *J. Ferment. Technol.* **65:**589–91.

Warner, P., and Higgins, I., 1977, Examination of obligate and facultative methylotrophs for plasmid DNA, *FEMS Microbiol. Lett.* **1:**339–42.

Whitta, S., Sinclair, M., and Holloway, B., 1985, Transposon mutagenesis in *Methylobacterium* AM1 (*Pseudomonas* AM1), *J. Gen. Microbiol.* **131:**1547–51.

Windass, J., Worsey, M., Pioli, E., Pioli, D., Barth, P., Atherton, K., Dart, E., Byrom, D., Powell, K., and Senior, P., 1980, Improved conversion of methanol to single-cell protein by *Methylophilus methylotrophus, Nature* **287:**396–401.

Methanol-Utilizing Yeasts 7

W. DE KONING and W. HARDER

1. INTRODUCTION

The ability of yeasts to grow on methanol as a source of carbon and energy has been discovered relatively recently. Whereas bacterial utilization of methanol was found before the end of the previous century, it was not until 1969 that growth of eukaryotic microorganisms at the expense of this one-carbon compound was reported (Ogata *et al.* 1969). Independent studies by several other research groups soon followed this first report and established that a variety of yeasts (Lee and Komagata, 1980a) and some filamentous fungi (Goncharova *et al.*, 1977) are able to utilize methanol as a source of carbon and energy for growth.

The enzymology of methanol dissimilation was readily elucidated (Fujii and Tonomura, 1972), but it took almost 10 years and considerable confusion before the assimilatory pathway was firmly established (van Dijken *et al.*, 1978; Kato *et al.*, 1979; Waites and Quayle, 1980). The uniformity of methanol dissimilation and the employment of the xylulose-5-phosphate cycle for formaldehyde fixation in all methylotrophic yeasts investigated so far is probably reflected in the close taxonomic relationship of these organisms (Komagata, 1981).

The first step in the metabolism of methanol in yeasts is catalyzed by a hydrogen peroxide-producing oxidase (see below). In eukaryotes, such enzymes generally occur in subcellular compartments collectively called microbodies (Veenhuis *et al.*, 1983) and, due to the crystalline nature of the oxidase in these organelles, yeasts possess a unique ultrastructure during

---

W. DE KONING • Department of Microbiology, University of Groningen, 9751 NN Haren, The Netherlands. W. HARDER • TNO-Institute of Environmental Sciences, 2600 AE Delft, The Netherlands.

Methane and Methanol Utilizers, edited by J. Colin Murrell and Howard Dalton. Plenum Press, New York, 1992.

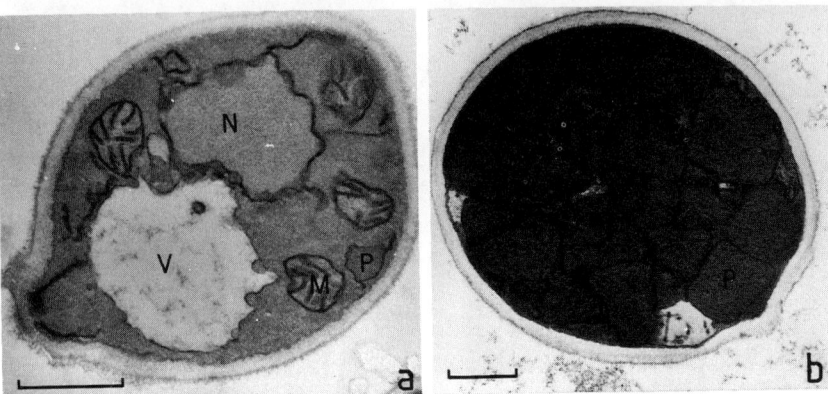

Figure 1. Survey of cells of *Hansenula polymorpha* showing the overall cell morphology after growth on glucose (a) and methanol (b) as the sole source of carbon. In the glucose-grown cell (log-phase in batch culture) one small peroxisomal profile is observed. In the methanol-grown cell (chemostat culture, $D = 0.1\ h^{-1}$) approximately 20 peroxisomal profiles are visible. The cells were fixed and postfixed with potassium permanganate. Abbreviations: M, mitochondria; N, nucleus; P, peroxisome; V, vacuole. The bar marker represents 0.5 µm. (Reproduced from Veenhuis et al., 1983.)

growth on methanol (Fig. 1). This aspect of methylotrophic yeasts has been reviewed recently (Veenhuis and Harder, 1987) and will not be reiterated here. The possibility to manipulate the proliferation of microbodies in these organisms by a simple change in media composition has been instrumental in the study of the biogenesis and functioning of these organelles (for a general review see Borst, 1989).

During the early 1970s there was considerable interest in producing single-cell protein from methanol using methylotrophic yeasts, and several companies developed methanol-yeast processes. This potential commercial application stimulated much of the subsequent research into the physiology, biochemistry, and genetics of these organisms, and during the past decade a considerable body of knowledge has become available. This in turn has sparked renewed interest in using these yeasts for novel applications. It is the purpose of this chapter to describe the developments in our understanding of these remarkable organisms and discuss their potential in applications. For further details concerning the use of methanol as a feedstock for biotechnological processes, the reader is referred to Dijkhuizen et al. (1985). Harder et al. (1987) recently reviewed methanol metabolism in yeasts, while Gleeson and Sudbery (1988a) and Sibirny et al. (1988) reviewed progress in the genetics of these organisms.

2. ISOLATION AND PROPERTIES

2.1. Enrichment and Isolation

The publication by Ogata *et al.* (1969), reporting the isolation of a yeast capable of using methanol as a sole carbon- and energy-source, initiated an extensive search for methylotrophic yeasts. Screening of yeast species from the CBS culture collection at Delft (Hazeu *et al.*, 1972) revealed 16 different methanol-utilizing species out of 422 tested. Other investigators were less successful; for instance, Oki *et al.* (1972) screened a different culture collection and found no methanol-utilizing representative among the 192 strains tested.

The isolation of novel methanol-utilizing yeasts has been described by several groups. The methods were based on different strategies for enrichment. Van Dijken and Harder (1974) isolated 26 *Candida boidinii* and 10 *Pichia pinus* strains using a mineral medium of pH 5 with methanol (0.1–0.5% v/v) as the sole carbon source. Bacterial growth was discouraged by addition of the antibiotics penicillin G and D-cycloserine. Incubation was at 28°C and 37°C, but only the incubations at 28°C were successful in about 20% of the cases. A low pH to prevent bacterial growth and 2% (v/v) methanol was used by Oki *et al.* (1972), who isolated 20 methanol-utilizing yeasts, which, according to the reclassification by Yarrow and Meyer (1978), all belong to the genus *Candida*. Fourteen of these isolates were classified as *C. boidinii* and six belonged to a new species, *C. methanolovescens*. The same techniques were used by Kato *et al.* (1974) and Lee and Komagata (1980b) to isolate new species of *Candida* and *Pichia*. Positive growth on methanol observed in these studies also yielded a large number of slowly growing yeasts, which were not identified or further investigated. The use of continuous culture methods for the enrichment of methylotrophic yeasts also was successful (Levine and Cooney, 1973; Pal and Hamdan, 1979). The strains of *Hansenula polymorpha* isolated by these groups showed a somewhat higher growth rate and higher temperature tolerance than the type strain.

The inocula used in the above studies were predominantly from soil, but also (rotting) fruits and vegetables were used. With the exception of *C. boidinii*, which turned up in all extended isolation studies, the source of the inoculum appears to determine, at least partly, the type of yeast that is isolated. Due to the usually inaccurate description of the habitats tested for the isolation of methylotrophic yeasts, the ecology of these organisms is largely unknown. The only ecological study is from Miller *et al.* (1976), who reported the occurrence of the yeast *Torulopsis sonorensis* (renamed *C. sonorensis;* Yarrow and Meyer, 1978) in necrotic tissue of various cactus species. They isolated also a small number of non-cactus-specific species, like *C. boidinii* and *H. polymorpha*. In general studies on the ecology of yeasts, *P. pastoris* was often found in exudates of trees in North America and Europe, but not in Japan (Phaff and Starmer, 1987). *P. pinus* and *H. capsulata* were among the principal species isolated from bark beetles associated with coniferous trees (Shifrine and Phaff, 1956) whereas *H. polymorpha* was frequently isolated from the crops of fruitflies (*Drosophila* and *Aulacigaster;* Phaff *et al.*, 1956a,b). The methylotrophic yeasts found in the CBS culture collection were isolated predominantly from the bark of trees or from the frass of insects living on trees (Hazeu *et al.*, 1972; Kreger-Van Rij, 1984). A possible explanation for the occurrence of methylotrophic yeasts in these habitats is that methanol can be derived from the methoxy groups present in wood lignin. Also, nearly all methylotrophic yeasts tested are able to grow on pectin (Lee and Komagata, 1980a), a polymer rich in methoxy groups and found especially in fruit.

2.2. Taxonomy

Yeast taxonomy is often based on both morphology and growth characteristics. In addition, information such as % G+C content of the DNA, DNA–DNA hybridization, coenzyme Q structure, cell wall mannan analyses, and immunological characteristics is used, especially to confirm classifications. DNA homology between strains is usually regarded as a sufficient criterion for their conspecificity; when there is less than 25% homology, strains are thought to belong to different species, whereas more than 65% homology places them in the same species (Kurzman and Phaff, 1987). The region in between is somewhat uncertain, but usually strains showing between 25 and 65% homology are regarded to belong to closely related species. These boundaries are based on studies with strains that are able to mate (Kurzman and Phaff, 1987). Mating followed by viable spore formation implicates conspecificity, but negative results do not exclude it. The separation into families, subfamilies, and genera is in large part based on morphological criteria, whereas growth characteristics are mainly used to discriminate between species.

To determine the morphological characteristics, both vegetative growth and sexual reproduction are studied, as well as the appearance of the colonies on agar plates. Vegetative growth is characterized by the size and shape of the cells, by the formation of (pseudo) mycelia, and by budding. All methylotrophic yeasts known at present show multilateral budding and thus belong to the subfamily of the Saccharomycetodes. Formation of pseudomycelia under standard conditions is observed only for *C. boidinii* and in a rudimentary form for *P. pinus* and *P. haplophila* (Kreger-van Rij, 1984). Sexual reproduction is characteristic (by definition) of species of the fungi perfecti. A scheme of the genetic cell cycle is shown in Figure 2. Methylotrophic yeasts are, in contrast to other species like *Saccharomyces*, predominantly haploid during vegetative growth. Under appropriate conditions, which can vary between different strains, formation of diploids, vegetative growth of the diploid cells, and sporulation can be observed (specific details of some strains are discussed in Section 4).

All methylotrophic yeasts belonging to the fungi perfecti form ascospores that are hat-shaped and homothallic. These yeasts are members of the genera *Hansenula* and *Pichia*. *Hansenula* differs from *Pichia* only in its ability to assimilate nitrate, and Kurzman (1984) recently suggested the transfer of those strains that form hat-shaped spores from *Hansenula* to the genus *Pichia*. One of the arguments is that the methylotrophic strains *H. minuta* and *P. lindneri* possess a DNA homology of 75% and should be considered to belong to one species (Kurzman *et al.*, 1980a,b). Thus, nitrate assimilation varies within one species and is therefore not a proper

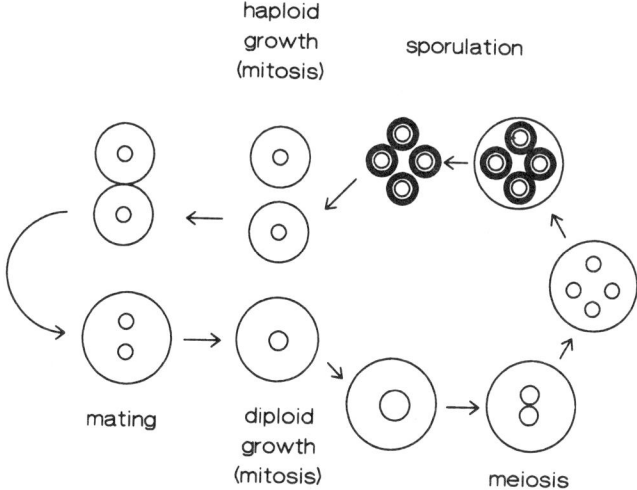

Figure 2. Life cycle of a (methylotrophic) yeast.

criterion for discrimination at the genus level. One significant drawback of the proposed change is that the name of one of the most widely investigated methylotrophic yeasts, *Hansenula polymorpha*, has to be changed to *Pichia angusta*, which may cause confusion to researchers. This is presumably the reason that most workers still use the old name.

The imperfect yeasts encompass strains in which sporulation has not (yet) been observed. On the basis of similarities in assimilation patterns and cell morphology, several imperfect strains are regarded as sterile counterparts of perfect strains; for example, *C. molischiana* is the anamorph of *H. capsulata* (Kreger-van Rij, 1984, but see Lee and Komagata, 1983). All imperfect methylotrophs described so far are included in the genus *Candida* after the transfer of most of the species of the former genus *Torulopsis* to the *Candida* group (Yarrow and Meyer, 1978).

The distribution of the methylotrophic yeasts over only two genera could suggest that these organisms are all closely related. However, the genera *Pichia (Hansenula)* and *Candida* contain more than 50% of the species described in *The Yeasts* (Kreger-van Rij, 1984). Moreover, both genera are very heterogeneous (Kurzman, 1984; Nakase and Komagata, 1971a,c) and should not be considered phylogenetic units. On the other hand, the metabolic pathways for methanol utilization are identical in the methylotrophic yeasts investigated so far and the enzymes involved are usually similar.

Lee and Komagata (1980a,b, 1983) investigated the relationship between a variety of methylotrophic yeasts taking into account carbon assimilation patterns, G+C content of DNA, properties of cell wall mannan, coenzyme Q characteristics, and the electrophoretic mobility of 13 different enzymes. They suggested the placement of these yeasts in four major groups. Two of these groups contain only a few species; the two other groups were further divided into different subgroups on the basis of mannan properties and coenzyme Q differences. Six of the eight resulting (sub-)groups contained only one or two species; the other two (sub-)groups, with five and 14 species showed significant variation in % G+C content (34.3–43.3%, resp. 35.5–50.5%) indicating distant phylogenetic relationships. From these results and from other investigations (Nakase and Komagata, 1970, 1971a,b,c; Yamada et al., 1973; Campbell, 1973), it is concluded that further subdivision is in order. But this has to wait until a more detailed analysis of the various properties of these yeasts is available. As a guide, the properties of the four species that have been most thoroughly investigated are listed in Table I. Three of the four main groups (Lee and Komagata, 1983) are included in the table; only the *H. capsulata* group is not represented.

TABLE I. Various Taxonomic Characteristics of Important Methylotrophic Yeasts[a]

	C. boidinii	H. polymorpha	P. pastoris	P. pinus
Spore formation	−	+	+	+
Pseudomycelia formation	+	−	−	− or w
Glucose Fermentation	+	+	+	v
Assimilation of				
nitrate	+	+	−	−
trehalose	−	+	+	+
sucrose	−	+	−	−
xylose	+	v	−	v
ribose	+	+	−	v
erythritol	+	+	−	+
ribitol	+	+	−	+
succinic acid	−	v	+	v
GC content DNA	29.5–33.3	47.8–48.3	40.2–41.0	44.4
coenzyme-Q system	Q7	Q7	Q8	Q7
Mannan type	18-p	9-b	6-b	9-b
Group	1	2B	4B	2B

[a] All yeasts assimilate glucose, ethanol, glycerol, and mannitol. Data from *The Yeasts* (Kreger-van Rij, 1984), Mannan type, CoQ, GC content, and group division from Lee and Komagata, 1983, w = weak, v = variable.

3. PHYSIOLOGY AND BIOCHEMISTRY

3.1. Enzymology and Compartmentation of Methanol Metabolism

A schematic representation of the pathways involved in the metabolism of methanol in yeasts and their compartmentation is shown in Figure 3. The initial reaction is an oxidation of methanol to formaldehyde, which is catalyzed by alcohol oxidase. This enzyme has been purified from various yeasts and studied extensively. It is a homo-octameric protein with a molecular mass of around 600 kDa, contains FAD as a prosthetic group, and is capable of oxidizing a variety of primary aliphatic alcohols, their halogen derivatives, and formaldehyde. Alcohol oxidase possesses a variable affinity for methanol (K_m 0.2–3.1 mM) and O_2 (K_m 0.2–1.0 mM) (van Dijken *et al.*, 1976; Kato *et al.*, 1976; Couderc and Baratti, 1980), and it is possible that this is not only strain-dependent but is also related to growth conditions. Recent evidence (Sherry and Abeles, 1985; Bystrykh *et al.*, 1989) suggests that alcohol oxidase possesses two chemically distinct forms of FAD. One form is identical to the classical FAD as found in many

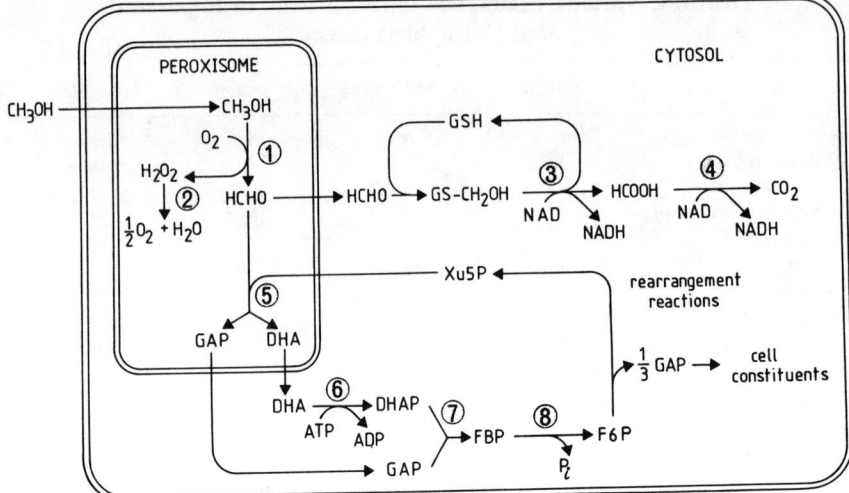

Figure 3. Compartmentation of methanol metabolism in *Hansenula polymorpha*. 1, alcohol oxidase; 2, catalase; 3, formaldehyde dehydrogenase; 4, formate dehydrogenase; 5, dihydroxyacetone synthase; 6, dihydroxyacetone kinase; 7, aldolase; 8, fructose-1,6-bis phosphate phosphatase. (After Douma *et al.*, 1985.)

flavoproteins, while the structure of the other has not yet been established. It is of interest to note that the ratio of the two species of FAD in alcohol oxidase is dependent on growth conditions (Bystrykh *et al.*, 1989), and this might explain in part the observed variable affinity of alcohol oxidase for methanol. For further details on alcohol oxidase see Hopkins and Muller (1987) and Bystrykh *et al.* (1989).

Hydrogen peroxide that is produced in the alcohol oxidase–catalyzed reaction is decomposed by catalase. It has been postulated (Roggenkamp *et al.*, 1974, but see also van Dijken *et al.*, 1981) that this peroxisomal enzyme may also play a role in the oxidation of methanol by way of its peroxidative activity using the H_2O_2 generated by alcohol oxidase. However, recent kinetic studies performed with catalase purified to homogeneity from *Candida boidinii* 2201 (Mozaffar *et al.*, 1986) suggest that this is unlikely.

Formaldehyde, generated in the peroxisomes, is at a crossroad in the metabolism of methanol. The study of reactions involving formaldehyde is, however, complicated by its hydration in aqueous solution. The product is called methylene glycol, and in a solution of formaldehyde in water, more than 99.9% is present in this form (Bieber and Trümpler, 1947). The establishment of the equilibrium, however, is rather slow (the dehydration reaction takes minutes to complete; W. de Koning, unpublished). *In vivo*, formaldehyde is produced in the nonhydrated form, and this form is also the substrate for the assimilation reaction and the dissimilatory pathway. Hydrated formaldehyde molecules can be oxidized by alcohol oxidase (Sahm, 1975), but the role of this reaction *in vivo* remains to be elucidated.

The enzymes catalyzing the oxidation of formaldehyde to CO_2 are localized in the cytosol, and in order to react with these enzymes, formaldehyde must pass the peroxisomal membrane. After entering the cytosol, it first condenses nonenzymically with reduced glutathione (GSH) to form S-hydroxymethylglutathione (GS-CH$_2$OH; Uotila and Koivusalo, 1974), which is the true substrate for formaldehyde dehydrogenase (Schütte *et al.*, 1976; Allais *et al.*, 1983a). The product S-formylglutathione is either hydrolyzed into formate and oxidized to CO_2 by formate dehydrogenase (Schütte *et al.*, 1976; Allais *et al.*, 1983b) or oxidized directly (van Dijken *et al.*, 1981). Representative properties of the dehydrogenases involved in the oxidation of formaldehyde are shown in Table II. A further enzyme, formaldehyde reductase, which reduces formaldehyde to methanol has been detected in methylotrophic yeasts. This enzyme is probably important when the cytosolic formaldehyde concentration is significantly increased over the levels normally found in growing cells (Trotsenko *et al.*, 1984). Under these conditions the activity of formaldehyde dehydrogenase *in vivo* is presumably limited by the concentration of cytosolic NAD, which can be replenished by the action of formaldehyde reductase.

As stated above, both dehydrogenases involved in the oxidation of formaldehyde are located in the cytosol, and therefore most of the NADH required for the generation of ATP is produced outside the mitochondria. This cytosolic NADH is oxidized by a mitochondrial NADH dehydrogenase that is localized at the outer face of the inner membrane (Moore and

TABLE II. Some Properties of Formaldehyde and Formate Dehydrogenases Purified from *Candida boidinii*[a]

Property	Formaldehyde dehydrogenase	Formate dehydrogenase
Molecular mass (kDa)	81	74
Subunits	2 (identical)	2 (identical)
pH optimum	7.5–7.9	7.5–8.5
Apparent K_m (mM)	0.25 (HCHO)	13 (HCOOH)
	0.13 (GSH)	
	0.09 (NAD)	0.09 (NAD)
Inhibitors	NADH, ATP	NADH, ATP

[a]Data of Schütte *et al.*, 1976; for comparison see also Allais *et al.*, 1983a,b.

Rich, 1980), whereas the citric acid cycle is used mainly for biosynthetic purposes (van Dijken et al., 1981; Egli and Lindley, 1984).

Assimilation of formaldehyde is accomplished by the xylulose monophosphate (XuMP) cycle. Recent studies (Douma et al., 1985; Goodman, 1985) have demonstrated that the first enzyme involved in the assimilatory sequence, dihydroxyacetone (DHA) synthase, is located inside the peroxisomes. This enzyme is a special transketolase that catalyzes the glycolaldehyde transfer from xylulose-5-phosphate as a donor to the acceptor molecule formaldehyde in a reversible reaction (K_{eq} = 622) by way of a ping-pong mechanism (Kato et al., 1982). Representative properties of the enzyme are summarized in Table III (the K_m for formaldehyde and the equilibrium constant are not corrected for hydration of this substrate; if this is taken into account, these values decrease or increase, respectively, by at least 1000-fold). DHA synthase differs from the classical transketolase in several respects, and studies with a DHA synthase minus mutant of *H. polymorpha* (de Koning et al., 1987a; L. V. Bystrykh and Y. A. Trotsenko, personal communication) have shown that the latter enzyme cannot substitute for the *in vivo* function of the former. However, de Koning et al. (1990a) recently observed assimilation of methanol carbon in a DHA synthase minus mutant of *H. polymorpha* when it was grown in a carbon-limited chemostat on a mixture of xylose and methanol. This indicates that under special conditions the transketolase may function in the assimilation of

TABLE III. Some Properties of Key Enzymes of the Xylulose Monophosphate Cycle Purified from *Candida boidinii* KD1[a]

Property	Dihydroxyacetone synthase	Dihydroxyacetone kinase
Molecular mass (kDa)	155	140
Subunits	2 (identical)	2 (identical)
pH optimum	6.8–7.1	7.5–8.2
Cofactors	TPP, Mg^{2+}	Mg^{2+}
Apparent K_m (mM)	0.5 (HCHO)	0.006 (DHA)
	0.35 (xylulose-5P)	
	0.4 (glycolaldehyde)	0.006 (glyceraldehyde)
	0.75 (ribose-5P)	0.35 (Mg ATP)
	0.75 (hydroxypyruvate)	Also active with MgITP, MgCTP, MgUTP, MgGTP
	0.90 (fructose-6P)	
	1.0 (sedoheptulose-7P)	
Inhibitors	GSH, HCHO	ADP, excess ATP
Localization	Peroxisome	Cytosol

[a]DHA, dihydroxyacetone; TPP, thiamin pyrophosphate; GSH, glutathione. Data of Trotsenko et al., 1984, and Bystrykh, 1985.

formaldehyde, as was also shown by *in vitro* experiments with purified transketolase (de Koning *et al.*, 1990a; Waites and Quayle, 1981).

In vitro DHA synthase is an unstable enzyme (Bystrykh *et al.*, 1981), but purified preparations can be stabilized with GSH or dodecanthiol (L. V. Bystrykh and Y. A. Trotsenko, personal communication) and inactivated by sulfhydryl reagents like *p*-chloromercuribenzoate (Kato *et al.*, 1982). This response to these reagents indicates that -SH groups are essential for activity. Inhibition of DHA synthase by GSH under *in vivo* conditions, as reported for the purified enzyme (Bystrykh *et al.*, 1981), is a result of the reaction of (nonhydrated) formaldehyde with GSH, which lowers the concentration of the substrate in the assay. It is not likely that this reaction plays an important role *in vivo* due to the limited exchange of GSH over the peroxisomal membrane. DHA synthase is inhibited by higher concentrations of CH_2O (K_i 8.2 mM), but because it is not known if this effect is due to formaldehyde itself, or to its hydrated adduct, the *in vivo* significance of this finding remains to be elucidated.

The peroxisomal localization of DHA synthase requires that its substrate xylulose-5-phosphate and its two products dihydroxyacetone and glyceraldehyde-3-phosphate are translocated across the peroxisomal membrane. Little is known about the transport properties of these organelles. Recent work (Nicolay *et al.*, 1987) using ^{31}P nuclear magnetic resonance spectroscopy has shown that *in vivo* these organelles are acidic inside (pH 5.8–6.0) and that a Δ pH of 1.2–1.4 units exists over their membranes that disappears upon addition of an ATPase inhibitor. Subsequent biochemical and immunohistochemical studies revealed the presence of a membrane-associated, proton-translocating ATPase (Douma *et al.*, 1987, 1990). The significance of this electrochemical proton gradient in the translocation across the organellar membrane of various low-molecular-weight (xylulose-5-phosphate, GSH, FAD, TPP, heme, dihydroxyacetone, glyceraldehyde-3-phosphate) and/or high-molecular-weight (subunits of alcohol oxidase, DHA synthase, catalase) compounds required for sustained growth on methanol remains to be established.

A second enzyme specifically involved in the XuMP cycle is DHA kinase. This enzyme is localized in the cytosol (Douma *et al.*, 1985), and mutant studies (de Koning *et al.*, 1987a) have indicated that it is essential for growth on methanol. The enzyme has a high affinity for both dihydroxyacetone and glyceraldehyde and requires MgATP as a second substrate (Table III). Significantly, excess ATP inhibits its activity, but it is doubtful whether this inhibition plays any role *in vivo* since the cellular Mg^{2+} pool measured in *C. boidinii* KD1 exceeds that of ATP by almost an order of magnitude (L. V. Bystrykh and Y. A. Trotsenko, personal communication).

Fructose-1,6-bisphosphatase (FBPase) is also induced during growth of methylotrophic yeasts on methanol (van Dijken *et al.*, 1981) and plays an indispensable role in the regeneration of the xylulose-5-phosphate required for formaldehyde assimilation. Attwood and van Dijken (1982) demonstrated that the properties of the FBPase purified from methanol- or ethanol-grown cells of *H. polymorpha* were identical. Comparable results have been obtained in Trotsenko's laboratory (L. V. Bystrykh and Y. A. Trotsenko, personal communication), where it was found that the enzyme purified from similarly grown *C. boidinii* KD1 was composed of four identical subunits of 46 kDa molecular mass and that it was inhibited by AMP. In contrast, FBPase purified from glucose-grown cells had a 30-fold lower specific activity and a reduced subunit molecular mass (40 kDa). The significance of these findings remains to be established.

3.2. Regulation of the Synthesis and Activity of Enzymes Involved in Methanol Metabolism

During exponential growth of a variety of methylotrophic yeasts on methanol in batch culture, the activities of all the C_1-specific enzymes described above (see also Fig. 3) are increased significantly above the levels observed during growth on glucose or ethanol (Sahm, 1977; van Dijken *et al.*, 1981; Egli *et al.*, 1982b). Detailed studies on the regulation of the synthesis of these enzymes, conducted mainly in strains of *C. boidinii* and *H. polymorpha*, have shown that although the fine tuning of the metabolic controls involved may be different, synthesis of the C_1-specific enzymes in none of the strains is completely dependent on the presence of methanol (induction). For example, in *H. polymorpha* alcohol oxidase, formaldehyde and formate dehydrogenases were all virtually absent during exponential growth on glucose, but their synthesis occurred after depletion of the substrate in the stationary phase of growth (Eggeling and Sahm, 1978). In this organism expression of these enzymes was strongly influenced by the growth substrate; in cells growing with xylose, glycerol, ribose, or sorbitol, specific activities as high as 60% of the values detected in methanol-grown cells were reported. Addition of methanol to a culture growing on ribose (but not glucose or ethanol) enhanced the activity of alcohol oxidase to a value close to that found in methanol-grown cells (Eggeling and Sahm, 1980). Although the reported evidence for the assimilatory enzymes is not as detailed, there are several indications that their synthesis is similarly controlled (see below).

This (partial) catabolite repression, modified by methanol induction, type of regulation, although encountered in several methylotrophic yeasts, does not always mediate expression of C_1-specific enzymes to the same extent in different yeasts when they are placed under identical growth conditions. Generally glucose and ethanol are most effective as sources of catabolite repression; for other compounds, control by catabolite repression in some strains is tighter than in others, while methanol may have a small or a significant inducing effect. For example, in *C. boidinii* sp. 2201 the derepression of alcohol oxidase synthesis following glucose exhaustion does not occur as readily as in *H. polymorpha* (Egli *et al.*, 1982a). Also, in *C. boidinii* (Eggeling and Sahm, 1980) as in *P. pastoris* (Cregg, 1987) alcohol oxidase is not detected in glycerol-grown cells and these organisms are more dependent on methanol for a high level of expression. This means that the sensitivity to catabolite repression by different carbon sources and the effectiveness of methanol as an inducer varies among different strains. Methanol also functions as an inducer of catalase and formaldehyde- and formate-dehydrogenase in an alcohol oxidase minus mutant of *H. polymorpha* (Eggeling and Sahm, 1980), indicating that the compound itself rather than a metabolite derived from it exerts this effect. On the other hand, formaldehyde also acts as an inducer for all the methanol-dissimilatory enzymes in *H. polymorpha* (Zwart and Harder, 1983; Giuseppin *et al.*, 1988a) and the same is true to a lesser extent for formate.

The molecular details of the regulation of the synthesis of the C_1-specific enzymes in methylotrophic yeasts remain to be established. There is compelling evidence, at least for two enzymes, that their synthesis is primarily controlled at the level of transcription (Cregg, 1987, but see also Giuseppin *et al.*, 1988b); however, studies of the regulation of mRNA synthesis at the promoter level have not yet been reported. In view of the recent advances made in the molecular genetics of these yeasts, it should now be possible to investigate the occurrence of promoter sequences specifically involved in catabolite repression and methanol induction as predicted by physiological studies.

Studies of the activity levels of the enzymes of methanol metabolism during growth on mixtures of glucose and methanol in carbon-limited continuous cultures further detailed the regulatory phenomena as found in batch cultures (Egli *et al.*, 1980, 1982a,b,c, 1983). For example, glucose repression diminished at lower dilution rates, and addition of methanol to the feed in increasing amounts resulted in increasing induction of enzymes until a plateau was reached, showing that both phenomena are gradually and carefully regulated. The regulation of the other enzymes involved in methanol metabolism differed somewhat from alcohol oxidase: apart from induction and repression, their activities are kept at least at a level required to metabolize the formaldehyde formed by alcohol oxidase.

Since alcohol oxidase, DHA synthase, and catalase are located in peroxisomes, synthesis of these enzymes in the cell is associated with proliferation of these organelles (Veenhuis et al., 1983). When cultures of *H. polymorpha* grown on methanol are transferred to media containing glucose or ethanol, the peroxisomes and the enzymes contained within them are actively destroyed. This active degradation is due to fusion of peroxisomes with vacuolar vesicles followed by proteolysis and has been called degradative inactivation (see Veenhuis et al., 1983). This inactivation is also found when a glucose- and methanol-limited culture is changed to nitrogen limitation shortly after nitrogen in the culture becomes undetectable, long before any accumulation of glucose or methanol is found (Egli, 1982). In contrast to these peroxisomal enzymes, the cytosolic C_1-specific dissimilatory enzymes, although their synthesis was completely repressed, were not degraded but were diluted out over newly formed cells. Compared to *H. polymorpha*, two of the peroxisomal enzymes in *Kloeckera* sp. 2201 (re-identified as *C. boidinii*; Lee and Komagata, 1980a) are not as severely affected by degradative inactivation (Egli, 1982; Yasuhara et al., 1976). This again suggests that different organisms may behave differently.

Because the inducible FAD-containing alcohol oxidase of methylotrophic yeasts may form a considerable portion (over 30% under certain conditions) of the total protein, an increased rate of FAD synthesis is required during growth on methanol. Studies of Sahm (1977) and Brooke et al. (1986) have shown that this is brought about by increased levels of enzymes involved in FAD biosynthesis, whose synthesis is controlled by intracellular levels of free FAD.

3.3. Regulation of the Flux of Formaldehyde over Dissimilatory and Assimilatory Pathways

Before methanol carbon can enter the dissimilatory or assimilatory pathways in yeasts, it must first be converted into formaldehyde by alcohol oxidase. As described in the previous section, the presence of this enzyme is controlled by a number of mechanisms, but there is little evidence for regulation of the activity of existing and properly packaged alcohol oxidase. It must therefore be assumed that the flux of methanol through reaction [1] (Fig. 3) is controlled by the concentration of enzyme [1] in the cells and the concentration of its substrates. One condition is known in which alcohol oxidase is inactivated; this is generally seen when methanol-limited cells, which contain a high concentration of enzyme, are suddenly exposed to excess methanol. Then the rate of methanol oxidation far exceeds that of formaldehyde utilization and as a result formaldehyde rapidly accumulates in the cells. Concomitantly, inactivation of alcohol oxidase occurs due to dissociation of FAD from the enzyme (see Veenhuis et al., 1983) and the cells eventually die.

In growing cells, partitioning of formaldehyde over assimilatory and dissimilatory pathways (Fig. 3) appears to be carefully regulated because the yield of cells on methanol is remarkably constant over a wide range of growth conditions (van Dijken *et al.*, 1976; Sahm, 1977; Egli *et al.*, 1986). Under steady-state conditions, when the appropriate enzymes are all present, this partitioning can be most readily accounted for on the basis of control by the energy status of the cells because it affects the rate at which xylulose-5-phosphate becomes available for the DHA synthase-catalyzed reaction (Douma *et al.*, 1985). After its synthesis in the cytosol, xylulose-5-phosphate must be translocated across the peroxisomal membrane. This process might involve the electrochemical proton gradient that exists over this membrane (Nicolay *et al.*, 1987), which is maintained by an ATPase with a low affinity for ATP (Douma *et al.*, 1987), making it sensitive to changes in the cytosolic ATP level. A high ATP level therefore may result in an increase in xylulose-5-phosphate transport leading to an enhanced assimilation rate. In this model, formaldehyde that cannot be assimilated will diffuse out of the organelles where it is trapped by GSH that is present in significant concentrations in the cytosol (Ubiyvovk *et al.*, 1983; W. de Koning, unpublished) and is dissimilated. Thus energy limitation increases the amount of formaldehyde available for dissimilation. A second control of the assimilation rate by the ATP level is possibly exerted via DHA kinase. The K_m of this enzyme for ATP is in the same order of magnitude as the cytosolic ATP concentration.

Methylotrophic yeasts grow generally much faster on multicarbon compounds than on methanol, and the question must therefore be asked whether this is due to a limited capacity of alcohol oxidase in supplying formaldehyde or to a limited capacity of either the assimilatory or dissimilatory pathways (or both). Several lines of evidence are available to suggest that the flux of formaldehyde over the assimilatory pathway may reach a maximal value above which synthesis of C_1-specific enzymes becomes repressed. During growth of *H. polymorpha* in a methanol-limited chemostat at different dilution rates, constant cell yields and a linear increase in the specific methanol consumption rates (q_{C1}) with increasing values of D were recorded up to $D = 0.15$ h^{-1}. At higher dilution rates where q_{C1} exceeded 0.42 g methanol.g dry wt^{-1}.h^{-1}, this linear relationship between q_{C1} and D no longer existed because a higher proportion of the methanol consumed was now oxidized to CO_2 (Egli *et al.*, 1986). The maximum rate of formaldehyde assimilation that can be calculated from these data (approximately 6 mmol.g dry wt^{-1}.h^{-1}) is close to the maximum activity found in cell-free extracts for DHA synthase. This, as well as another study (Egli *et al.*, 1983), indicated that under these conditions DHA synthase (and also DHA kinase and FBPase) operated close to its maximum capacity, whereas formaldehyde dehydrogenase activity is always found in excess. A maximum rate of methanol consumption of 0.42 g methanol.g dry wt^{-1}.h^{-1} was also found during growth of *H. polymorpha* on glucose/methanol mixtures of various composition at D values in excess of 0.20 h^{-1} (Egli *et al.*, 1986). Apparently there is an upper limit to the flux of methanol carbon that a cell can handle, and attempts to push the cells over this limit by increasing either the methanol/glucose ratio in the mixture or D invariably led to repression of the synthesis of C_1-specific enzymes. In the strain used by Egli *et al.* (1986) there was little flexibility in the sense that a switch of metabolism toward the use of methanol as an energy source was very limited. In other organisms, however, and under other conditions, such a switch has been observed (Eggeling and Sahm, 1981; Müller *et al.*, 1985), although methanol was never used for energy generation only, as is the case in several prokaryotes (Egli and Harder, 1984).

4. GENETICS AND MOLECULAR BIOLOGY

4.1. Mutagenesis and Auxotrophic Mutants

As mutants are essential in the study of genetics, some aspects of the methods frequently used for mutagenesis of methylotrophic yeasts are discussed briefly. Mutant selection in these yeasts is facilitated by their haploid nature, which enables direct expression of recessive mutations. The alkylating agents ethylmethane sulfonate (EMS) and *N*-methyl-*N*-nitro-*N*-nitrosoguanidine (NTG) have frequently been used to obtain mutants (for protocols see Gleeson *et al.*, 1984). When positive selection is not possible, mutagenesis is usually followed by enrichment to obtain a higher percentage of the desired mutants. Nystatin is the preferred agent for enrichment; it is an antibiotic directed against growing yeasts and a 400-fold enrichment was found when the method was tested (Gleeson, 1986).

Auxotrophic mutants are generally readily isolated. For instance, they can make up 1–3% of the colonies obtained after NTG mutagenesis or EMS treatment followed by nystatin enrichment. Adenine-requiring mutants usually make up the largest part of such mutants, followed by methionine-requiring strains; this is independent of the strain or the method used. Mutations in other biosynthetic pathways are much less frequent (Gleeson *et al.*, 1984; Tolstorukov *et al.*, 1977). Preferential selection, different reversion frequencies, mutational hot spots, and large numbers of duplicate genes have been suggested as possible explanations. Whatever the mechanism, these observations indicate that the frequencies of other mutations, like those in methanol metabolism, should be interpreted with care.

4.2. Classical Genetics

Several characteristics of mutants can be studied via the formation of diploids and their meiotic segregation resulting in spore formation (see Section 2). For three methylotrophic yeasts, methods for classical genetics are available. The purpose of the experiment and the mating frequency of the strain used determine which of the various techniques is most appropriate.

For *H. polymorpha*, Gleeson *et al.* (1984) found that mating was strongly dependent on the strain used; strain NCYC 495 was easy to cross, whereas the type strain from the CBS culture collection (CBS 4732) was semisterile. Crossings between the two strains were possible at a low frequency of diploid formation and revealed that the semisterility is apparently located on a single gene (Gleeson, 1986). Using strain NCYC 495, Gleeson *et al.* (1984) isolated marker mutants and optimized mating and sporulation procedures. Tetrad analysis using a micromanipulator to separate the spores of single tetrads was possible, although relatively difficult. Spore viability was high, 70–95%, and a normal distribution of the genetic markers was observed. For handling large numbers of mutants a mass-mating procedure and a random spore analysis method were described. Poor mating and sporulation is not uncommon among strains of *H. polymorpha*; see *H. polymorpha* DL-1 (Levine and Cooney, 1973) and *H. polymorpha* ML-3, KT-2, and Z (Savchenko *et al.*, 1983) were reported to sporulate very poorly or not at all. Bodunova *et al.* (1986a,b) described how the genetic performance of a strain of *H. polymorpha* was improved by repeating the cycle of diploidization and sporulation several times.

In *P. pastoris*, mating and sporulation are easily performed in mutants derived from the wild-type NRRL Y-11430, but markers frequently do not segregate in the normal $2^-:2^+$ distribution (Digan and Lair, 1986). One of the histidine marker mutations can be complemented by the *HIS4* genes from *S. cerevisiae* using molecular genetic techniques (Cregg *et al.*, 1985).

The most promising classical system is that developed for *P. pinus* by Tolstorukov and co-workers (1977, 1984). For this organism a large number of auxotrophic markers are available. They are distributed over four linkage groups, as established with induced mitotic haploidization of diploids, heterozygote for the marker mutations investigated (Tolstorukov *et al.*, 1983). Tetrad analysis of 26 different auxotrophic mutants resulted in a partial genomic map of four different chromosomes (Tolstorukov and

Efremov, 1984), and mutations linked with the centromeres of these four chromosomes were found.

Also, mutants defective in switching of the mating type were isolated; strains with opposite mating type showed that the homothallic nature of the type strains is due to the presence of a nuclear gene responsible for the mutational conversion between the α and a mating type (Benevolenskii and Tolstorukov, 1980). The conversion is induced during nitrogen limitation and growth including a division cycle is essential for the switching (Tolstorukov and Benevolenskii, 1980).

For asporogenous yeasts, transfer of genes can be achieved using molecular genetic techniques, whereas properties of mutated genes like the characteristic dominant or recessive and complementation groups can be studied using protoplast fusion (Lahtchev and Tuneva, 1986; Savchenko *et al.*, 1983; Fujii *et al.*, 1988). The protocols are usually too tedious to expect large-scale use for mutant characterization.

4.3. Methanol Pathway Mutants

Mutant studies are usually essential in establishing that a proposed metabolic pathway is not only present but also essential for the metabolism. Mutants blocked in activity of alcohol oxidase (Eggeling and Sahm, 1980), formaldehyde dehydrogenase (J. M. Cregg, personal communication), formate dehydrogenase (Bystrykh *et al.*, 1988), catalase (Eggeling and Sahm, 1980), DHA synthase (Gleeson and Sudbery, 1988b), and DHA kinase (de Koning *et al.*, 1987a) have been isolated. These mutants were all unable to grow on methanol, showing that all enzymes mentioned are essential for growth on methanol and confirming the pathway shown in Figure 3. Not all mutants described in the literature were investigated in detail and some of the mutants were possibly affected in a regulatory gene rather than in the structural gene coding for the enzyme.

Systematic investigations of 18 methanol-negative mutants by Gleeson and Sudbery (1988b) revealed five different complementation groups. In four of these groups no enzymatic defect was detected; they may be impaired in one or more regulatory genes or as yet unknown genes of methanol metabolism. The remaining complementation group contained a DHA synthase-negative mutant. Tolstorukov *et al.* (1988) found that a large number of alcohol oxidase-negative mutants of *P. pinus*, selected for their resistance against allyl alcohol, could be divided in at least four different complementation groups. This means that a minimum of four genes are involved in the production of active alcohol oxidase. The ease at which methanol nongrowing mutants are isolated and the occurrence of many mutants without a detectable enzyme defect as mentioned by various investigators indicate that a large number of genes are involved in growth on methanol.

Among the regulatory mechanisms, catabolite repression and inactivation have been established. In *H. polymorpha* (Gleeson, 1986), *P. pinus* (Sibirny *et al.*, 1986, 1987, 1988), and *C. boidinii* (Sakai *et al.*, 1987), glucose and ethanol repress the synthesis of methanol-metabolizing enzymes via different mechanisms.

A preliminary analysis of mutants blocked in catabolite repression and mutants with constitutive catabolite repression revealed that in *H. polymorpha* several dominant and recessive genes are involved, acting in a complex way (Gleeson, 1986).

A relationship between the citric acid cycle and the synthesis of alcohol oxidase in *H. polymorpha* was found by Sibirny *et al.* (1988). A glutamate auxotrophic mutant, lacking a key enzyme of the citric acid cycle, citrate synthase, was unable to grow on methanol in the presence of glutamate. The level of alcohol oxidase activity in this mutant was low in contrast to that of the other enzymes involved in methanol metabolism; observations on various revertants revealed a possibly direct relationship between the conformation of citrate synthase and the level of alcohol oxidase synthesis.

The regulation of alcohol oxidase activity has been investigated in more detail in two mutants of *P. pinus* defective in glucose- and ethanol-catabolite repression, *gcr1* and *ecr1*, respectively (Sibirny *et al.*, 1986, 1987). The way in which the repressive effect of various carbon sources was influenced by the *gcr1* and *ecr1* mutations led to the classification of these carbon sources into four different groups. Glucose and some related compounds exerted their effect via *gcr1*, while ethanol and some others acted via *ecr1*. The *ecr1* and *gcr1* mutations did not effect repression by dihydroxyacetone or malate. In the fourth group, containing various different compounds, the repression is partly reduced by both mutations. As in some similar *S. cerevisiae* mutants, the levels of phosphofructokinase activity were reduced in the *gcr1* mutant, resulting in higher glucose-6-phosphate and fructose-6-phosphate levels in the cell and a reduction of the fructose-1,6-bisphosphate concentration. In the *ecr1* mutant, reduced levels of 2-oxoglutarate were found during growth on ethanol. Although both mutants are reasonably well characterized, the actual mechanisms for the different forms of catabolite repression remain to be resolved.

4.4. Properties of Methanol-Regulated Genes

The two most dominant proteins related to methanol metabolism found in cell-free extract of methanol-grown cells are alcohol oxidase and dihydroxyacetone synthase. For *H. polymorpha* the genes coding for both enzymes (*AOX* and *DAS,* respectively) have been cloned and sequenced (Ledeboer *et al.,* 1985; Janowicz *et al.,* 1985). Neither of the genes contained intervening sequences, and the calculated monomeric molecular masses were in close agreement with results obtained by SDS polyacrylamide gel electrophoresis of the respective proteins. A similar bias in codon usage for *AOX* and *DAS* was found for several amino acids, although the bias was not as strong as found for abundant proteins in *S. cerevisiae*. A comparison of the amino acid sequence of the N-terminus of the protein and the first part of the coding DNA sequence revealed that no cleavable leader sequence is present for these peroxisomal enzymes, a property found for all investigated peroxisomal enzymes (Borst, 1989). The noncoding 5' and 3' regions of the genes were also investigated. Several general structures were recognized, like the TATA box (Goldberg–Hogness sequence) and regions of dyad symmetry at the 5' noncoding region, but a common sequence likely to be involved in regulation of gene expression by methanol was not found.

4.5. Transformation Systems and Vectors

Methods for transformation have been described for *H. polymorpha* and *P. pastoris*. In *H. polymorpha* comparable methods have been developed for various strains by different groups. They are all based on plasmids containing selectable genes from *S. cerevisiae*. Gleeson *et al.* (1986) selected leucine-negative mutants of *H. polymorpha* NCYC 495 and tested them for absence of β-isopropylmalate dehydrogenase. One stable mutant was isolated and used for transformation with YEp13, a construct based on the 2-μm plasmid containing the *S. cerevisiae* gene for β-isopropylmalate dehydrogenase *(LEU2)* and an ARS (autonomous replication sequence). Relatively low frequencies of 30–60 transformants per μg DNA were found using either a protoplasting method or LiCl treatment. Roggenkamp *et al.* (1986) found up to 1.5×10^3 transformants per μg DNA using the protoplast method and a plasmid containing a 0.5-kb fragment with a HARS (*Hansenula* ARS) and the *S. cerevisiae URA3* gene, which codes for the enzyme orotidine 5'-decarboxylase. A 5-fluoro-orotic acid–resistant mutant of *H. polymorpha* ATCC 34438 blocked in the activity of this enzyme served as the host in these transformation experiments. The transformation of *H. polymorpha* DL-1 was described by Tikhomirova *et al.* (1986, 1988). Small modifications in the protoplast method resulted in transformation frequencies up to 3×10^4 per μg DNA, using shuttle plasmids containing the *Saccharomyces LEU2* gene and a HARS.

The strategy used by Cregg *et al.* (1985) to develop a transformation system for *P. pastoris* did not differ essentially from that used for *H. poly-*

morpha. In this case the *HIS4* of *S. cerevisiae* was used to complement a mutation in the histidinol dehydrogenase gene in *P. pastoris* NRRL Y-11430. Using a PARS (*Pichia* ARS) cloned into the plasmid pBR325, transformation frequencies of $1-2 \times 10^5$ per µg DNA were obtained, both with *HIS4* and the equivalent gene isolated from *Pichia*. Integration was observed when the *Pichia HIS4* equivalent was used in combination with a *Saccharomyces* ARS. A combination of PARS and the *HIS4* led to stable autonomous replication during 50 generations. The integration at homologous genes opens the way to methods like gene disruption, gene replacement, and site-directed mutagenesis in this yeast.

These methods were used in combination with classical genetics to study the role of the two different alcohol oxidase genes *(AOX)*, present in this organism. The promoter of the *AOX1* gene is very strong; this can be deduced from the fact that alcohol oxidase can make up 30% of the total protein in this yeast, but it was also shown by cloning the *E. coli lacZ* gene behind this promotor (Tschopp *et al.*, 1987a). Although the second gene, *AOX2*, closely resembles the *AOX1* gene, its gene product was only present in small amounts under the conditions investigated. The *AOX2* gene, cloned behind the *AOX1* promoter, supported growth on methanol at a growth rate indistinguishable from the wild type. The presence of only one *AOX* gene *(1* or *2)* behind the *AOX2* promotor led to a strong reduction of the growth rate (Cregg *et al.*, 1989). The remaining growth proved, however, to be very useful in obtaining maximal expression when the *AOX1* gene was exchanged for a heterologous gene, coding for commercially interesting proteins. The physiological function of the *AOX2* gene in this organism remains to be elucidated. In other methylotrophic yeasts no evidence was found for the presence of more than one *AOX* gene.

4.6. Expression of Heterologous Genes

Expression of heterologous genes in methylotrophic yeasts has been reported for *P. pastoris* by cooperating groups at Sibia (San Diego, CA) and Phillips Petroleum Company (Bartlesville, OK).

Invertase from *S. cerevisiae (SUC2)* was cloned behind the *AOX1* promotor in *P. pastoris* using a plasmid containing the *Pichia HIS4* gene and a PARS. The plasmid integrated at the *his4* or *AOX1* locus and gave the yeast the ability to grow on sucrose. The amino-terminus of the glycosylated invertase was identical with the one found in *S. cerevisiae*, indicating that a 19-amino-acid N-terminal sequence is removed in the proper way. The excreted protein differed from the invertase found in the periplasmic space of *S. cerevisiae* in its glycosylation: The *S. cerevisiae* glycoprotein is 100–140 kDa, whereas the enzyme excreted by *P. pastoris* was only 85–90 kDa, identical to invertase produced by a mutant of *S. cerevisiae* blocked in the endoplasmic reticulum secretory pathway *(sec18)*. Glycosylation of the 56-kDa protein in *P. pastoris* was slow, taking hours to complete. The highest excretion (2.5 g/liter) was found in a strain growing only very slowly on methanol due to the lack of the *AOX1* gene (Tschopp *et al.*, 1987b).

High expression of hepatitis B surface antigen (HBsAg) was obtained by Cregg *et al.* (1987) in the same organism. The vector for these transformations was especially designed for integration at the *AOX1* locus. The expression cassette consisted of the *Pichia HIS4* gene and the HBsAg gene behind the *AOX1* promoter flanked by DNA fragments from the 5' and 3'noncoding regions of the *AOX1* gene. Transformants with a properly integrated expression cassette were obtained by selecting for slow growth on methanol. About 30% of the transformants showed this phenotype. After induction on methanol, HBsAg appeared in the cell. In contrast to *E. coli*, in *P. pastoris* the protein was found mainly in 22-nm particles, a configuration also found in human sera and essential for high response. After 250 hr of induction, the 22-nm particles made up about 2.5% of the cell protein. Slow growth was essential for the particle formation; a transformant in which the *AOX1* was not deleted yielded about the same amount of HBsAg protein, but only about 10% was found in the desired 22-nm particle form (Cregg *et al.*, 1987).

A similar expression cassette was used by Sreekrishna *et al.* (1988) to produce human tumor necrosis factor (TNF). In some transformants, after induction on methanol, more than 30% of the cell protein was found to be TNF. The protein had a biological activity comparable to TNF obtained from *E. coli* transformed with the TNF gene. The *Pichia* TNF was more stable upon storage, presumably due to a lower proteolytic activity of this organism.

5. APPLICATIONS

5.1. General Economics

In this section a description of some industrial applications of methylotrophic yeasts is given. The economic feasibility of some of the applications is strongly dependent on the market price of methanol and the costs of alternative feedstocks (Linton and Niekus, 1987). At the time of writing (1989), the price of methanol is low due to significant overcapacity in production facilities and a moderate price of natural gas, the raw material for its production. In some other applications, advantage is taken of one or more specific properties of methylotrophic yeasts (e.g., availability of a strong promotor for heterologous gene expression) which can alter the economic considerations.

The application of methylotrophic yeasts in industry may be considered in three different areas:

- Production of single-cell protein (SCP). Here the price of the carbon source (i.e., methanol) is very important. It has to be compared with other feedstocks for producing SCP, and when issued as fodder protein, the price of interchangeable protein sources like soya bean should also be considered.
- Production of bulk and fine chemicals. Depending on the added value of the product, the price of methanol is more or less important. Other features of the process, such as yield, productivity, recovery, and purity of the end-product, play important roles.
- Production of pharmaceuticals and specialties. The price of the carbon source is usually not of major importance. Other properties of methanol, such as its strong inducing effect and its chemical purity, may be important in certain cases.

When compared with other feedstocks for fermentation, methanol has some characteristics that can be advantageous: It is a pure chemical, easy to store and handle; even crude methanol contains only minor amounts of formaldehyde and higher alcohols, which are readily consumed during most fermentations. Because it does not support growth of many organisms, contaminations supposedly occur less frequently, especially in concentrated stocks.

5.2. Single-Cell Protein

Shortly after the discovery of methylotrophic yeasts, the possibility of using these organisms for SCP production was investigated by a number of research groups. Compared with methylotrophic bacteria, their direct competitors, methylotrophic yeasts have some advantages but also some drawbacks, both of which are considered below.

Yeasts are generally larger than bacteria and are therefore easier to separate from the culture broth. When centrifugation is used, this can be done at lower centrifugal forces and/or a higher throughput of broth per hour. For filtration, a larger pore diameter can be used and clogging of the filter is usually less severe.

The chemical composition of yeasts differs in some important aspects from that of bacteria: yeasts usually have a lower nucleic acid content (typically 7–12% of the dry weight compared to 10–20% for bacteria), which is an advantage. However, the lower protein content (45–50% of the dry weight compared to 50–70% in the case of bacteria) is a drawback. When used for human (or pet) food, addition of yeast protein is psychologically more easily accepted by the consumer.

The pathway of methanol metabolism in yeasts is energetically less favorable for SCP production than that of their bacterial rivals. The first step, the oxidation of methanol to formaldehyde in yeasts, is catalyzed by an oxidase yielding no biologically useful energy, whereas bacteria employ cofactor-dependent dehydrogenases, which are thought to deliver metabolic energy (Anthony and Jones, 1987). Energy generation in yeast takes place by the NAD-dependent oxidation of formaldehyde to CO_2, a process occurring in the cytosol. The efficiency of energy generation from cytosolic NADH is only about 2/3 of the usual mitochondrial efficiency. Also, the assimilation of carbon in methylotrophic yeasts via the XuMP cycle requires more energy in the form of ATP than via the bacterial RuMP cycle. A consequence of these biochemical differences is not only a lower biomass yield (0.36–0.42 kg/kg methanol for yeasts compared with 0.54 kg/kg methanol for bacteria, Anthony, 1982), but also a higher heat production, thereby increasing the demand for cooling.

Other potential drawbacks of methylotrophic yeasts when compared with bacteria are the usually lower growth rate and the low affinity for oxygen of yeasts due to the high K_m for oxygen of the enzyme alcohol oxidase (0.4–1 mM in *P. pastoris* and *H. polymorpha*), which can easily result in a reduction of the growth rate in large fermenters.

Most problems and drawbacks were solved or minimized by researchers at the Phillips Petroleum Company (Bartlesville, OK). After extensive screening they selected a strain of *P. pastoris* that could be grown stably in a large-scale continuous fermentor (with improved oxygen and heat transfer properties) up to a density of more than 130 g/liter (dry weight). Although the strain had some potential drawbacks, such as a low optimum growth temperature (30°C, compared with 42°C for some strains of *H. polymorpha* and 55°C for certain methylotrophic bacilli) and relatively low growth rate (doubling time 5–6 hr, *H. polymorpha* 3 hr, *Methylophilus methylotrophus* 1 hr), the extremely high cell density obtained with the *Pichia* strain allowed direct spray drying of the culture, thereby significantly reducing the downstream processing costs. At this cell density the growth rate is limited by fermentor characteristics, implicating that a faster growing strain would not give an immediate advantage. The productivity of the fermentor, 12 g/hr per liter is comparable with the productivity obtained for baker's yeast. Also advantageous is the option to use alternative carbon sources, such as molasses or whey, for the process (Linton and Niekus, 1987).

5.3. Enzymes

Enzymes have found use in the industry for different purposes. Certain chemical conversions are preferably carried out using specific enzymes, while certain enzymes are also added to various products such as food and detergents. At least two enzymes involved in the metabolism of methanol are of industrial interest.

Formate dehydrogenase has been used for cofactor regeneration (NADH) in enzymatic reductions of various compounds by NAD-dependent dehydrogenases. Because formate is oxidized to CO_2, which is easily removed as a gas from the reaction mixture, the reaction can continue without reaching an equilibrium.

Alcohol oxidase is a promising enzyme, and various applications have been patented (Giuseppin, 1988). For some of these applications the production of hydrogen peroxide is an essential aim. Due to the large amount of alcohol oxidase present in methanol-grown yeasts, the price of this enzyme can be kept reasonably low. Alcohol oxidase from *P. pastoris* is sold by the Phillips Petroleum Company in a kit for the determination of alcohol (Alcoscan). Alcohol oxidase oxidizes the alcohol present in the test sample, thereby producing an equivalent amount of hydrogen peroxide. In the following reaction, horseradish peroxidase is used to oxidize a dye, resulting in a color development.

Production of peroxide associated with the oxidation of an alcohol can also be used in detergents to obtain a bleaching effect at low temperature. For this purpose the enzyme should be free of catalase. Because removal of catalase through purification of the enzyme is expensive, the production of catalase-free alcohol oxidase using a catalase-negative mutant of the thermotolerant yeast *H. polymorpha* was studied by the Unilever Company (Giuseppin *et al.*, 1988a). Although this mutant had lost its ability to grow on methanol as the only carbon and energy source, methanol utilization was possible during growth on mixtures of methanol and glucose in continuous culture. A maximum induction of alcohol oxidase (3 U/mg protein) was obtained by using an approximately equimolar methanol/glucose mixture. The hydrogen peroxide produced *in vivo* during these fermentations was removed by the mitochondrial cytochrome *c* peroxidase (Giuseppin *et al.*, 1988c; Verduyn *et al.*, 1988). Although some problems with the application of alcohol oxidase in detergents (such as the required presence of proteases) have to be solved, the potential market for alcohol oxidase for this purpose could be very large. Other potential uses of alcohol oxidase are in the synthesis of various aldehydes from alcohols. For this purpose it is sometimes not essential to work with a purified enzyme. Tani and co-workers studied the production of formaldehyde from methanol by *Candida boidinii* (see next section).

Oxidases can also be used as oxygen scavengers in certain products; the absence of oxygen can elongate the shelf life and eliminate changes in taste in food.

5.4. Bulk and Fine Chemicals

A variety of chemicals can be produced using methylotrophic yeasts. A division can be made into three groups: chemicals with no obvious relationship to methanol metabolism, chemicals more or less directly related to methanol metabolism, and chemicals for which cofactor regeneration is essential; in the latter case a precursor of the product is usually added and the product is formed by a reduction by NADH or by phosphorylation using ATP. The production of chemicals with no direct relationship to the metabolism of methanol, such as amino acids (Denenu and Demain, 1981a,b; Sanchez and Demain, 1978), citric acid (Sahm, 1977) and ketones from alcohols, based on a secondary alcohol dehydrogenase reaction (Patel *et al.*, 1979, 1981), will not be discussed in detail. Usually, the use of methanol is not essential for the process, and none of the processes has been developed into commercial production.

Chemicals related to the metabolism of methanol are potentially more attractive to produce. The relationship with methanol metabolism can be either direct, as in the case of formaldehyde or dihydroxyacetone, or more indirect, as in the case of riboflavin or FAD, the cofactor found in alcohol oxidase.

Production of formaldehyde from methanol using a methylotrophic yeast was studied for *H. polymorpha* (Baratti *et al.*, 1978) and for *C. boidinii* (Tani *et al.*, 1985a). Formaldehyde is used in various chemical processes and is presently produced chemically. Using resting cells in an atmosphere of pure oxygen, formaldehyde was produced from methanol by *C. boidinii* S2 at 4°C, the optimum temperature for the process (Tani *et al.*, 1985a). Although it is not very likely that these conditions can be used in an economically attractive process, a comparison of the influence of various cultivation parameters was made. Not surprisingly, formaldehyde production increased when the amount of alcohol oxidase present in the cells was increased either by mutation (Tani *et al.*, 1985b) or by using chemostat-grown cells (Sakai and Tani, 1986). Using heat-treated cells, the maximum amount produced in these experiments was more than 1 M (3%) formaldehyde (Sakai and Tani, 1987).

Production of dihydroxyacetone and glycerol using a mutant of *H. polymorpha* lacking dihydroxyacetone kinase was studied by Kato *et al.* (1986) and de Koning *et al.* (1987a, 1990b). Using only methanol in the incubation mixture, a significant amount of dihydroxyacetone was excreted, although, on the basis of the cyclical nature of the assimilation reaction, a fast exhaustion of the acceptor molecule xylulose-5-phosphate is expected. Labeling experiments showed that most of the carbon in the dihydroxyacetone originated from methanol (Kato *et al.*, 1986). A bypass via the phosphorylation of glycerol was shown (de Koning *et al.*, 1987b). A double mutant blocked in glycerol kinase and dihydroxyacetone kinase did not accumulate significant amounts of trioses unless a second carbon source was added that could replenish the xylulose-5-phosphate. After optimization of a number of parameters, in this mutant accumulation of trioses (mainly glycerol) up to 70 g/liter was found with intermittent feeding of mixtures of xylose and methanol (de Koning *et al.*, 1990b). The overall efficiency and the rate of production were, however, low (50% of the xylose and 20% of the methanol were recovered as triose and a maximum rate of 1 g product per g biomass per day was found).

The most abundant protein (up to 30%) in methanol-limited cells is the enzyme alcohol oxidase, which contains eight FAD molecules per octamer. This implies that methylotrophic yeasts must have a significant capacity to produce FAD and its precursors FMN and riboflavin. The production of riboflavin (vitamin B_2) by methylotrophic yeasts was mentioned in patents in 1974, but as far as we are aware, no commercial process based on methylotrophic yeasts is operating at present. The amounts of riboflavin, published up to now are too low to expect a quick change in this situation (75 mg/liter, Patent AS USSR Biochem Mic, 1978). Investigations on the metabolic pathway of riboflavin, FMN, and FAD synthesis showed a strict control on the last three enzymes of this pathway (riboflavin synthetase, riboflavin kinase, FMN adenylyltransferase) by the level of free intracellular FAD (Brooke *et al.*, 1986). Isolation of mutants deregulated in this control, using, for example, analogs, have not yet been reported.

A large number of chemicals may be produced using methanol for the supply of energy and/or reducing power (NADH). At present, two processes have been studied in detail: the production of sorbitol from fructose or glucose and the production of ATP from AMP or adenosine.

Sorbitol is a well-known sugar substitute in food products. In yeasts, growth on sorbitol is possible by its oxidation to fructose, a reaction catalyzed by the enzyme sorbitol dehydrogenase. The reaction requires NAD, although (in *C. boidinii*) some activity was found with NADP (Vongsuvanlert and Tani, 1988a). Because the reaction is reversible, sorbitol can be produced from fructose, provided that the NADH/NAD ratio is kept high. Tani and Vongsuvanlert (1987) selected *C. boidinii* no. 2201 out of more than 100 methanol-utilizing yeasts for sorbitol-production experiments. Growth on mixtures of xylose and methanol provided cells with good methanol-oxidizing capacity and high levels of sorbitol dehydrogenase. Beside fructose, which could be converted with 95% efficiency, glucose could also be used. Glucose was converted by the cells into fructose by way of xylose isomerase (Vongsuvanlert and Tani, 1988b). Although during the initial screening, glucose provided a higher rate of sorbitol production, after optimization of the reaction conditions (e.g., pH, aeration, addition of polyethylene glycol 6000 and divalent cations), fructose was the better substrate.

C. *boidinii* no. 2201 was also used to study ATP formation from precursors like AMP, adenosine, and adenine by Tani and co-workers (1984a,b,c, 1987). In this case the NADH, produced by the oxidation of formaldehyde via formate to CO_2, is oxidized in the mitochondria, which results in the conversion of ADP into ATP. Adenylate kinase, converting AMP + ATP into 2 ADP, catalyzes the second step in these transformations. For optimal production, cells were treated with sorbitol and heated for 1 hr at 37°C. This resulted in permeabilization of the cells without disrupting the organelles (Tani *et al.*, 1984b). Addition of NAD, glutathione, pyrophosphate, $MgSO_4$, and sorbitol improved the production and decreased the unwanted production of IMP. Using intermittent feeding of adenosine and K_2HPO_4, 200 mM ATP was produced with 77% conversion of the adenosine (Yonehara and Tani, 1987). Also, with respect to energy conservation of the added methanol and productivity per g of cells, reasonable figures were obtained (30% of the theoretical maximum and, 4 g/g, respectively, Tani *et al.*, 1987). Apart from the production of ATP itself, the possibility of using this system for regeneration of ATP or NADH in enzymatic conversions of various compounds can also be of interest.

5.5. Hosts for Heterologous Gene Expression

As already mentioned in the section on genetics, methylotrophic yeasts are good host organisms for the expression of heterologous genes. At the moment, only *P. pastoris* and *H. polymorpha* transformation systems are available, and so far the *Pichia* system has been considered for large-scale production of some heterologous proteins. To evaluate the potential of the *Pichia* system, a comparison with other competing systems should be made (Thill *et al.*, 1987). *E. coli* is the organism of choice for basic genetic manipulation. It can also be grown in high density on a large scale, and strong inducible promoters are available. Proteins of several heterologous genes, when brought to high expression in *E. coli*, are found in so-called inclusion bodies. Recovery of biologically active enzyme from these particles is generally difficult. Also, when eukaryotic proteins have to be glycosylated, *E. coli* is not the organism of choice.

Less problems with insolubility and glycosylation are usually encountered when mammalian cells are used. However, stable high-level expression in these cells is more difficult to achieve than in *E. coli* and, most important, cultivation of mammalian cells is slow and expensive. Yeasts offer an intermediate between *E. coli* and mammalian cells, being both eukaryotic and simple to cultivate. *S. cerevisiae* is by far the most investigated yeast, and a well-developed genetic system for this organism exists. The system developed for *P. pastoris* shows, when compared with *S. cerevisiae*, some significant advantages. In *S. cerevisiae*, strong expression is usually achieved when a multicopy plasmid is used based on the 2-µm plasmid. Without selective pressure this plasmid is slightly unstable, which can cause problems during large-scale cultivation. In *P. pastoris* genomic integration can be easily ensured, and the alcohol oxidase or dihydroxyacetone synthase promoters are strong enough to achieve high expression levels. The possibility of cultivating a large amount of biomass in the absence of expression of the heterologous gene by using repressing substrates like glucose or glycerol further reduces the risk that cells loose the heterologous gene. Other advantages of the use of *P. pastoris* are the stable, very high cell densities that can be obtained and the low proteolytic activity, compared with *S. cerevisiae* (and also *E. coli*).

The production of HBsAg (see Section 4.6) has shown that laboratory experiments with *P. pastoris* are relatively easily scaled up to a 250-liter fermentor with cell densities of 50–100 g/liter without losing much of the productivity (Cregg *et al.*, 1987). In these experiments the cells were grown on mineral media using sufficient glycerol to reach the desired cell density. As soon as the glycerol is depleted, methanol is fed to the culture for a period of ca. 200 hr to obtain maximal expression of the protein. Although 250-liter scale is small for industrial fermentation, the amount of HBsAg produced in one fermentation run is enough to vaccinate 3,000,000 people. For TNF, even higher production levels were obtained in a 5-liter fermentor: this protein made up 25% of the soluble protein in cell-free extracts, resulting in production of 6–10 g/liter, or about 5×10^{11} Units (Sreekrishna *et al.*, 1988). For pharmaceutical applications of the highly active proteins like the ones mentioned above, significant efforts in further scaling up will be economically unattractive since the output of a few fermentors will suffice the world market.

6. CONCLUDING REMARKS

This chapter discusses several scientific and applied aspects of methylotrophic yeasts. These eukaryotic organisms differ from their prokaryotic counterparts in the manner that methanol is metabolized. Whereas in yeasts a flavin-containing alcohol oxidase is used to convert methanol into formaldehyde, bacteria employ dehydrogenases, which require the presence of special electron acceptors (see other chapters in this volume) and contribute to the generation of useful energy for the cell. Assimilation of C_1 carbon in bacteria can occur via three pathways: the Calvin cycle, the serine pathway, and the ribulose monophosphate pathway. Although the initial reactions are different, the xylulose monophosphate pathway in yeasts resembles the latter, both in overall energy costs of carbon assimilation and in the enzymes involved in the regeneration of the acceptor molecule. Compared with bacteria, the metabolic regulation in yeasts is complicated by the compartmentation of alcohol oxidase and DHA synthase. This increased complexibility might give the yeasts a better control over methanol metabolism, which may result in a usually broader range of metabolizable carbon sources, when compared with RuMP pathway-utilizing bacteria. The availability of peroxisome-negative mutants (J. M. Cregg and G. J. Sulter, personal communication) is expected to stimulate future studies of both the biogenesis and the physiological functioning of these organelles that are so important in the metabolism of methanol.

The use of methylotrophic yeasts in biotechnological processes has been studied by various groups, but at present, applications in industry are rare. Because yeasts differ from methylotrophic bacteria, the selection of a particular organism will depend on the application. Further studies of the genetics and physiology of methylotrophic yeasts and bacteria are expected to facilitate this choice and the development of processes that exploit the specific properties of these organisms.

REFERENCES

Allais, J. J., Louktibi, A., and Baratti, J., 1983a, Oxidation of methanol by the yeast, *Pichia pastoris*, purification and properties of the formaldehyde dehydrogenase, *Agric. Biol. Chem.* **47**:1509–1516.

Allais, J. J., Louktibi, A., and Baratti, J., 1983b, Oxidation of methanol by the yeast *Pichia pastoris*. Purification and properties of the formate dehydrogenases, *Agric. Biol. Chem.* **47**:2547–2554.

Anthony, C., 1982, *Biochemistry of Methylotrophs*, Academic Press, London.

Anthony, C., and Jones, C. W., 1987, Energy metabolism of aerobic, methylotrophic bacteria, in: *Microbial Growth on C_1 Compounds* (H. W. van Verseveld and J. A. Duine, eds.), Martinus Nijhoff, Dordrecht, pp. 195–202.

Attwood, M. M., and Dijken, J. P. van, 1982, Characteristics of fructose-1,6-bisphosphatase from the methanol-utilizing yeast *Hansenula polymorpha*, *J. Gen. Microbiol.* **128**:2313–2317.

Baratti, J., Couderc, R., Cooney, C. L., and Wang, D. I. C., 1978, Preparation and properties of immobilized methanol oxidase, *Biotechnol. Bioeng.* **20**:333–348.

Benevolenskii, S. V., and Tolstorukov, I. I., 1980, Study of the mechanisms of mating and self-diploidization in haploid yeasts *Pichia pinus*. III. Study of heterothallic mutants, *Genetika* **16**:1342–1349.

Bieber, R., and Trümpler, G., 1947, Angenäherte spektrographische Bestimmung der Hydratationsgleichgewichtskonstanten wäßriger Formaldehydlösungen, *Helv. Chim. Acta* **30**:1860–1865.

Bodunova, E. N., Donich, V. N., Nesterova, G. F., and Soom, Y. O., 1986a, Genetic lines of *Hansenula polymorpha* yeast. Communication I. Preparation and characterization of genetic lines, *Genetika* **22**:741–747.

Bodunova, E. N., Donich, V. N., and Nesterova, G. F., 1986b, Genetic lines of *Hansenula polymorpha* yeast. II. Inheritance of abnormalities in meiotic segregation, *Genetika* **22**:939–950.

Borst, P., 1989, Peroxisome biogenesis revisited, *Biochim. Biophys. Acta* **1008**:1–13.

Brooke, A. G., Dijkhuizen, L., and Harder, W., 1986, Regulation of flavin biosynthesis in the methylotrophic yeast *Hansenula polymorpha*, *Arch. Microbiol.* **145**:62–70.

Bystrykh, L. V., 1985, Kinetic properties of dihydroxyacetone kinase of the methylotrophic yeast *Candida boidinii*, *Biochemistry* (USSR). **40**:1611–1616.

Bystrykh, L. V., Sokolov, A. P., and Trotsenko, Y. A., 1981, Purification and properties of dihydroxyacetone synthase from the methylotrophic yeast *Candida boidinii*, *FEBS Lett.* **132**:324–328.

Bystrykh, L. V., Aminova, L. R., and Trotsenko, Y. A., 1988, Methanol metabolism in mutants of the methylotrophic yeast *Hansenula polymorpha*, *FEMS Microbiol. Lett.* **51**:89–94.

Bystrykh, L. V., Romanov, V. P., Steczko, J., and Trotsenko, Y. A., 1989, Catalytic variability of alcohol oxidase from the methylotrophic yeast *Hansenula polymorpha*, *Biotechnol. Appl. Biochem.* **11**:184–192.

Campbell, I., 1973, Numerical analysis of *Hansenula*, *Pichia* and related yeast genera, *J. Gen. Microbiol.* **77**:427–441.

Couderc, R., and Baratti, J., 1980, Oxidation of methanol by the yeast, *Pichia pastoris*. Purification and properties of the alcohol oxidase, *Agric. Biol. Chem.* **44**:2279–2289.

Cregg, J. M., 1987, Genetics of methylotrophic yeasts, in: *Microbial Growth on C_1 Compounds* (H. W. van Verseveld and J. A. Duine, eds.), Martinus Nijhoff, Dordrecht, pp. 158–167.

Cregg, J. M., Barringer, K. J., Hessler, A. Y., and Madden, K. R., 1985, *Pichia pastoris* as a host system for transformations, *Mol. Cell. Biol.* **5**:3376–3385.

Cregg, J. M., Tschopp, J. F., Stillman, C., Siegel, R., Akong, M., Craig, W. S., Buckholz, R. G., Madden, K. R., Kellaris, P. A., Davis, G. R., Smiley, B. L., Cruze, J., Torregrossa, R., Velicelebi, G., and Thill, G. P., 1987, High-level expression and efficient assembly of hepatitis B surface antigen in the methylotrophic yeast *Pichia pastoris*, *Bio/Technology* **5**:479–485.

Cregg, J. M., Madden, K. R., Barringer, K. J., Thill, G. P., and Stillman, C. A., 1989, Functional characterization of the two alcohol oxidase genes from the yeast *Pichia pastoris*, *Mol. Cell. Biol.* **9**:1315–1323.

Denenu, E. O., and Demain, A. L., 1981a, Enzymatic basis for overproduction of tryptophan and its metabolites in *Hansenula polymorpha* mutants, *Appl. Environm. Microbiol.* **42**:497–501.

Denenu, E. O., and Demain, A. L., 1981b, Relationship between genetic deregulation of *Hansenula polymorpha* and production of tryptophan metabolites, *Eur. J. Appl. Microbiol. Biotechnol.* **13**:202–207.

Digan, M. E., and Lair, S. V., 1986, Genetic methods for the methylotrophic yeast *Pichia pastoris*, Thirteenth International Conference on Yeast Genetics and Molecular biology, Banff, Alberta, Canada, Book of abstracts, p. 589.

Dijken, J. P. van, and Harder, W., 1974, Optimal conditions for the enrichment and isolation of methanol-assimilating yeasts, *J. Gen. Microbiol.* **84**:409–411.

Dijken, J. P. van, Otto, R., and Harder, W., 1976, Growth of *Hansenula polymorpha* in a methanol-limited chemostat. Physiological responses due to the involvement of methanol oxidase as a key enzyme in methanol metabolism, *Arch. Microbiol.* **111**:137–144.

Dijken, J. P. van, Harder, W., Beardsmore, A. J., and Quayle, J. R., 1978, Dihydroxyacetone: an intermediate in the assimilation of methanol by yeasts? *FEMS Microbiol. Lett.* **4**:97–102.

Dijken, J. P. van, Harder, W., and Quayle, J. R., 1981, Energy transduction and carbon assimilation in methylotrophic yeasts, in: *Microbial Growth on C_1 Compounds* (H. Dalton, ed.), Heyden, London, pp. 191–201.

Dijkhuizen, L., Hansen, T. A., and Harder, W., 1985, Methanol, a potential feedstock for biotechnological processes, *Trends Biotechnol.* **3**:262–267.

Douma, A. C., Veenhuis, M., de Koning, W., Evers, M., and Harder, W., 1985, Dihydroxyacetone synthase is localized in the peroxisomal matrix of methanol-grown *Hansenula polymorpha*, *Arch. Microbiol.* **143**:237–243.

Douma, A. C., Veenhuis, M., Sulter, G. J., and Harder, W., 1987, A proton-translocating adenosine triphosphatase is associated with the peroxisomal membrane of yeasts, *Arch. Microbiol.* **147**:42–47.

Douma, A. C., Veenhuis, M., Waterham, H. R., and Harder, W., 1990, Immunological demonstration of the peroxisomal ATPase of yeasts, *Yeast* **6**:45–52.

Eggeling, L., and Sahm, H., 1978, Derepression and partial insensitivity to carbon catabolite repression of the methanol dissimilating enzymes in *Hansenula polymorpha*, *Eur. J. Appl. Microbiol. Biotechnol.* **5**:197–202.

Eggeling, L., and Sahm, H., 1980, Regulation of alcohol oxidase synthesis in *Hansenula polymorpha*: oversynthesis during growth on mixed substrates and induction by methanol, *Arch. Microbiol.* **127**:119–124.

Eggeling, L., and Sahm, H., 1981, Enhanced utilization-rate of methanol during growth on a mixed substrate: a continuous culture study with *Hansenula polymorpha*, *Arch. Microbiol.* **130**:362–124.

Egli, T., 1982, Regulation of protein synthesis in methylotrophic yeasts: Repression of methanol dissimilating enzymes by nitrogen limitation, *Arch. Microbiol.* **131**:95–101.

Egli, T., and Harder, W., 1984, Growth of methylotrophs on mixed substrates, in: *Microbial Growth on C_1 Compounds* (R. L. Crawford and R. S. Hanson, eds.), American Society for Microbiology, Washington, DC, pp. 330–337.

Egli, T., and Lindley, N. D., 1984, Mitochondrial activities in the methylotrophic yeast *Kloeckera* sp. 2201 during growth with glucose and/or methanol, *J. Gen. Microbiol.* **130**:3239–3249.

Egli, T., Dijken, J. P. van, Veenhuis, M., Harder, W., and Fiechter, A., 1980, Methanol metabolism in yeasts: regulation of the synthesis of catabolic enzymes, *Arch. Microbiol.* **124**:115–121.

Egli, T., Käppeli, O., and Fiechter, A., 1982a, Regulatory flexibility of methylotrophic yeasts in chemostat cultures: Simultaneous assimilation of glucose and methanol at a fixed dilution rate, *Arch. Microbiol.* **131**:1–7.

Egli, T., Käppeli, O., and Fiechter, A., 1982b, Mixed substrate growth of methylotrophic yeasts in chemostat culture: Influence of the dilution rate on the utilization of a mixture of glucose and methanol, *Arch. Microbiol.* **131**:8–13.

Egli, T., Haltmaier, T., and Fiechter, A., 1982c, Regulation of the synthesis of methanol oxidizing enzymes in *Kloeckera* sp. 2201 and *Hansenula polymorpha*, a comparison, *Arch. Microbiol.* **131**:174–175.

Egli, T., Lindley, N. D., and Quayle, J. R., 1983, Regulation of enzyme synthesis and variation of residual methanol concentration during carbon-limited growth of *Kloeckera* sp. 2201 on mixtures of methanol and glucose, *J. Gen. Microbiol.* **129**:1269–1281.

Egli, T., Bosshard, C., and Hamer, G., 1986, Simultaneous utilization of methanol-glucose mixtures by *Hansenula polymorpha* in chemostat: influence of dilution rate and mixture composition on utilization pattern, *Biotechnol. Bioeng.* **28**:1735–1741.

Fujii, T., and Tonomura, K., 1972, Oxidation of methanol, formaldehyde and formate by a *Candida* species, *Agric. Biol. Chem.* **36**:2297–2306.

Fujii, T., Yamamoto, H., Takenaka, E., Fujinami, K., Ando, A., and Yabuki, M., 1988, Intraspecific hybridization of a methanol-utilizing yeast, *Candida* sp. N-16, through protoplast fusion, *Agric. Biol. Chem.* **52**:1661–1667.

Giuseppin, M. L. F., 1988, Optimization of methanol oxidase production by *Hansenula polymorpha*: an applied study on physiology and fermentation, Ph.D. thesis, Technical University of Delft, The Netherlands.

Giuseppin, M. L. F., van Eijk, H. M. J., Verduyn, C., Bante, I., and van Dijken, J. P., 1988a, Production of catalase-free methanol oxidase (MOX) by *Hansenula polymorpha*, *Appl. Microbiol. Biotechnol.* **28**:14–19.

Giuseppin, M. L. F., van Eijk, H. M. J., and Bes, B. C. M., 1988b, Molecular regulation of methanol oxidase activity in continuous cultures of *Hansenula polymorpha*, *Biotechnol. Bioeng.* **32**:577–583.

Giuseppin, M. L. F., van Eijk, H. M. J., Bos, A., Verduyn, C., and van Dijken, J. P., 1988c, Utilization of methanol by a catalase-negative mutant of *Hansenula polymorpha*, *Appl. Microbiol. Biotechnol.* **28**:286–292.

Gleeson, M. A., 1986, The genetic analysis of the methylotrophic yeast *Hansenula polymorpha*, Ph.D. thesis, University of Sheffield, U.K.

Gleeson, M. A., and Sudbery, P. E., 1988a, The methylotrophic yeasts, *Yeast* **4**:1–15.

Gleeson, M. A., and Sudbery, P. E., 1988b, Genetic analysis in the methylotrophic yeast *Hansenula polymorpha*, *Yeast* **4**:293–303.

Gleeson, M. A., Waites, M. J., and Sudbery, P. E., 1984, Development of techniques for genetic analysis in the methylotrophic yeast *Hansenula polymorpha*, in: *Microbial Growth on C_1 Compounds* (R. L. Crawford and R. S. Hanson, eds.), American Society for Microbiology, Washington, DC, pp. 228–243.

Gleeson, M. A., Ortori, G. S., and Sudbery, P. E., 1986, Transformation of the methylotrophic yeast *Hansenula polymorpha*, *J. Gen. Microbiol.* **132**:3459–3465.

Goncharova, I. A., Babitskaya, V. G., and Lobanok, A. G., 1977, Growth and formation of protein biomass by fungi *Trichoderma* and *Penicillium* on methanol, in: *Microbial Growth on C_1 Compounds* (G. K. Skryabin, M. V. Ivanov, E. N. Kondratjeva, G. A. Zavarzin, Y. A. Trotsenko, and A. I. Nesterov, eds.), USSR Academy of Sciences, Moscow, pp. 187.

Goodman, J. M., 1985, Dihydroxyacetone synthase is an abundant constituent of the methanol-induced peroxisome of *Candida boidinii*, *J. Biol. Chem.* **260**:7108–7113.

Harder, W., Trotsenko, Y. A., Bystrykh, L. V., and Egli, T., 1987, Metabolic regulation in methylotrophic yeasts, in: *Microbial Growth on C_1 Compounds* (H. W. van Verseveld and J. A. Duine, eds.), Martinus Nijhoff, Dordrecht, pp. 139–149.

Hazeu, W., de Bruin, J. C., and Bos, P., 1972, Methanol assimilation by yeasts, *Arch. Mikrobiol.* **87:**185–188.

Hopkins, T. R., and Muller, F., 1987, Biochemistry of alcohol oxidase, in: *Microbial Growth on C_1 Compounds* (H. W. van Verseveld and J. A. Duine, eds.), Martinus Nijhoff, Dordrecht, pp. 150–157.

Janowicz, Z. A., Eckart, M. R., Drewke, C., Roggenkamp, R. O., Hollenberg, C. P., Maat, J., Ledeboer, A. M., Visser, C., and Verrips, C. T., 1985, Cloning and characterization of the DAS gene encoding the major methanol assimilatory enzyme from the methylotrophic yeast *Hansenula polymorpha*, *Nucl. Acids Res.* **13:**3043–3062.

Kato, K., Kurimura, Y., Makiguchi, N., and Asai, Y., 1974, Determination of methanol strongly assimilating yeasts, *J. Gen. Appl. Microbiol.* **20:**123–127.

Kato, N., Omory, Y., Tani, Y., and Ogata, K., 1976, Alcohol oxidases of *Kloeckera* sp. and *Hansenula polymorpha*, Catalytic properties and subunit structures, *Eur. J. Biochem.* **64:**341–350.

Kato, N., Nishizawa, T., Sakazawa, C., Tani, Y., and Yamada, H., 1979, Xylulose 5-phosphate dependent fixation of formaldehyde in a methanol-utilizing yeast *Kloeckera* sp. no. 2201, *Agric. Biol. Chem.* **43:**2013–2015.

Kato, N., Higuchi, T., Sakazawa, C., Nishizawa, T., Tani, Y., and Yamada, H., 1982, Purification and properties of a transketolase responsible for formaldehyde fixation in a methanol-utilizing yeast, *Candida boidinii (Kloeckera* sp.) no. 2201, *Biochim. Biophys. Acta* **715:**143–150.

Kato, N., Kobayashi, H., Shimao, M., and Sakazawa, C., 1986, Dihydroxyacetone production from methanol by a dihydroxyacetone kinase deficient mutant of *Hansenula polymorpha*, *Appl. Microbiol. Biotechnol.* **23:**180–186.

Komagata, K., 1981, Taxonomic studies of methanol-utilizing yeasts, in: *Microbial Growth on C_1 Compounds* (H. Dalton, ed.), Heyden, London, pp. 301–311.

Koning, W. de, Gleeson, M. A. G., Harder, W., and Dijkhuizen, L., 1987a, Regulation of methanol metabolism in the yeast *Hansenula polymorpha:* isolation and characterization of mutants blocked in methanol assimilatory enzymes, *Arch. Microbiol.* **147:**375–382.

Koning, W. de, Harder, W., and Dijkhuizen, L., 1987b, Glycerol metabolism in the methylotrophic yeast *Hansenula polymorpha:* phosphorylation as the initial step, *Arch. Microbiol.* **148:**314–320.

Koning, W. de, Bonting, K., Harder, W., and Dijkhuizen, L., 1990a, Classical transketolase functions as the formaldehyde-assimilating enzyme during growth of a dihydroxyacetone synthase-negative mutant of the methylotrophic yeast *Hansenula polymorpha* on mixtures of xylose and methanol in continuous cultures, *Yeast* **6:**117–125.

Koning, W. de, Weusthuis, R. A., Harder, W., and Dijkhuizen, L., 1990b, Methanol-dependent production of dihydroxyacetone and glycerol by mutants of the methylotrophic yeast *Hansenula polymorpha* blocked in dihydroxyacetone kinase and glycerol kinase, *Appl. Microbiol. Biotechnol.* **32:**693–698.

Kreger-van Rij, N. J. W. (ed.), 1984, *The Yeasts*, Elsevier, Amsterdam.

Kurzman, C. P., 1984, Synonomy of the yeast genera *Hansenula* and *Pichia* demonstrated through comparisons of deoxyribonucleic acid relatedness, *Antonie van Leeuwenhoek* **50:**209–217.

Kurzman, C. P., and Phaff, H. J., 1987, Molecular Taxonomy, in: *The Yeasts, Vol. 1, Biology of Yeasts* (A. H. Rose and J. S. Harrison, eds.), Academic Press, London, pp. 63–94.

Kurzman, C. P., Smiley, M. J., and Johnson, C. J., 1980a, Emendation of the genus *Issatchenkia* Kudriavzev and comparison of species by deoxyribonucleic acid reassociation, mating reaction, and ascospore ultrastructure, *Int. J. Syst. Bacteriol.* **30:**503–513.

Kurzman, C. P., Smiley, M. J., Johnson, C. J., Wickerham, L. J., and Fuson, G. B., 1980b, Two new and closely related heterothallic species, *Pichia amylophila* and *Pichia mississippiensis:* characterization by hybridization and deoxyribonucleic acid reassociation, *Int. J. Syst. Bacteriol.* **30:**208–216.

Lahtchev, K., and Tuneva, D., 1986, Mitotic segregation in hybrid of methylotrophic yeast *Candida pelliculosa, Curr. Microbiol.* **14:**121–125.

Ledeboer, A. M., Edens, L., Maat, J., Visser, C., Bos, J. W., Verrips, C. T., Janowicz, Z., Eckart, M., Roggenkamp, R., and Hollenberg, C. P., 1985, Molecular cloning and characterization of a gene coding for alcohol oxidase in *Hansenula polymorpha, Nucl. Acids Res.* **13:**3063–3082.

Lee, J. D., and Komagata, K., 1980a, Taxonomic study of methanol-assimilating yeasts, *J. Gen. Appl. Microbiol.* **26:**133–158.

Lee, J. D., and Komagata, K., 1980b, *Pichia cellobiosa, Candida cariosilignicola* and *Candida succiphila,* new species of methanol-assimilating yeasts, *Int. J. Syst. Bacteriol.* **30:**514–519.

Lee, J. D., and Komagata, K., 1983, Further taxonomic study of methanol-assimilating yeasts with special references to electrophoretic comparison of enzymes, *J. Gen. Appl. Microbiol.* **29:**395–416.

Levine, D. W., and Cooney, C. L., 1973, Isolation and characterization of a thermotolerant methanol-utilizing yeast, *Appl. Microbiol.* **26:**982–990.

Linton, J. D., and Niekus, H. G. D., 1987, The potential of one-carbon compounds as fermentation feedstocks, in: *Microbial Growth on C_1 Compounds* (H. W. van Verseveld and J. A. Duine, eds.), Martinus Nijhoff, Dordrecht, pp. 263–271.

Miller, M. W., Phaff, H. J., Miranda, M., Heed, W. B., and Starmer, W. T., 1976, *Torulopsis sonorensis,* a new species of the genus *Torulopsis, Int. J. Syst. Bacteriol.* **26:**88–91.

Moore, A. L., and Rich, P. R., 1980, The bioenergetics of plant mitochondria, *Trends Biochem. Sci.* **5:**284–287.

Mozaffar, S., Ueda, M., Kitatsuji, K., Shimizu, S., Osumi, M., and Tanaka, A., 1986, Properties of catalase purified from a methanol-grown yeast, *Kloeckera* sp. 2201, *Eur. J. Biochem.* **155:**527–531.

Müller, R. H., Uhlenhut, G. J., and Babel, W., 1985, Flow of ^{14}C-methanol via assimilatory and dissimilatory sequences with yeast in presence of glucose, *Arch. Microbiol.* **143:**77–81.

Nakase, T., and Komagata, K., 1970, Significance of DNA base composition in the classification of yeast genus *Pichia, J. Gen. Appl. Microbiol.* **16:**511–521.

Nakase, T., and Komagata, K., 1971a, Further investigation on the DNA base composition of the genus *Hansenula, J. Gen. Appl. Microbiol.* **17:**77–84.

Nakase, T., and Komagata, K., 1971b, Significance of DNA base composition in the classification of yeast genus *Torulopsis, J. Gen. Appl. Microbiol.* **17:**161–166.

Nakase, T., and Komagata, K., 1971c, Significance of DNA base composition in the classification of yeast genus *Candida, J. Gen. Appl. Microbiol.* **17:**259–279.

Nicolay, K., Veenhuis, M., Douma, A. C., and Harder, W., 1987, A ^{31}P NMR study of the internal pH of yeast peroxisomes, *Arch. Microbiol.* **147:**37–41.

Ogata, K., Nishikawa, H., and Ohsugi, M., 1969, A yeast capable of utilizing methanol, *Agr. Biol. Chem.* **33:**1519–1520.

Oki, T., Kouno, K., Kitai, A., and Ozaki, A., 1972, New yeasts capable of assimilating methanol, *J. Gen. Appl. Microbiol.* **18:**295–305.

Pal, H. S., and Hamdan, I. Y., 1979, Growth of a methanol-utilizing yeast, *Enzyme Microbiol. Technol.* **1:**265–268.

Patel, R. N., Hou, C. T., Laskin, A. I., Derelanko, P., and Felix, A., 1979, Oxidation of secondary alcohols to methyl ketones by yeasts, *Appl. Environm. Microbiol.* **38:**219–223.

Patel, R. N., Hou, C. T., Laskin, A. I., and Derelanko, P., 1981, Microbial production of methylketones: Properties of purified yeast secondary alcohol dehydrogenase, *J. Appl. Biochem.* **3**:218–226.

Phaff, H. J., and Starmer, W. T., 1987, Yeasts associated with plants, insects and soil, in: *The Yeasts, Vol. 1, Biology of Yeasts* (A. H. Rose and J. S. Harrison, eds.), Academic Press, London, pp. 123–180.

Phaff, H. J., Miller, M. W., and Shifrine, M., 1956a, The taxonomy of yeasts isolated from *Drosophila* in the Yosemite region of California, *Antonie van Leeuwenhoek* **22**:145–161.

Phaff, H. J., Miller, M. W., Recca, J. A., Shifrine, M., and Mrak, E. M., 1956b, Studies on the ecology of *Drosophila* in the Yosemite region of California. II. Yeasts found in the alimentary canal of *Drosophila*, *Ecology* **374**:533–538.

Roggenkamp, R., Sahm, H., and Wagner, F., 1974, Microbial assimilation of methanol, induction and function of catalase in *Candida boidinii*, *FEBS Lett.* **41**:283–286.

Roggenkamp, R., Hansen, H., Eckart, M., Janowicz, Z., and Hollenberg, C. P., 1986, Transformation of the methylotrophic yeast *Hansenula polymorpha* by autonomous replication and integration vectors, *Mol. Gen. Genet.* **202**:302–308.

Sahm, H., 1975, Oxidation of formaldehyde by alcohol oxidase of *Candida boidinii*, *Arch. Microbiol.* **105**:179–181.

Sahm, H., 1977, Metabolism of methanol by yeasts, *Adv. Biochem. Eng.* **6**:77–103.

Sakai, Y., and Tani, Y., 1986, Formaldehyde production by cells of a mutant of *Candida boidinii* S2 grown in methanol-limited chemostat culture, *Agric. Biol. Chem.* **50**:2615–2620.

Sakai, Y., and Tani, Y., 1987, Formaldehyde production with heat-treated cells of methanol yeast, *J. Ferment. Technol.* **65**:489–491.

Sakai, Y., Sawai, T., and Tani, Y., 1987, Isolation and characterization of a catabolite repression-insensitive mutant of a methanol yeast, *Candida boidinii* A5, producing alcohol oxidase in glucose-containing medium, *Appl. Environm. Microbiol.* **53**:1812–1818.

Sanchez, S., and Demain, A. L., 1978, Tryptophan excretion by a bradytroph of *Hansenula polymorpha* growing on methanol, *Appl. Environm. Microbiol.* **35**:459–461.

Savchenko, G. V., Kapul'tsevich, Y. G., Temina, A. V., and Nikitina, I. A., 1983, Hybridization of the asporogenic strains of *Hansenula polymorpha* by protoplast fusion, *Microbiologiya* **52**:449–452.

Schütte, H., Flossdorf, J., Sahm, H., and Kula, M. R. 1976, Purification and properties of formaldehyde dehydrogenase and formate dehydrogenase from *Candida boidinii*, *Eur. J. Biochem.* **62**:151–160.

Sherry, B., and Abeles, R. H., 1985, Mechanism of action of methanol oxidase, reconstitution of methanol oxidase with 5-deazaflavin, and inactivation of methanol oxidase by cyclopropanol, *Biochemistry* **24**:2594–2605.

Shifrine, M., and Phaff, H. J., 1956, The association of yeasts with certain bark beetles, *Mycologia* **48**:41–55.

Sibirny, A. A., Titorenko, V. I., Benevolenskii, S. V., and Tolstorukov, I. I., 1986, Differences in the mechanisms of ethanol and glucose catabolite repression of the enzymes of methanol metabolism in the yeast *Pichia pinus*, *Genetika* **22**:584–592.

Sibirny, A. A., Titorenko, V. I., Efremov, B. D., and Tolstorukov, I. I., 1987, Multiplicity of mechanisms of carbon catabolite repression involved in the synthesis of alcohol oxidase in the methylotrophic yeast *Pichia pinus*, *Yeast* **3**:233–241.

Sibirny, A. A., Titorenko, V. I., Gonchar, M. V., Ubiyvovk, V. M., Ksheminskaya, G. P., and Vitvitskaya, O. P., 1988, Genetic control of methanol utilization in yeasts, *J. Basic Microbiol.* **28**:293–319.

Sreekrishna, K., Potenz, R. H. B., Cruze, J. A., McCombie, W. R., Parker, K. A., Nelles, L., Mazzaferro, P. K., Holden, K. A., Harrison, R. G., Wood, P. J., Phelps, D. A., Hubbard, C. E., and Fuke, M., 1988, High level expression of heterologous proteins in methylotrophic yeast *Pichia pastoris, J. Basic Microbiol.* **28**:265–278.
Tani, Y., and Vongsuvanlert, V., 1987, Sorbitol production by a methanol yeast, *Candida boidinii (Kloeckera* sp.) no. 2201, *J. Ferment. Technol.* **65**:405–411.
Tani, Y., Mitani, Y., and Yamada, H., 1984a, ATP production by protoplasts of a methanol yeast, *Candida boidinii (Kloeckera* sp.) no. 2201, *Agric. Biol. Chem.* **48**:431–437.
Tani, Y., Mitani, Y., and Yamada, H., 1984b, Preparation of ATP-producing cells of a methanol yeast *Candida boidinii (Kloeckera* sp.) no. 2201, *J. Ferment. Technol.* **62**:99–101.
Tani, Y., Yonehara, Y., Mitani, Y., and Yamada, H., 1984c, ATP production by sorbitol-treated cells of a methanol yeast, *Candida boidinii (Kloeckera* sp.) no. 2201, *J. Biotechnol.* **1**:119–127.
Tani, Y., Sakai, Y., and Yamada, H., 1985a, Production of formaldehyde by a mutant of methanol yeast, *Candida boidinii* S2, *J. Ferment. Technol.* **63**:443–449.
Tani, Y., Sakai, Y., and Yamada, 1985b, Isolation and characterization of a mutant of a methanol yeast *Candida boidinii* S2, with higher formaldehyde productivity, *Agric. Biol. Chem.* **49**:2699–2706.
Tani, Y., Yonohara, T., Sakai, Y., and Yoon, B. D., 1987, Microbiological synthesis from C_1-compounds: application of some methylotrophic functions to synthesis of useful chemicals, in: *Microbial Growth on C_1 Compounds* (H. W. van Verseveld and J. A. Duine, eds.), Martinus Nijhoff, Dordrecht, pp. 282–288.
Thill, G., Davis, G., Stillmann, C., Tschopp, J. F., Graig, W. S., Velicelebi, G., Greff, J., Akong, M., Stroman, D., Torregrossa, R., and Siegel, R. S., 1987, The methylotrophic yeast *Pichia pastoris* as a host for heterologous protein production, in: *Microbial Growth on C_1 Compounds* (H. W. van Verseveld and J. A. Duine, eds.), Martinus Nijhoff, Dordrecht, pp. 289–296.
Tikhomirova, L. P., Ikonomova, R. N., and Kuznetsova, E. N., 1986, Evidence for autonomous replication and stabilization of recombinant plasmids in the transformants of yeast *Hansenula polymorpha, Curr. Genet.* **10**:741–747.
Tikhomirova, L. P., Ikonomova, R. N., Kuznetsova, E. N., Fodor, I. I., Bystrykh, L. V., Aminova, L. R., and Trotsenko, Y. A., 1988, Transformation of methylotrophic yeast *Hansenula polymorpha:* Cloning and expression of genes, *J. Basic Microbiol.* **5**:343–351.
Tolstorukov, I. I., and Benevolenskii, S. V., 1980, Study of the mechanism of mating and self-diploidization in haploid yeasts *Pichia pinus.* II. Mutations in the mating type locus, *Genetika* **16**:1335–1341.
Tolstorukov, I. I., and Efremov, B. D., 1984, Genetic mapping of the yeast *Pichia pinus.* II. Mapping by tetrad analysis, *Genetika* **20**:1099–1107.
Tolstorukov, I. I., Dutova, T. A., Benevolenskii, S. V., and Soom, Y. O., 1977, Hybridization and genetic analysis of the methanol-utilizing yeasts *Pichia pinus, Genetika* **13**:322–329.
Tolstorukov, I. I., Efrimov, B. D., and Bliznik, K. M., 1983, Construction of a genetic map of the yeast *Pichia pinus.* I. Determination of linkage groups using induced mitotic haploidization, *Genetika* **19**:897–902.
Tolstorukov, I. I., Motruk, O. M., and Efrimov, B. D., 1988, genetic control of alcohol oxidase activity in methylotrophic yeast *Pichia pinus* MH4, in: *14th Int. Conf. on Yeast Genetics and Molecular Biology,* Wiley, London, p. 375.
Trotsenko, Y. A., Bystrykh, L. V., and Ubiyvovk, V. M., 1984, Regulatory aspects of methanol metabolism in yeasts, in: *Microbial growth on C_1 compounds* (R. L. Crawford and R. S. Hanson, eds.), American Society for Microbiology, Washington, DC, pp. 118–122

Tschopp, J. F., Burst, P. F., Cregg, J. M., Stillman, C. A., and Gingeras, T. R., 1987a, Expression of the *lacZ* gene from two methanol-regulated promotors in *Pichia pastoris*, *Nucl. Acids Res.* **15:**3859–3876.

Tschopp, J. F., Sverlow, G., Kosson, R., Craig, W., and Grinna, L., 1987b, High-level secretion of glycosylated invertase in the methylotrophic yeast, *Pichia pastoris*, *Bio/Technology* **5:**1305–1308.

Ubiyvovk, V. M., Bystrykh, L. V., and Trotsenko, Y. A., 1983, Participation of glutathione in regulation of methanol metabolism in yeast, *Mikrobiologiya* **52:**383–387.

Uotila, L., and Koivusalo, M., 1974, Formaldehyde dehydrogenase from human liver. Purification, properties and evidence for the formation of glutathione thioesters by the enzyme, *J. Biol. Chem.* **249:**7653–7663.

Veenhuis, M., and Harder, W., 1987, Metabolic significance and biogenesis of microbodies in yeasts, in: *Peroxisomes in biology and medicine* (H. D. Fahimi and H. Sies, eds.), Springer-Verlag, Berlin, Heidelberg, pp. 436–458.

Veenhuis, M., Dijken, J. P. van, and Harder, W., 1983, The significance of peroxisomes in the metabolism of one-carbon compounds in yeasts, *Adv. Microbial Physiol.* **24:**1–82.

Verduyn, C., Giuseppin, M. L. F., Scheffers, W. A., Dijken, J. P. van, 1988, Hydrogen peroxide metabolism in yeasts, *Appl. Environm. Microbiol.* **54:**2086–2090.

Vongsuvanlert, V., and Tani, Y., 1988a, Characterization of D-sorbitol dehydrogenase involved in D-sorbitol production of a methanol yeast, *Candida boidinii (Kloeckera* sp.) no. 2201, *Agric. Biol. Chem.* **52:**419–426.

Vongsuvanlert, V., and Tani, Y., 1988b, Purification and characterization of xylose isomerase of a methanol yeast, *Candida boidinii*, which is involved in sorbitol production from glucose, *Agric. Biol. Chem.* **52:**1817–1824.

Waites, M. J., and Quayle, J. R., 1980, Dihydroxyacetone: a product of xylulose 5-phosphate-dependent fixation of formaldehyde by methanol-grown *Candida boidinii*, *J. Gen. Microbiol.* **118:**321–327.

Waites, M. J., and Quayle, J. R., 1981, The interrelation between transketolase and dihydroxyacetone synthase activities in the methylotrophic yeast *Candida boidinii*, *J. Gen. Microbiol.* **124:**309–316.

Yamada, Y., Okada, T., Ueshima, O., and Kondo, K., 1973, Coenzyme Q system in the classification of the ascosporogenous yeast genera *Hansenula* and *Pichia*, *J. Gen. Appl. Microbiol.* **19:**189–208.

Yarrow, D., and Meyer, S. A., 1978, Proposal for amendment of the diagnosis of the genus *Candida* Berkhout nom. cons., *Int. J. Syst. Bacteriol.* **28:**611–615.

Yasuhara, S., Kawamoto, S., Tanaka, A., Osumi, M., and Fukui, S., 1976, Induction of catalase activity in a methanol-utilizing yeast, *Kloeckera* sp. no. 2201, *Agr. Biol. Chem.* **40:**1771–1780.

Yonehara, T., and Tani, Y., 1987, Highly efficient production of ATP by a methanol yeast, *Candida boidinii (Kloeckera* sp.) no. 2201, *J. Ferment. Technol.* **65:**255–260.

Zwart, K. B., and Harder, W., 1983, Regulation of the metabolism of some alkylated amines in the yeasts *Candida utilis* and *Hansenula polymorpha*, *J. Gen. Microbiol.* **129:**3157–3169.

Biotechnological and Applied Aspects of Methane and Methanol Utilizers

8

DAVID J. LEAK

1. INTRODUCTION

The rapid developments in our understanding of many aspects of methylotrophy have, in no small part, resulted from the perceived industrial applications of these organisms, particularly the aerobic methanol and methane utilizers. The ICI Pruteen process stands as a landmark of technical achievement in industrial fermentation. Yet few of these applications have reached commercial reality, at least in Western economies, or withstood the competition from other quarters. This is not to say that C_1 biotechnology has no future. Indeed, the relative maturity of C_1 research and technology might in itself count in favor of adopting new C_1-based processes. However, it is probably true to say that C_1 biotechnology is at a crossroads. The successful processes of the future may well be very different from those envisaged 15–20 years ago but will no doubt benefit from the knowledge base already established. It is therefore timely to review C_1 biotechnology with the benefit of hindsight and perhaps establish the ground rules for future developments.

Within the spectrum of activities encompassed by the term "biotechnology," the current or potential applications of methylotrophs can be considered in three broad, but not exclusive, categories.

DAVID J. LEAK • Centre for Biotechnology, Imperial College of Science, Technology and Medicine, London SW7 2AZ, England.

Methane and Methanol Utilizers, edited by J. Colin Murrell and Howard Dalton. Plenum Press, New York, 1992.

1. Processes highly dependent on substrate economics
2. Processes based on C_1-specific enzymes/metabolism
3. Processes or products coincidentally associated with C_1 metabolism

2. BIOTECHNOLOGY AND ECONOMICS

Like a number of products from the chemical industry, the products of biotechnology exhibit a typical inverse relationship (log–log plot) of production volume and price (Fig. 1). Three aspects of this should be highlighted. First, the cost of the carbon substrate forms an increasingly important component of total production costs with increasing operational volume and is thus considerably more important for bulk products such as ethanol or single-cell protein (SCP) than for specialty products, e.g., pharmaceuticals. Second, biotechnology products are subject to the usual laws of supply and demand. Thus the economies of scale are lost if increased production results in a glut and a consequent reduction in selling price. However, improvements in product yield and volumetric productivity (which would allow smaller production units to be used) or the utilization

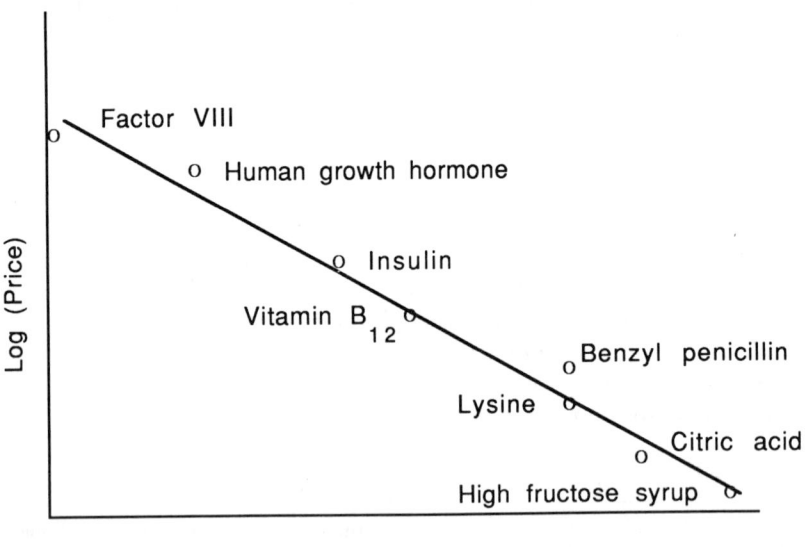

Figure 1. Relationship between worldwide production volume and selling price for a range of products of biotechnology.

of cheaper substrates with similar yield characteristics serves to reduce the process costs and thus increase profitability. Third, where biotechnology products have direct chemical or agricultural competition, the relationship between price and quantity sold (the demand curve) is highly "elastic" (Fig. 2); i.e., a small increase in price can result in a significant drop in demand as customers switch to alternative suppliers. At the other end of the scale, specialized products such as pharmaceuticals, which may have no effective substitute, exhibit highly inelastic demand curves (Fig. 2). As a generalization, products that fall between the two volumetric extremes exhibit an intermediate level of elasticity of demand (for further discussion of these points see Hacking, 1986).

3. PROCESSES HIGHLY DEPENDENT ON SUBSTRATE ECONOMICS

Although Foster (1962) had previously noted the industrial potential of cooxidation by methanotrophs, the main factor that propeled methylotrophs into the biotechnology limelight was the comparatively low cost of methane and methanol as carbon substrates for the production of microbial biomass, i.e., single-cell protein (SCP). The reasons behind the drive to make high-protein feedstuffs for animal or human consumption and the development (and subsequent demise) of various processes are well documented elsewhere (Solomons, 1983; Goldberg, 1985). Suffice it to say that the combination of increasing substrate costs (particularly for long-chain hydrocarbons derived from oil) and the stable, if not decreasing, price of soya-derived protein as a direct competitor to SCP have made

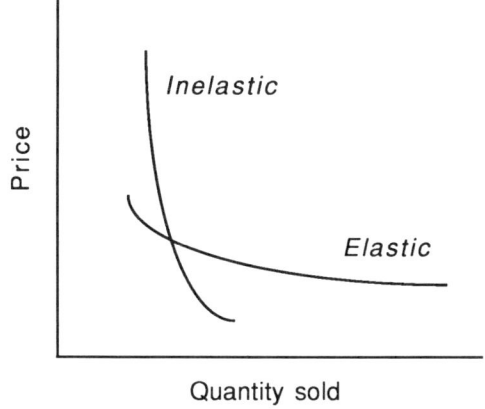

Figure 2. Demand curves displaying the extremes of elasticity of product sales in response to price fluctuations in products with significant competition (Elastic) and products for which there is no effective substitute (Inelastic).

SCP uncompetitive as a self-contained basic commodity,* at least in Western economies. However, where SCP is produced as a by-product from a process waste (e.g., whey), the prognosis is more favorable and such processes are in commercial operation.

Despite the demise of most C_1-based SCP processes (except perhaps in the Soviet Union, with some interest being retained in the oil-rich nations of the Middle East), the information gained on the production of biomass provides a valuable basis for the assessment of any commodity (high volume) product derived from C_1 substrates.

3.1. Single-Cell Protein

Historically, the first intensive SCP processes were based on yeast fermentations using long-chain (C_{10}–C_{18}) alkanes as substrates. Apart from these oil-derived substrates being relatively inexpensive (at that time), being highly reduced they should also be capable of giving high biomass yields (g. dry wt. cells/g. substrate). With a wide range of substrates there is a close correlation between the degree of reduction of the substrate and maximum aerobic growth yield (Roels, 1980; Heijnen and Roels, 1981). In biochemical terms, this implies that the yield of biomass is dependent primarily on the ATP yield from substrate oxidation. In the case of long-chain alkanes, this arises almost entirely by the coupled respiration of NADH/FADH produced during the process of β oxidation. Examples where the biomass yield from a relatively oxidized substrate falls below that predicted reflect a lower than expected yield of ATP resulting either from inefficient coupling of respiration and ATP synthesis (e.g., Ackrell and Jones, 1971; Hardy and Dawes, 1985) or from an imbalance of metabolism resulting in product formation or storage polymer biosynthesis. Above a substrate reduction value of 4 [i.e., substrates more reduced than $(CH_2O)n$] with ammonia as the nitrogen source, the relationship breaks down. One reason for this is that the relationship between ATP yield and biomass yield breaks down; i.e., yield is no longer energy limited, but becomes carbon limited or possibly NADH limited (Heijnen and Roels, 1981; Anthony, 1982). This is observed in cases where a significant amount of energy-yielding substrate oxidation occurs before the divergence of assimilatory and dissimilatory metabolic pathways and is exemplified by the case of long-chain hydrocarbons. A second reason is that although the oxidation of saturated hydrocarbons to their respective alcohols is an exothermic reaction, the energy released is not harnessed in biological

---

*This does not include products like Mycoprotein (Quorn™), which has additional properties, particularly texture, which give it added value as a meat substitute.

systems. In contrast, the biological process requires an initial, energy-dependent (usually NADH requiring) activation of oxygen. The effects of this are most evident in the case of methane oxidation, where reported yields for methanotrophs are significantly lower than those expected from the reduction state of the substrate (Drozd et al., 1978; Leak and Dalton, 1986).

Although SCP from C_1 substrates was being considered by a number of companies in the late 1960s, attention focused on these substrates after the dramatic oil price rises in the 1970s. Methane, as natural gas, is available in large amounts and, where it is associated with oil recovery, is effectively available at zero cost. Additionally, other major deposits of methane have been discovered in recent years (Kvenvolden, 1988), but these have yet to be exploited. However, although SCP processes based on methane are still being investigated to its use, which have generally resulted in methanol becoming the substrate of choice. These include explosive hazards, poor compressibility (dictating use at source or piping to the fermentation plant), and the low aqueous solubility of methane. Additionally, the oxygen demand for growth on methane is very high, primarily due to the involvement of a monooxygenase in methane oxidation. Together with the oxidation of methanol coupled via a respiratory chain to a terminal oxidase, a minimum ration of $O_2:CH_4 = 1.5$ would be expected, compared to $O_2:(CH_2O) = 0.4$ (approximately) for glucose. This also results in higher heat output, implying increased cooling costs or the use of thermotolerant strains. As a consequence, high-density fermentations with methane as substrate require highly efficient aeration, possibly using O_2-enriched air, and efficient cooling systems. Because of the energy requirement for methane oxidation, molar growth yields (g cells/mol substrate) are lower for methane oxidation than for methanol with comparable strains (Leak and Dalton, 1986). Furthermore, the broad substrate range of methane monooxygenase (MMO) results in the oxidation of higher hydrocarbons (e.g., ethane, propane), which are often present in significant amounts in natural gas. With further metabolism by broad-specificity alcohol and aldehyde dehydrogenases, this can result in the accumulation of toxic levels of the respective carboxylic acids. One solution, adopted in the Shell process (Wilkinson et al., 1974), was to establish a stable mixed culture in which the nonmethanotrophic organisms utilized the cooxidation products from higher alkanes.

3.2. SCP from Methanol

Methanol is produced commercially in a low-pressure catalytic process from synthesis gas (Fig. 3), which may be derived from methane or

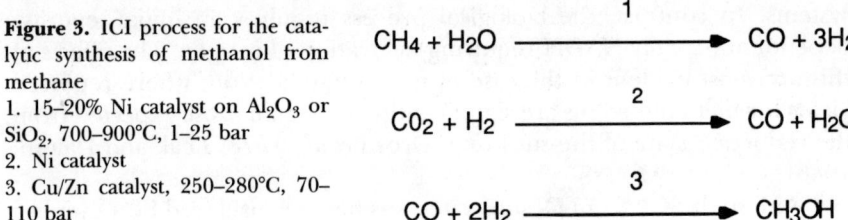

Figure 3. ICI process for the catalytic synthesis of methanol from methane
1. 15–20% Ni catalyst on Al_2O_3 or SiO_2, 700–900°C, 1–25 bar
2. Ni catalyst
3. Cu/Zn catalyst, 250–280°C, 70–110 bar

from a variety of fossil fuels (e.g., fuel oil, coal) or potentially from renewable resources (e.g., cellulose). Currently there is considerable overcapacity for methanol production (Linton and Niekus, 1987), primarily because the predicted use of methanol as a petroleum additive has not been realized. The use of methanol as carbon source avoids most of the fermentation and handling problems highlighted for methane, and a number of SCP processes have been developed to pilot or full-scale production (Table I).

3.2.1. Selection of Organism

A primary factor in selection of suitable organisms is nutritional balance, including the content of essential dietary amino acids, lack of toxic components, and general digestibility. To this should be added lack of pathogenicity and consistency (i.e., genetic stability) of the product. Most of these criteria can be met by selected methanol-utilizing bacteria or yeast strains. However, one crucial factor, excluding public acceptability, that determines the potential market for the product is the level of contained nucleic acid. Specified limits for total nucleic acid intake (4 g/day) and

TABLE I. Full or Pilot-Scale SCP Processes Using Methanol as Substrate

Company	Capacity	Product (organism)	Reference
Imperial Chemical Industries PLC, UK	1500 m^3	Pruteen (*Methylophilus methylotrophus*)	Senior and Windass (1980)
Hoechst-Uhde, FRG	20 m^3	Probion (*Methylomonas clara*)	Faust (1979)
Phillips Petroleum Co. (Provesta Corp.) USA	25 m^3	Provesteen (*Pichia pastoris*)	Shay et al. (1987)
Norsk-Hydro, Sweden	45 m^3	Norprotein (*Methylomonas methanolica*)	Mogren (1979)
Mitsubishi Gas-Chemical Co., Japan	2 m^3	— (*Pseudomonas sp.*)	Solomons (1983) Kuraishi et al. (1979)

SCP-derived RNA (2 g/day) in human nutrition have been recommended by the World Health Organization. This means that an RNA reduction step is necessary at the end of the fermentation if the product is to be acceptable as a human foodstuff. The problem is particularly acute with bacterial SCP, as nucleic acid in rapidly growing bacteria can comprise 20–25% of the dry weight of the organism, with the major part being RNA. Because of their lower growth rates, yeasts and fungi have lower nucleic acid levels, but these still require further postfermentation reduction. Although methods are available for reduction of nucleic acids in whole organisms (usually by stimulation of endogenous nucleases), these have not been entirely successful on a large scale, except with the nonmethylotrophic Mycoprotein process using the fungus *Fusarium graminearum* (Solomons, 1985). Thus, without resorting to cell disruption and protein recovery (see later), SCP based on methylotrophic bacteria is only suitable as an animal feedstuff, animals being generally more tolerant of the high nucleic acid levels. However, except for certain niche areas where the nutritional balance of SCP may give improved growth performance in high-intensity processes (e.g., ICI's Pruteen was used in the rearing of veal calves and in fish farming), microbial SCP based on methanol as a feedstock is not competitive as a commodity feed protein.

The alternative to nucleic acid reduction is cell disruption and physical separation of the protein and nucleic acids. While this adds additional steps, and therefore additional expense to the process, the potential added value of the products, which may then be suitable for human consumption, would justify this. In this respect it should be noted that the Probion process (Hoechst), based on the growth of the bacterium *Methylomonas clara*, was developed with this in mind from the outset, using a proprietary disruption and separation technology (Faust, 1979). As a semipurified source of protein the product could then be more useful as a source of "functional" protein, i.e., protein that influences the physical properties of a foodstuff such as foaming or gelling. Additionally, the separation into component parts yields a nucleic acid fraction, which, through the action of appropriate nucleases, could be hydrolyzed to yield flavor-enhancing 5'nucleotides [5' guanosine monophosphate (GMP) and 5' inosine monophosphate (IMP)]. McNairney (1984) has discussed such a process developed for Pruteen. However, as the Pruteen plant is no longer operating, it is evident that even this prospect was not going to make the process economical. Along the same lines as increasing added value by separating the component parts is the recognition that, having developed the fermentation technology, the organism could be used as a host for the expression of heterologous proteins. This path has been followed as an extension to the Pruteen (Hennam *et al.*, 1982), Probion (de Vries *et al.*,

1990), and Provesteen (Thill et al., 1987) processes. However, the economic logic of producing proteins with pharmaceutical applications is not clear. The problems associated with production of heterologous proteins are those of efficient expression, protein export, and posttranslational modification. Given the expected market value of these products (Fig. 1), the effect of carbon substrate costs on the economics of the process is minimal. If methylotrophs are to find application as hosts for expression of foreign proteins, then economics would again point to the high-volume commodities, e.g., food-related applications (particularly with an organism approved for food use) or possibly in biotransformations. *Pichia pastoris*, the yeast used in the Provesteen process, may prove to be the exception, not because of C_1 economics but because, as a facultative methylotroph, the alcohol oxidase (AOX) system is highly regulated and can be induced to high levels of expression (Thill et al., 1987). If this organism achieves commercial exploitation, then the growth substrate would probably be glucose or glycerol!

Two processes, the ICI Pruteen process based on the bacterium *Methylophilus methylotrophus* and the Phillips Petroleum Provesteen process based on the yeast *P. pastoris*, have been developed to full-scale production levels, although in the latter case the plant has been used primarily to produce specialty products derived from nonmethylotrophic yeasts. While unfavorable economics means that neither are currently producing SCP based on C_1 substrates, the technological development of the processes warrants further discussion as they impinge on any C_1-based process.

3.2.2. Pruteen

M. methylotrophus is an obligate methylotroph that was selected from a screen of methanol-utilizing bacteria as the most suitable for SCP applications. In addition to having a favorable nutritional status, *M. methylotrophus* grows optimally at 40°C with a μ_{max} of 0.55 hr^{-1} and a molar growth yield of approximately 16 g dry weight. (mol. methanol)$^{-1}$. The relatively high growth temperature reduces the cooling requirements, as outlined previously. Although this organism was selected empirically, it was perhaps not surprising to discover that it uses the energetically more efficient ribulose monophosphate pathway (RuMP) of carbon assimilation (Beardsmore et al., 1982a) rather than the serine pathway. A significant (4–7%) improvement in growth yield was obtained by replacing the glutamine synthetase/glutamate synthase (GS/GOGAT) pathway of ammonia assimilation with glutamate dehydrogenase cloned from *Escherichia coli* (Windass et al., 1980), thus saving 1 ATP per NH_4^+ assimilated. However, the improved strain was never put into full-scale production, partly because the construct proved unstable under production conditions, but also because of the prohibitive cost of repeating toxicological testing. A report in the patent literature (Beardsmore et al., 1982b) also suggests that greater carbon conversion efficiency could be obtained by replacing methanol dehydrogenase from M. methylotrophus with an alcohol dehydrogenase cloned from Bacillus stearothermophilus, but there are no further reports of the use of this construct.

Given the economies of scale outlined earlier, the production process required the development of novel technology and resulted in the construction of the largest aerobic aseptic fermentor in the world, built in 1979 at a cost of approximately U.S. $80 × 10^6. This comprised a single 1500-m^3 pressure cycle fermentor with associated upstream and downstream processing equipment (Fig. 4). Mixing was achieved on the airlift principle with pressurized air and nutrients fed into the riser. To avoid high, potentially toxic, local concentrations of methanol, the carbon substrate was sparged via numerous outlets to achieve a fairly even distribution. The high aspect ratio of the fermentor (42 m high × 7 m wide) meant that air entered the fermentor under a considerable head of pressure, and hence good mass transfer was obtained, which was required to satisfy the relatively high oxygen demand for growth on methanol. Economics dictated continuous operation, and runs in excess of 100 days without contamination problems have been reported (Powell and Rodgers, 1984). No doubt the lack of mechanical stirring with associated sterile seals reduced the contamination risk. The fermentation typically ran at a dilution rate of 0.16–0.19 hr^{-1} with a cell density of 30 g dry weight $liter^{-1}$ and a projected annual production capacity of 50–70,000 tonnes. In addition to introducing the E. coli glutamate dehydrogenase gene into M. methylotrophus, the potential for heterologous gene expression in this organism was demonstrated (Hennam et al., 1982) by expressing the representative eukaryotic genes chicken ovalbumin and mouse dihydrofolate reductase from lac and β-lactamase promoters, respectively. Expression of ovalbumin was significantly increased using the strong lac UV5 promoter construct, although not to the level found in E. coli Byrom (1984) subsequently confirmed that the lac and lac UV5 promoters and also the trp promoter from E. coli function in M. methylotrophus. Additionally, de Maeyer et al. (1982) have produced α interferon in this organism. However, advantages of lower substrate costs and use of a food-grade organism are not sufficient to warrant switching from established host systems for this type of product.

3.2.3. Provesteen

Yeast already has a well-established place in human nutrition and therefore, within the context of SCP, is likely to gain more public acceptance. Thus there may be good commercial grounds for adopting a yeast-

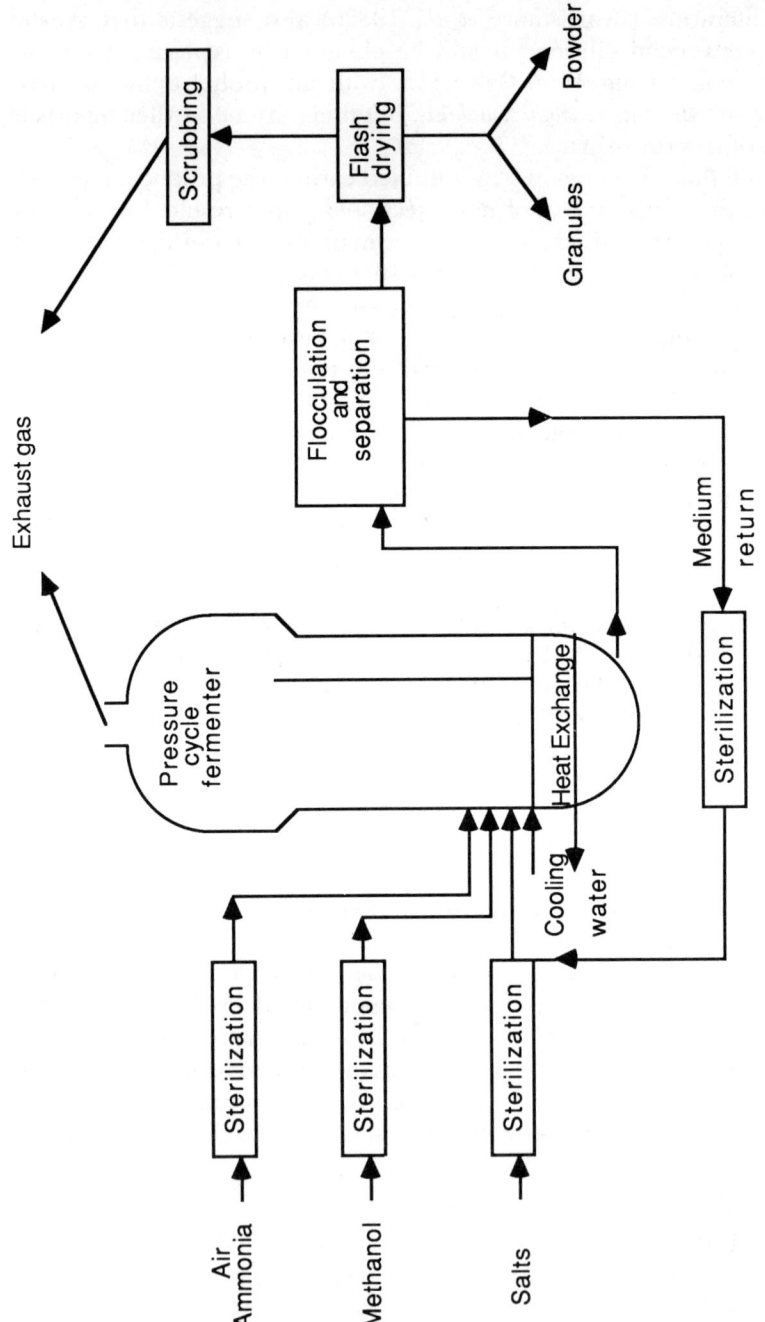

Figure 4. Flow diagram of the ICI single-cell protein (Pruteen) process. [Adapted from McNairney (1984).]

based process even though the productivity and yield may be less than for a bacterial fermentation. With this in mind, researchers at Phillips Petroleum initiated studies in the late 1960s to isolate methanol-utilizing yeasts suitable for SCP production, settling finally on a strain of *P. pastoris* with relatively high biomass yields and protein content. The inefficient first step in the metabolism of methanol by yeasts, involving an alcohol oxidase coupled with catalase, was not appreciated until some years later (Sahm and Wagner, 1973). This compounds the traditional problems of oxygen demand and heat dissipation in large-scale aerobic yeast fermentations. *P. pastoris* grows optimally at 30°C with a doubling time of 5.5–6 hr. Although seemingly a disadvantage, such a slow growth rate could be tolerated if the fermentation was run at very high cell densities, thus maintaining a respectable volumetric productivity.

Research on medium optimization and fermentation design at Provesta Corporation, a subsidiary of Phillips Petroleum, led to the development of the Phillips/Provesta continuous high-cell-density, direct dry process (Shay *et al.*, 1987), which, by achieving oxygen transfer rates of 0.8–1 mol liter^{-1} hr^{-1}, can produce a continuous slurry containing 125–150 g dry weight/liter of cells, which after heat treatment can be spray-dried directly without cell concentration. A commercial plant based a 25-m^3 fermentor came into operation in 1988, but, as stated, is currently used for monmethylotrophic yeast fermentations.

Although the envisaged SCP process is currently uneconomical, research on *P. pastoris* at Phillips Petroleum and the Salk Institute Biotechnology and Industrial Associates (SIBIA) has opened up two other potential avenues of exploitation. One is the provision of alcohol oxidase, which under appropriate conditions may constitute up to 35% of the soluble cell protein (Couderc and Baratti, 1980) and may have diverse applications. The second exploits the regulatory sequences for alcohol oxidase (AOX) in the expression of heterologous eukaryotic proteins. Although baker's yeast, *Saccharomyces cerevisiae*, has been shown to have a number of advantages compared with bacteria in the expression of eukaryotic proteins, it is by no means the ideal host. Hence there is a strong precedent for examining other simple eukaryotes (yeast and fungi) for improved performance.

In *P. pastoris* most of the alcohol oxidase activity derives from a single gene AOX 1, which is subject to catabolite repression but, in the absence of repression, is induced to high levels by the presence of methanol (see also Chapter 7). This has been exploited to construct an expression cassette (Thill *et al.*, 1987), which after introduction of a foreign gene can be integrated into the chromosome specifically at the AOX locus using an integrative vector. In batch fermentations this may be controlled by growth first on 5–10% glycerol, a nonrepressing substrate, and after exhaustion of the glycerol expression is switched on by feeding methanol. In continuous fermentations glycerol–methanol mixtures are used (Wegner, 1990). The tight regulation of expression from AOXI and use of a chromosomally integrated cassette means that in the absence of methanol there is no metabolic burden arising from either expression or maintenance of plasmids in high copy number and hence no selection against the construct. This system has been used to express a range of proteins from different sources (Table II). Of particular note is the production of tumor necrosis factor (TNF) at 6–10 g.liter^{-1} from high-density fermentations and expression of hepatitis B surface antigen (HBsAg) at levels equal to *S. cerevisiae* but with 10 times higher concentrations of the antigenically active assembled 22-nm particles (Cregg *et al.*, 1987). *P. pastoris* recognizes both yeast and mammalian secretion signals, and recent evidence (Grinna and Tschopp, 1989) indicates a posttranslational glycosylation pattern more similar to that of higher animals than that of *S. cerevisiae*.

Although this expression system exploits elements of methylotrophic metabolism, its principal attractions put it into the third category listed in the introduction.

4. METABOLIC PRODUCTS FROM METHYLOTROPHS

It is evident from the previous chapters that C_1 primary metabolism is sufficiently distinct from glycolysis to offer new possibilities for the accumulation of intracellular and extracellular metabolites. This includes both traditional microbial products, such as amino acids, and specialized

TABLE II. Heterologous Proteins Expressed in *Pichia pastoris*

Intracellular proteins	Secreted proteins
β-Galactosidase	Bovine lysozyme
Hepatitis B surface antigen	Human serum albumin
HIV antigens	Invertase
Human serum albumin	Tissue plasminogen activator
γ interferon	
IL-2 (mature)	
Salmon growth hormone	
Streptokinase	
Tumor necrosis factor (TNF)	
TNF analogs	

products, such as $^{13}C/^{14}C$-labeled metabolites. In the latter context, Cambridge Isotope Laboratories market 1-, 3-, and U-^{13}C-labeled lactate and pyruvate, ^{13}C serine, and U-^{13}C glucose and galactose produced from ^{13}C methanol by Celgene Corporation (D. Stirling, personal communication).

Amino acids, nucleotides, and vitamins occupy the middle ground in the economic table of biotechnology products. Many of these compounds are already produced by established fermentation processes. Thus in considering a C_1-based process the competition is likely to come from fermentation or chemical synthesis. In this respect, the outcome of long-term research will be less vulnerable to fluctuations in agricultural or oil prices. Economies in the use of carbon substrates will have some bearing on the price of the product, but because the established processes are relatively mature, it is doubtful whether investment in research based on utilization of C_1 substrates is prudent, except by the existing manufacturers, unless some aspect of C_1 metabolism makes the process particularly attractive. Therefore, apart from the odd cases where high levels of extracellular metabolites have been fortuitously associated with methylotrophs, this topic falls mainly into the second category listed in the introduction.

4.1. Amino Acids

Enzymes catalyzing steps in C_1 primary metabolism should be expressed in significantly higher amounts than those of peripheral pathways and subject to different patterns of control. In particular, the central role of serine hydroxymethyltransferase (STHM) in serine pathway organisms and C_1 carriers in both C_1 assimilatory routes makes serine and methionine obvious targets. RuMP pathway organisms and yeasts using the XuMP route may also find application in the production of aromatic amino acids that require erythrose-4-phosphate and phosphoenol pyruvate as precursors.

4.1.1. Serine

Serine is used primarily as a precursor in the biosynthesis of tryptophan and is produced from glycine by glycine-resistant heterotrophs (Kubota *et al.*, 1972; Tanaka *et al.*, 1980) expressing high levels of STHM. Fermentative production from carbohydrates has proved impossible because of the high rates of serine turnover. Together with the possibility that STHM may be expressed at high levels in some serine pathway organisms, methylotrophs also have the potential to give stoichiometric conversion from glycine because the C_1 unit can be produced from methanol, unlike the situation with heterotrophic bacteria, in which the C_1 unit is produced by cleavage of a second molecule of glycine. With wild-type *Pseudomonas* 3ab, Keune *et al.* (1976) produced 4.7 g. liter^{-1} serine with resting cells incubated at pH 8.5 with glycine (20 g. liter^{-1}) and methanol (8 g. liter^{-1}) supplements. This was subsequently improved to 7 g. liter^{-1} from 10 g. liter^{-1} glycine and 20 g. liter^{-1} methanol by optimization in a study (Behrendt *et al.*, 1984) in which the serine produced was converted directly and stoichiometrically to tryptophan by incubation of the unprocessed serine production broth, supplemented with indole, with a tryptophan-producing strain of *E. coli.*

Serine production from wild-type strains appears to be favored by high pH (Keune *et al.*, 1976; Izumi *et al.*, 1982), possibly because this reduces the CO_2 availability to PEP carboxylase and consequently the degradation of serine, although other explanations have been suggested. An alternative approach adopted by Morinaga *et al.* (1981b) was to use a mutant of the serine pathway organism *Pseudomonas* MS31 that was temperature sensitive (T^s) for serine degradation. Cells grown on methanol at the permissive temperature (30°C) then switched to a nonpermissive temperature (38–42°C) accumulated 6.9 g. liter^{-1} serine from 12 g. liter^{-1} glycine. Metal chelators and some divalent cations such as cobalt and nickel also appear to block serine degradation as well as stimulating STHM (Morinaga *et al.*, 1981a).

As with carbohydrate-utilizing, amino acid–producing strains, increased productivity may be obtained by relieving end-product inhibition and repression. Thus a further mutant of the T^s strain, resistant to the serine analog *o*-methyl D,L-serine (Morinaga *et al.*, 1983), was found to produce 12 g. liter^{-1} serine from 15 g. liter^{-1} glycine. Release of repression by methionine (by using methionine auxotrophs) also increased serine production in a strain of *Arthrobacter globiformis* (Tani *et al.*, 1978). In both instances higher productivity correlated with higher levels of STHM. Higher productivities have also been obtained in glycine-resistant strains, glycine being inhibitory to growth of a wide range of bacteria. Thus Yamada *et al.* (1986) increased productivity of an already high-yielding *Hyphomicrobium methylovorum* KM146 from 24 g. liter^{-1} to 34 g. liter^{-1} serine from 100 g. liter^{-1} glycine and 48 g. liter^{-1} methanol at pH 9.2 using a glycine-resistant strain GM2, which was shown to have higher levels of methanol dehydrogenase (MDH) and STHM. Resting cells of KM146 and a glycine-resistant mutant of a *Pseudomonas* sp. KM193 have also been shown to produce modest yields of threonine (Yamada *et al.* 1985) by replacing the methanol for serine production with ethanol. This results from the ability of MDH to oxidize ethanol and STHM to assimilate the resulting acetaldehyde.

The highest reported yield (Sirirote et al., 1986) of 54 g. liter^{-1} serine from a methylotroph was obtained with resting cells of a strain of *Pseudomonas extorquens* NR-1 by increasing the permeability of cells to serine and glycine with a freeze–thaw regime. However, this pales into insignificance compared to the claimed 400 g/liter of serine produced in a enzyme bioreactor containing STHM cloned initially from *Escherichia coli* and expressed in *Klebsiella pneumoniae* during growth on starch. This process, developed by Genex Corporation (Hamilton et al., 1985; Hsiao et al., 1986), takes serine production by STHM to its logical conclusion, ironically using a gene cloned from a nonmethylotroph. Furthermore, because the equilibrium between tetrahydrofolate (THF) and formaldehyde is strongly in favor of the product methylene-THF, it is not even necessary to use C_1 metabolism to provide the C_1 unit. Thus if an enzymic route to serine production is adopted, it is unlikely to be based on methylotrophic organisms unless alcohol oxidase is adopted for the *in situ* production of formaldehyde.

4.1.2. Methionine

Methionine, an essential amino acid used in animal feed supplements, is currently produced chemically from methanethiol and acrolein. Biosynthesis is convergent, requiring L-homoserine (from aspartate), serine, sulfide, and the C_1 carrier 5-methyl-THF. Yamada et al. (1982) demonstrated that production of methylotrophs was feasible using an obligate methylotroph (RuMP) strain OM33 that excreted 70 mg. liter^{-1} L-methionine from L-homoserine (5 g. liter^{-1}) and sodium sulfide (2 g. liter^{-1}) during growth on a methanol medium. A mutant resistant to ethionine, a methionine analog, and consequently desensitized to feedback inhibition accumulated 420 mg. liter^{-1} without any homoserine additions (Yamada et al., 1982). Higher yields (800 mg. liter^{-1}) have been reported for an ethionine-resistant mutant of the serine pathway organism *Pseudomonas* FM518 (Morinaga et al., 1982) grown in batch culture on a methanol medium. Studies showed that methionine production was highest in the early exponential phase, and cells harvested under optimal conditions could be used to produce L-methionine from D,L-homocysteine and methanol by transmethylation (Morinaga et al., 1984).

Given that the major requirement for methionine is as an animal feed supplement, an alternative strategy to extracellular accumulation and purification might be to use methionine-enriched whole cells. Lim and Tani (1988) have shown that an ethionine-resistant mutant of *Candida boidinii* no. 2207 accumulated 282 mg intracellular pool L-methionine (in

38.5 g dry weight cells) per liter of culture under optimal conditions. However, it is unlikely that methionine production *per se* would warrant the costly toxicological testing necessary for an SCP product.

4.1.3. Aromatic Amino Acids (L-Tryptophan, L-Tyrosine, and L-Phenylalanine)

Erythrose-4-phosphate, a precursor in the biosynthesis of the aromatic amino acids, occupies a central position in the metabolism of C_1 substrates by the RuMP and XuMP pathways, offering the possibility that C_1 metabolism might present some advantage for aromatic amino acid overproduction. Suzuki and co-workers (1977) demonstrated that a mutant strain of *Methylomonas methanophila* resistant to the aromatic amino acid analog β-2-thienylalanine, 5-methyltryptophan, and 3-aminotyrosine, and therefore presumably resistant to feedback repression, accumulated 4, 0.2, and 1.1 g. liter^{-1} of L-phenylalanine, L-tryptophan, and L-tyrosine, respectively, in a methanol medium. Following the work of Terui (1972) on the conversion of anthranilic acid or indole to tryptophan by *Hansenula anomola*, Demain's group (Longin *et al.*, 1982) demonstrated tryptophan production at low levels from a mutant of the methylotrophic yeast *H. polymorpha*. More recently, Dijkhuizen and co-workers (Dijkhuizen *et al.*, 1985; de Boer *et al.*, 1987) have been studying the regulation of L-phenylalanine biosynthesis in a gram-positive facultative methylotroph (RuMP) *Nocardia* sp. 239 with a view to overproduction. The use of a gram-positive organism could give some advantage in improved excretion of the amino acid compared with the more commonly encountered gram-negative methylotrophs, while the use of a facultative organism will enable exploitation of methanol–glucose mixtures to counteract any imbalance in precursor requirements. Overproduction of L-phenylalanine, which is used primarily in the artificial sweetener Aspartame (Enei and Hirose, 1985), has been obtained with analog-resistant mutants, and attempts to increase yield through genetic manipulation are in progress.

4.1.4. Other Amino Acids

C_1 metabolism offers no obvious advantage for the overproduction of the other amino acids currently produced by fermentation or bioconversion, namely, L-lysine (Nakayama, 1985), L-aspartate (Chibata *et al.*, 1985), and L-glutamate (Hirose *et al.*, 1985). Nevertheless, the production of L-lysine by a methanol-grown *Protaminobacter* sp. has been patented by Kyowa Hakko Kogyo Co., and yields of glutamate up to 12 g. liter^{-1} have been claimed (Oki *et al.*, 1973; Nakayama *et al.*, 1975) for thiamin-requiring and auxotrophic strains of methylotrophs. In the latter case, it is foreseeable that continuing demand (worldwide production now exceeds 400,000 tonnes/year) and the decreasing unit price may warrant reinvestigation of C_1 route solely on the basis of substrate costs. The branched-chain amino acids valine and leucine have also been overproduced in modest amounts (Izumi et al., 1977) by an analog (valine hydroxamate) resistant mutant of *Methylomonas aminofaciens*.

4.2. Vitamins and Coenzymes

Vitamins and coenzymes occupy the higher price/lower volume end of the midrange biotechnology products. Thus the differences in growth substrate costs are generally secondary in importance to volumetric yield and processing costs. However, methylotrophs might be expected to be good sources of certain compounds in this class. Cytochrome c, for instance, the electron acceptor from MDH, is a potential product especially as at least some of the cytochrome is soluble within the periplasmic space and therefore could be released by osmotic shock; *M. methylotrophus* surprisingly releases approximately 40% of the soluble cytochrome c into the culture medium during growth on methanol (Cross and Anthony, 1980). In facultative methylotrophs this soluble cytochrome c is considerably enriched during growth on methanol compared to growth on heterotrophic substrates. Tani *et al.* (1985a) found that a cyanide-resistant mutant of the obligate methanol utilizer *Methylomonas* YK1 contained 3 times more cytochrome c than the parent strain and that a glycine-resistant mutant of this strain, AR 67, produced 103.5 mg. liter^{-1} (in 6.6 g. dry weight of cells) of intracellular cytochrome c in 40 hr under optimized growth conditions (Yoon *et al.*, 1987). However, as the current source of cytochrome c is from spent yeast or bovine/equine heart, neither of which incurs specific production costs, yields from methylotrophs would have to be outstanding to justify the capital investment needed to adopt this route.

By analogy, methanol-grown yeasts might be expected to be a ready source of FAD, the cofactor of methanol oxidase. FAD concentration in methanol-grown *Kloeckera* sp. 2207 were 3–5 times higher (Shimizu *et al.*, 1977a) than on other substrates (glucose, ethanol, or glycerol). However, most of this is enzyme-bound (noncovalently), and the problems of downstream processing for FAD recovery have not been addressed. An alternative strategy might be to use the coincident higher levels of FMN adenylyltransferase for the conversion of riboflavin or FMN to FAD (Shimizu *et al.*, 1977b).

The high oxygen demands for methylotrophic growth imply a high respiratory capacity and an increased level of electron transfer components, although this is not necessarily the case in practice. However, coenzyme Q_{10}, which like cytochrome *c* has applications in cardiotherapy, is sufficiently abundant in mutant strains of facultative methylotrophs (e.g., Natori and Nagasaki, 1981) to persuade Mitsubishi Gas Chemical Co. to develop a methanol-based route as a commercial process (Urakami *et al.*, 1981).

The production of vitamin B_{12} is another area where C_1-utilizing organisms may find application given the involvement of coenzyme B_{12} in C_1 transfer reactions. However, although many facultative methylotrophs do yield increased levels of vitamin B_{12} when grown on methanol, these are still disappointing, even after physiological optimization (Nishio *et al.*, 1977). A better source may be the methanol-utilizing methanogens, e.g., *Methanosarcina barkeri*, which, although slow growing, accumulated relatively large amounts of corrinoids (Mazumder *et al.*, 1986), including vitamin B_{12}, of which the majority was extracellular. Despite the low yields of methanogens on methanol, this may still be an attractive route.

4.3. Carboxylic Acids

The relatively recent discovery of acidogenic methylotrophic anaerobes has opened up the possibility of producing low-molecular-weight carboxylic acids, in particular acetic (Ljungdahl, 1983) and butyric acids, by fermentation from methanol. The low growth yields and high product yields of anaerobic bacteria, together with a favorable energy balance (the conversion is not highly exothermic), makes this look attractive as a process that is highly dependent on substrate costs. However, in the case of acetic acid production, the chemical (Monsanto) process, which also starts from methanol, presents formidable competition. Linton and Niekus (1987) estimated that, on the basis of product selling price, fermentation productivities need to be increased to 25–80 g. liter^{-1}hr.$^{-1}$, which implies developing efficient continuous extraction methods to avoid the problem of product toxicity. However, the economics of the process could be significantly improved if vitamin B_{12} was obtained as a by-product.

5. MICROBIAL POLYMERS FROM C_1 SUBSTRATES

The production of microbial polymers such as polysaccharide gums, while ostensibly a bulk process, cannot be considered in the same light as SCP with regard to economics, but falls into the category of "coincidental association". Decisions to adopt an SCP process based on cheap C_1 substrates were based on the reasonable assumption that with sufficient screening and physiological optimization, a product with the required nutritional properties (nucleic acid levels excepted) could be obtained. However, with the exception of the common intracellular storage compound poly-β-hydroxybutyrate, the discovery of useful, functional polymers is less predictable. Many methylotrophic bacteria, in particular the RuMP pathway organisms, have been shown to produce extracellular polysaccharides at reasonable yields (based on carbon substrate utilization) and rates. However, Linton and co-workers (1986) pointed out that yields of polysaccharide based on oxygen consumption (g product/g O_2) are likely to be 10 times lower than on glucose. Given that in viscous fermentations oxygen transfer is a major problem, this indicates that in the unlikely situation of a methylotroph and hexose utilizer producing an identical useful polymer, process considerations would favor the latter. Thus for an extracellular polymer to be produced in a C_1-based process, that polymer will need to have significant functional advantages over existing microbial polysaccharides.

5.1. Polysaccharides

Poly 54

Despite this gloomy prognosis, Celgene Corporation is currently marketing an exopolysaccharide (Poly 54 or Methylophilan) produced by continuous culture of *Methylophilus viscogenes,* which has a number of useful features (Table III). The polymer is comprised of glucose (46%), galactose (27%), mannose (9%), glucuronic acid (10%), and *O*-acetyl groups (8%) and is supplied in native (N) and deacetylated (D) forms. Envisaged uses are as viscosifiers (where excellent properties are claimed for Poly 54-D), suspension and emulsion stabilizers, water binders, gelling agents, and enhancers

TABLE III. Properties of Poly 54 (Methylophilan) from *Methylophilus viscogenes*

High apparent viscosity
Pseudoplastic
Variable thixotropy
Limited sensitivity to salt
Stable over a wide pH range (pH 1–11)
Superior suspending properties
Good biological stability
Heat stable (>130°C)
Synergy with other biopolymers
Cationic compatibility

of other water-soluble polymers. Poly 54-N displays an atypical temperature response in which the viscosity, starting at an intermediate level, actually increases irreversibly with an increase in temperature up to 80°C, at which point it is more viscous than xanthan gum. Furthermore, the high viscosity at 80°C is stably maintained for extended periods, in contrast to xanthan gum where the viscosity decreases with time at 80°C. This and its gelling properties with multivalent cations suggest potential applications in enhanced oil recovery (EOR) and in oil-drilling muds. Typical production runs are operated at a dilution rate of 0.03 hr^{-1} with a steady-state polysaccharide concentration of 18–20 g. $liter^{-1}$, representing 45% of the cell yield. Under these conditions oxygen demand is apparently not a problem.

5.2. Poly-β-hydroxybutyrate (PHB)

PHB, a polyester bacterial storage compound produced in relatively large amounts by some methylotrophic bacteria, forms the basis of the biodegradable, biocompatible thermoplastic Biopol developed by ICI plc and marketed through a subsidiary, Marlbrough Biopolymers (Byrom, 1987). In those bacteria which form it, it usually accumulates under conditions of nitrogen or oxygen limitation. Asenjo and Suk (1986) and Suzuki et al. (1986) confirmed that nitrogen limitation was the key to maximizing PHB yields in methylotrophs, although Suzuki et al. (1986) showed that PHB production from methanol in their selected strain *Pseudomonas* sp. K was also enhanced by sulfate, magnesium, iron, or manganese deficiencies. In a computer-controlled, fed-batch fermentation, they achieved PHB yields of 136 g. $liter^{-1}$ corresponding to 66% of the cell dry weight. Asenjo and Suk (1986) also obtained high yields (as a percentage of cell dry weight) from methane with the type II methane utilizer *Methylocystis parvus* OBBP. However, large-scale production of PHB from methane is unattractive for the same reasons as SCP from methane. PHB production from methanol, while a more attractive proposition, has the same problems as exopolysaccharide production, with the yield from oxygen being lower than with glucose-grown cells. Additionally, it is reported (Byrom, 1987) that although production of PHB from methanol was examined by ICI, difficulties were encountered in extracting the polymer from methylotrophs, and the resulting material was of low molecular weight, restricting its application. Hence Biopol is currently produced using *Alcaligenes eutrophs* in fed-batch culture with glucose as the main carbon source. Interestingly, the strategy adopted employs phosphate rather than nitrogen limitation, and concentrations of up to 75% of the cell dry weight have been achieved.

Poly-β-hydroxybutyrate is, in fact, the major representative of a family of poly-β-hydroxyalkanoates that can be accumulated by bacteria. Copolymers comprised of β-hydroxybutyrate and β-hydroxyvalerate monomers, which are tougher and more flexible than PHB, can be obtained by supplementing the culture with low concentrations of propionate (Byrom, 1987). The β-hydroxyvalerate content of the polymer and hence its mechanical properties can be manipulated by varying the ratio of glucose to propionate. Given the generally competitive carbon substrate economics for methanol-based PHB and relative ease of polymer composition manipulation, it appears worthwhile continuing investigations of methylotrophs for the production of novel copolymers.

6. METABOLIC ACTIVITIES AND ENZYMES FROM METHYLOTROPHS

In addition to offering possibilities for the production of metabolites from a cheap substrate, methylotrophic bacteria and yeasts may be a valuable source of enzymes. Their applications may extend from the use of purified enzymes in analytical devices to the exploration of metabolic sequences in whole organisms for biotransformations or biodegradation. Ironically, in the case of single enzymes, or even short metabolic sequences, genetic engineering offers the possibility of high-level expression in heterologous hosts (as witnessed with STHM) and would thus remove exploitation from the bounds of C_1 biotechnology. For this reason alone, it is likely that the exploitation of enzyme activities in methylotrophs will concentrate on enzymes peculiar to C_1 metabolism and in particular the catabolic enzymes, either because of their unique metabolic activities/specificities or simply because they can utilize cheap substrates.

6.1. Cofactor Regeneration

In the latter context, Tani and co-workers (1982) have shown that cell extracts from *Candida boidinii* no. 2201 containing intact mitochondria, peroxisomes, and cytoplasmic enzymes produce ATP from AMP or, more economically, from adenine and adenosine by the oxidation of methanol to CO_2 coupled, via the reduction cycle of $NAD^+/NADH$, to oxidative phosphorylation. By using sorbitol-plasmolyzed cells (Tani *et al.*, 1984; Yonehara and Tani, 1988), activity was retained for up to 36hr, with a reported yield of 30 g. $liter^{-1}$. More recently, they have coupled the same system of NADH production with the reduction of sugars by sheep liver D-sorbitol (19.1 g. $liter^{-1}$) from fructose, and xylitol (48.5 g. $liter^{-1}$) from xylose. NADH and NADPH regeneration has also been demonstrated with methylotrophic bacteria. In the former case, freeze-thawed, air-dried, or acetone-dried cells of the facultative methylotroph *Arthrobacter* sp. reduced NAD^+ to NADH in the presence of formate (Izumi *et al.*, 1983) with 90% conversion (30 g. liter^{-1} NADH). Sixty percent conversion of $NADP^+$ to NADPH has also been reported (Eguchi *et al.*, 1983) using formate dehydrogenase in resting cells of a methanogenic bacterium permeabilized with 0.2% Triton X-100.

6.2. Analysis and Biosensors

In principle, any enzyme can constitute an analytical device if the course of its reaction can be followed, either directly or indirectly by a coupled analysis of the reaction products. The attractions of enzyme-based analyses compared to chemical methods lie primarily in the specificity of enzymes allowing detection of an analyte in a complex mixture without the need for prior separation. With the recent developments in biosensor technology in which the biological reaction can be coupled to the generation of an electrical signal, simplicity of analysis and portability of the biosensor might also be cited in favor of enzyme analysis.

In the case of purified enzymes, the costs of purification are likely to predominate over fermentation substrate costs such that enzymes not unique to C_1 metabolism are only likely to be derived from a methylotroph if they are particularly abundant, e.g., catalase from methylotrophic yeasts. Hence the considerations primarily revolve around enzymes unique to C_1 metabolism. Ironically, the wide substrate range of a number of methylotrophic enzymes, which makes them interesting as biocatalysts, makes them less useful for analytical purposes. The complexity and instability of methane monooxygenase (MMO) restrict its use to whole cells, but the high oxygen demand for methane oxidation makes methanotrophs good candidates for a whole-cell methane sensor linked to measurements of oxygen uptake. Karube *et al.* (1982) developed such a sensor using *Methylomonas flagellata* immobilized on acetylcellulose filters. The detection limit was approximately 5 mM and the sensor was stable for more than 10 days. However, the sensor will also respond to gaseous higher alkanes/alkenes.

6.2.1. Alcohol Oxidase and Methanol Dehydrogenase

Sensors for the determination of ethanol have a potentially huge market, and both alcohol oxidase (AOX) from methylotrophic yeasts and methanol dehydrogenase (MDH) from bacteria have been studied in this respect. Both enzymes have the problem that they will oxidize a range of primary alcohols, although the alcohol dehydrogenase from *Rhodopseudomonas acidophila* has a much higher affinity for ethanol than other primary alcohols (Bamforth and Quayle, 1978). However, MDH- or AOX-based sensors may be useful where no other primary alcohols are present. As with the methane sensor, AOX-based sensors generally rely on the measurement of oxygen uptake (Verduyn et al., 1983) in the coupled reaction of AOX and catalase (equations 1 and 2) and examples have been marketed commercially.

$$2CH_3CH_2OH + 2O_2 \xrightarrow{AOX} 2CH_3CHO + 2H_2O_2 \quad (1)$$

$$2H_2O_2 \xrightarrow{Catalase} 2H_2O + O_2 \quad (2)$$

MDH-based sensors have exploited the ability of MDH to reduce artificial electron acceptors which can subsequently be reoxidized at an electrode, thus acting as redox mediators. Such a sensor, which used 1,1′ dimethylferrocene as electron acceptor (Higgins et al., 1984), was shown to have a rapid response time (20 sec to steady state) and a useful range of 1–100 μM methanol.

6.2.2. Formate Dehydrogenase

Unlike the enzyme involved in anaerobic formate oxidation, formate dehydrogenase (FDH) from aerobic methylotrophs is generally oxygen stable. The enzyme from *C. boidinii* is commercially available and could be used in the analysis of formate. However, a more likely application is in the *in situ* regeneration of NADH, where the volatility of the end-product CO_2 is a particular asset.

6.3. Methylotrophs as Biocatalysts. 1. Synthesis

The potential of methane and methanol utilizers for the biotransformation of nongrowth substrates to more valuable products was originally noted by Foster (1962) but was not extensively investigated until the mid-1970s. More recently, the same catalytic activities have come under the spotlight for their potential in the biodegradation of toxic organic wastes.

6.3.1. Methane Monooxygenase

In synthetic chemistry the principal features that make a biological transformation attractive are the specificity (regio- and stereospecificity) of enzymes and the availability of reactions with no direct chemical counterpart. Monooxygenases, which catalyze the insertion of a single atom of dioxygen into the substrate, fall into the latter category, and their commercial importance is best exemplified by their key role in steroid transformations (Kieslich and Sebek, 1979). Although alkane monooxygenases from organisms grown on intermediate and long-chain alkanes had previously been shown to hydroxylate a range of alkanes and form epoxides from terminal alkenes, such as octene (Cardini and Jurtshuk, 1970; van Ravenswaay Claasen and van der Linden, 1971), they generally showed little activity toward the lower-molecular-weight substrates. Thus the feature that aroused commercial interest in MMO was its ability to hydroxylate and epoxidize a range of low-molecular-weight alkanes and alkenes (Colby *et al.*, 1977). At various times it has been suggested that MMO could be used for the conversion of alkanes to alcohols, including the conversion of methane to methanol (Ghisalba and Heinzer, 1982), reactions that are difficult to achieve chemically due to the further oxidation of the products. In the case of methanol production, the proposal is probably fanciful given the current low price of methanol and the inherent problems of further oxidation of the product by MDH (MMO is too complex and unstable to use other than in whole cells), removing the methanol sufficiently quickly to prevent further metabolism by MMO for which it is also a substrate (Colby *et al.*, 1977) and providing a source of reducing equivalents to drive the MMO. Examination of the substrate specificity of the soluble MMO from *Methylococcus capsulatus* or *Methylosinus trichosporium* reveals that, unlike many monooxygenases from long-chain alkane utilizers, it is essentially a subterminal and terminal monooxygenase (Dalton, 1980); see Table IV. Thus with alkanes longer than ethane, a mixture of subterminal and terminal alcohols is produced, with the former predominating owing to the greater reactivity of the subterminal position. This is, of course, an unattractive feature in any consideration of alcohol production using MMO, where high conversion rates to a single product are desirable.

This characteristic of subterminal hydroxylation probably accounts for the ability of MMO to monohydroxylate aromatic rings (Dalton *et al.*, 1981; Jezequel and Higgins, 1983), unlike many of the terminal alkane monooxygenases. However, exploitation of this potentially valuable attribute has again been limited by the observation of nonspecific oxidation of ring alkane and heteroatom substituents (Colby *et al.*, 1977; Dalton *et al.*, 1981).

TABLE IV. Selectivity of Oxygen Insertion into Linear Alkanes and Alkenes Observed with Soluble Methane Monooxygenase Derived from (1) *Methylococcus capsulatus* **(Bath), (2)** *Methylosinus trichosporium* **OB3b, and (3)** *Methylobacterium* **sp. CRL-26**

Substrate	Products	a (%)	b (%)	c (%)
Propane	Propan-1-ol	39	41	40
	Propan-2-ol	61	59	60
Butane	Butan-1-ol	55	59	63
	Butan-2-ol	45	41	37
Pentane	Pentan-1-ol	28	21	32
	Pentan-2-ol	72	79	68
Hexane	Hexan-1-ol	62	70	65
	Hexan-2-ol	38	30	35
Heptane	Heptan-1-ol	22	93	19
	Heptan-2-ol	78	7	81
trans-2-Butene	*trans*-2,3-epoxybutane	60	16	58
	trans-But-2-enol	40	84	42
cis-2-Butene	*cis*-2,3-epoxybutane	44	37	60
	cis-2,3-But-2-enol	41	44	40
	Butanone	15	19	—

From Dalton (1980) and Patel *et al.* (1982).

The biotransformation for which MMO (soluble and membrane associated) has received the most attention is the conversion of propene to propene oxide. Although racemic propene oxide, the form made by MMO (Weijers *et al.*, 1988), is a commodity, high-volume chemical and therefore, according to perceived wisdom, unsuitable for a biotransformation process, there are a number of favorable indications for a biological process. First, the chemical synthesis is multistep and the price of propene oxide is partly dependent on by-product sales (Hou, 1984). Second, this is one example of an MMO biotransformation where only a single product is formed from a substrate that is even better than methane when measured in terms of K_m (Green and Dalton, 1986). Finally, the product, which is not further metabolized, has a boiling point of 34°C and with thermotolerant organisms such as *M. capsulatus* (Bath) should be removed readily with the gas stream. In short-time-course assays in enclosed reactors, resting cells of a number of methanotrophs will produce propene oxide at rates of 50–100 nmoles.min.$^{-1}$ mg cell dry weight^{-1}. However, these rates generally fall fairly rapidly owing to toxicity of the product (usually at concentrations greater than 5 mM) or exhaustion of the endogenous supply of reductant. To have any chance of success, it has been estimated that bioconversion rates need to be increased 10- to 100-fold (i.e., 1–5 µmoles. min.$^{-1}$ mg cell dry weight^{-1}) and catalytic lifetimes need to be extended substantially. Hou (1984) at Exxon Research Laboratories and Dalton and co-workers have invested considerable effort into improving the process, but although catalytic rates in excess of 1 μmole.min.$^{-1}$ mg cell dry weight^{-1} have been obtained, these cannot be maintained for any length of time. Improvements have been made by optimizing growth physiology to maximize expression of the membrane-bound MMO, provision of a regulated supply of exogenous reductant, e.g., methanol or formate, and optimizing bioconversion conditions including product removal. The use of formate as an exogenous source of reducing equivalents suppresses the additional oxygen demand incurred by methanol oxidation via MDH, although the ATP supplied from methanol oxidation may perform a role in cell maintenance and potentially increase catalytic lifetime. However, a process operating in conventional stirred tank reactors faces limitations due to the conflicting requirements for product removal, favored by high temperatures, and efficient gas (propene and oxygen) transfer favored by lower temperatures.

Hou (1984) carried out some preliminary investigations of a gas–solid bioreactor in which the cells were physically absorbed onto the surface of glass beads packed into a column that was operated at 40°C. A 1:1 (v/v) mixture of propene and oxygen was introduced into the bottom of the column after passing through a water bottle to maintain approximately 70% humidity, and propene oxide was condensed from the outlet stream. Propene oxide production was apparently linear for 7hr at a rate of 18 mmoles. hr^{-1}.mg protein^{-1} (~9 mmoles.hr^{-1}.mg cell dry weight^{-1}), after which activity decreased. However, relatively high levels of activity could be regenerated by passing methanol vapor through the column for 30 min. Despite these improvements, there are still major technical hurdles to be overcome to make the process competitive.

6.3.2. Methanol Oxidase

The abundance of alcohol oxidase in methylotrophic yeasts and relative ease of purification make it into an enzyme searching for an application. Indeed, Provesta Corporation make no secret of their willingness to collaborate in the exploitation of this enzyme, which they can produce in large amounts via the Provesteen process. Couderc and Baratti (1980) and Tani and co-workers (Tani *et al.*, 1985b; Sakai and Tani, 1986) have demonstrated that whole yeast cells or isolated enzyme can be used to produce formaldehyde at conversion efficiencies of up to 70% and with remarkably high yields given the toxicity of formaldehyde. Kierstan (1982) suggested that it could be used to facilitate ethanol removal from fermentation broths by conversion to the more volatile acetaldehyde, and Mizumo and Imada (1986) described a system for the complete conversion of methanol to formic acid without requirement for cofactor by combining alcohol oxidase and catalase both from *H. polymorpha* with formaldehyde dismutase from *Pseudomonas putida*. Additionally, Unilever (1987) have patented a process to produce hydrogen peroxide as a bleaching agent from methanol using alcohol oxidase. However, at present none of these applications are close to commercialization.

6.3.3. Other Synthetic Biotransformations

Patel and co-workers (1980) proposed that the secondary alcohol dehydrogenases found in some methylotrophs could be exploited for the production of methyl ketones, possibly in a two-step reaction involving prior oxidation of the corresponding alkane. However, as pointed out earlier, MMO produces a mixture of terminal and subterminal alcohols from C_3 and higher alkanes. Thus a proportion of the substrate would be lost as the primary alcohol. The alternative suggestion of simply converting secondary alcohols to ketones (Hou *et al.*, 1979, Patel *et al.*, 1979; Huang *et al.*, 1985) is unlikely to be viable in its own right as it competes with a single-step chemical process. However, the use of secondary alcohol oxidation to supply reducing equivalents to the MMO, thus producing a ketone and, for example, an epoxide (Hou, 1984), is an interesting possibility, reducing the ketone to the level of a by-product.

6.4. Methylotrophs as Biocatalysts. 2. Biodegradation

A number of methylated compounds, e.g., methyl sulfides, methyl sulfates, and methylamines, are produced in industrial processes and constitute potential environmental contaminants. Methylotrophs capable of degrading those compounds by releasing the C_1 substituent for use as a carbon source are frequently isolated in enrichment culture with these contaminants as substrates (Ghisalba and Kuenzi, 1983a,b) and could form the basis of a waste treatment process, although this may not utilize methane or methanol and would thus fall outside the subject of this chapter. Methanol would, however, comprise the substrate if the proposed denitrification process employing Hyphomicrobia was adopted (Payne, 1981). These bacteria will oxidize methanol anaerobically using nitrate as the electron acceptor, converting the nitrate ultimately to N_2.

M. methylotrophus could also find a role in the increasingly regulated area of waste treatment as a catalyst for the degradation of the neurotoxin acrylamide (M. A. Carver, personal communication). This employs the aliphatic amidase activity of *M. methylotrophus*, which allows it to use short-chain aliphatic amides, including acrylamides, as a nitrogen source. As these amides only function as a nitrogen and not a carbon source, expression of the amidase is not repressed by organic acids or methanol. Hence unlike the situation with many other aliphatic amidases, catalytically active cells can readily be produced by growth on methanol with acetamide as the nitrogen source. The catalytic activity toward acrylamide has recently been improved by subjecting *M. methylotrophus* to a regime of "directed evolution" in continuous culture with selection for improved ability to use longer-chain aliphatic amides (Silman *et al.*, 1989).

Methanotrophic bacteria may also find a niche in the treatment of potential organic pollutants. This would exploit the broad substrate range of the soluble MMO to activate the substrate either to an unstable form that spontaneously degrades or alternatively to a form that can be degraded by other organisms, probably as a mixed culture. A number of reports have appeared recently describing the ability of methanotrophs to degrade trichloroethylene and other low-molecular-weight halogenated hydrocarbons (Mayer *et al.*, 1988; Moore *et al.*, 1989; Little *et al.*, 1988; Oldenhuis *et al.*, 1989; see also Chapter 1, Section 4) in a process catalyzed by the soluble MMO, which involves oxygen insertion followed by spontaneous chloride elimination. To be operated in a waste treatment process, catalytically active cells would have to be supplied with a continuous source of reductant such as methanol or formate.

7. THE FUTURE OUTLOOK

The driving force for C_1-based biotechnology has been, and will continue to be, the low cost of methane and methanol as potential fermentation feedstocks, and hence the primary players have been the major producers. On the basis of substrate costs, the "obvious" benefits from C_1 biotechnology are in the production of high-volume, commodity products (Linton and Niekus, 1987). However, with the exception of carboxylic acid production by methylotrophic anaerobes, which are still a subject of investigation, it is fair to surmise that the "obvious" applications have been tried, tested, and shelved for the reasons outlined. Some useful lessons have been learned and a lot of money has been spent along the way.

In the short term, those products which do reach the market on the back of C_1 biotechnology are more likely to be those coincidentally associated with C_1 metabolism or complex products (e.g., vitamin B_{12}) that have some specific association with C_1 metabolism. Methylotrophic yeasts could have an interesting future as eukaryotic expression systems, but as pointed out, they will most likely not be grown on methanol! There is also the possibility that a methylotrophic enzyme/organism could find a niche in a viable biotransformation, but this would depend more on the reaction specificity and the value of the transformation than any predictable feature of methylotrophs.

The long term could well be dictated by the current midrange biotechnology products, at least in terms of volume. As global demand and new applications for amino acids increase, the influence of substrate costs on increasingly large-scale fermentations will become more important. Unlike the case of SCP, where competition comes from a source with relatively independent economic determinants, the long-term economics for competitive fermentation processes should be easier to predict. In contrast to the current situation, it is worth noting that in this context, the main consumers of methanol are unlikely to be the major producers. Finally, we should not forget that genetic engineering is daily increasing our ability to mix and match enzymes and activities from different organisms. The activity at which aerobic methylotrophs really excel is the oxidation of methanol, thus releasing a considerable amount of reducing power (NADH). With the end-product of this oxidation being a volatile gas, the process should be able to continue indefinitely without end-product inhibition, assuming that the NADH is reoxidized. Even if MMO does not prove to be a useful biocatalyst, methylotrophs should be an ideal host for conducting other NADH-requiring biotransformations.

ACKNOWLEDGMENTS. I am grateful to Mark Carver, David Stirling, Yoshiki Tani, and Gene Wegner for communicating some of the material included in this chapter, and to Rosalind Chan for helping in the preparation of the manuscript.

REFERENCES

Ackrell, B. A. C., and Jones, C. W., 1971, The respiratory system of *Azotobacter vinelandii* 2. Oxygen effects, *Eur. J. Biochem.* **20**:29–35.

Anthony, C., 1982, *The Biochemistry of Methylotrophs*, Academic Press, London, pp. 251–260.

Asenjo, J. A., and Suk, S. S., 1986, Microbial conversion of methane into poly-beta-hydroxybutyrate (PHB): growth and intracellular accumulation in a type II methanotroph, *J. Ferment. Technol.* **64**:271–278.

Bamforth, C. W., and Quayle, J. R., 1978, The dye linked alcohol dehydrogenase of *Rhodopseudomonas acidophila*. Comparison with dye-linked methanol dehydrogenase, *Biochem. J.* **169**:677–686.

Beardsmore, A. J., Aperghis, P. N., and Quayle, J. R., 1982a, Characterization of the assimilatory and dissimilatory pathways of carbon metabolism during growth of *Methylophilus methylotrophus* on methanol, *J. Gen. Microbiol.* **128**:1423–1439.

Beardsmore, A. J., Collins, S. H., Powell, K. A., and Senior, P. J., 1982b, EPA 82,302,608.3.

Behrendt, U., Bang, W. G., and Wagner, F., 1984, The production of L-serine with a methylotrophic microorganism using the L-serine pathway and coupling with an L-tryptophan–producing process, *Biotechnol. Bioeng.* **26**:308–314.

de Boer, L., Vrijbloed, W., Van Rijssel, M., and Dijkhuizen, L., 1987, Regulation of phenylalanine metabolism in the facultative methylotroph *Nocardia* sp. 239 growing on a methanol culture medium, *Eur. Congr. Biotechnol.* **3**:466.

Byrom, D., 1984, Host-vector systems for *Methylophilus methylotrophus*, in: *Microbial Growth on C_1 Compounds. Proceedings of the 4th International Symposium* (R. L. Crawford and R. S. Hanson, eds.), American Society for Microbiology, Washington, DC, pp. 221–223.

Byrom, D., 1987, Polymer synthesis by micro-organisms: technology and economics, *Trends Biotechnol.* **5**:246–250.

Cardini, G., and Jurtshuk, P., 1970, Cytochrome P-450 involvement in the oxidation of n-octane by cell free extracts of *Corynebacterium* sp. strain 7E1C, *J. Biol. Chem.* **245**:2789–2796.

Chibata, I., Tosa, T., and Sato, T., 1985, Aspartic acid, in: *Comprehensive Biotechnology*, Vol. 3 (M. Moo-Young, ed.), Pergamon Press, Oxford, pp. 633–640.

Colby, J., Stirling, D. I., and Dalton, H., 1977, The soluble methane monooxygenase of *Methylococcus capsulatus* (Bath). Its ability to oxygenate n-alkanes, n-alkenes, ethers and alicyclic, aromatic and heterocyclic compounds, *Biochem, J.* **165**:395–402.

Couderc, R., and Baratti, J., 1980, Immobilized yeast cells with methanol oxidase activity: preparation and enzymatic properties, *Biotechnol. Bioeng.* **22**:1155–1173.

Cregg, J. M., Tschopp, J. F., Stillman, C., Siegel, R., Akong, M., Craig, W. S., Buckholz, R. G., Madden, K. R., Kellaris, P. A., Davis, G. R., Smiley, B. L., Cruze, J., Torregossa, R., Velicelebi, G., and Thill, G. P., 1987, High level expression and efficient assembly of hepatitis B surface antigen in the methylotrophic yeast, *Pichia pastoris*, *Bio/technology* **5**:479–485.

Cross, A. R., and Anthony, C., 1980, The electron-transport chains of the obligate methylotroph *Methylophilus methylotrophus*, *Biochem. J.* **192**:429–439.

Dalton, H., 1980, Oxidation of hydrocarbons by methane monooxygenase from a variety of microbes, *Adv. App. Microbiol.* **26**:71–87.

Dalton, H., Golding, B. J., Waters, B. W., Higgins, R., and Taylor, J. A., 1981, Oxidations of cyclopropane, methyl cyclopropane and arenes with the monooxygenase system from *Methylococcus capsulatus*, *J. Chem. Soc. Chem. Commun.* **189**:482–483.

de Vries, G. E., Kues, U., and Stahl, U., 1990, Physiology and genetics of methylotrophic bacteria, *FEMS Microbiol. Rev.* **75**:57–102.

Dijkhuizen, L., Hansen, T. A., and Harder, W., 1985, Methanol: a potential feedstock for biotechnological processes, *Trends Biotechnol.* **3**:262–267.

Drozd, J. W., Linton, J. D., Downs, J., and Stephenson, R. J., 1978, An *in situ* assessment of the specific lysis rate in continuous cultures of *Methylococcus* sp. (NCIB 11083) grown on methane, *FEMS Microbiol. Lett.* **4**:311–314.

Eguchi, S. Y., Nishio, N., and Nagai, S., 1983, NADPH production from $NADP^+$ by a formate-utilizing methanogenic bacterium, *Agric. Biol. Chem.* **47**:2941–2943.

Enei, H., and Hirose, Y., 1985, Phenylalanine, in: *Comprehensive Biotechnology*, Vol. 3 (M. Moo-Young, ed.), Pergamon Press, Oxford, pp. 601–605.

Faust, U., 1979, Process results from SCP-pilot plant based on methanol in: *Microbiology Applied to Biotechnology*, Proceedings of 12th International Congress of Microbiology, Verlag Chemie, Weinheim, pp. 125–133.

Foster, J. W., 1962, Hydrocarbons as substrates for microorganisms, *Ant. van Leeuw. J. Microbiol. Serol.* **28**:241–274.

Ghisalba, O., and Heinzer, F., 1982, Methanol from methane–a hypothetical microbial conversion compared to the chemical process, *Experientia* **38**:218–223.

Ghisalba, O., and Kuenzi, M., 1983a, Biodegradation of monomethyl sulfate by specialized methylotrophs, *Experientia* **39**:1257–1263.

Ghisalba, O., and Kuenzi, M., 1983b, Biodegradation and utilization of quaternary alkylammonium compounds by specialized methylotrophs, *Experientia* **39**:1264–1271.

Goldberg, I., 1985, *Single Cell Protein*, Springer Verlag, Berlin.

Green, J., and Dalton, H., 1986, Steady-state kinetic analysis of soluble methane monooxygenase from *Methylococcus capsulatus* (Bath) *Biochem. J.* **236**:155–162.

Grinna, L. S., and Tschopp, J. F., 1989, Size distribution and general structural features of N-linked oligosaccharides from the methylotrophic yeast, *Pichia pastoris*, *Yeast* **5**:107–115.

Hacking, A. J., 1986, *Economic Aspects of Biotechnology*, Cambridge University Press, Cambridge.

Hamilton, B. K., Hsiao, H-Y., Swann, W. E., Anderson, M., and Delente, J., 1985, Manufacture of L-amino acids with bioreactors, *Trends Biotechnol.* **3**:64–68.

Hardy, G. A., and Dawes, E. A., 1985, Effect of oxygen concentration on the growth and respiratory efficiency of *Acinetobacter calcoaceticus*, *J. Gen. Microbiol.* **131**:855–864.

Heijnen, J. J., and Roels, J. A., 1981, A macroscopic model describing yield and maintenance relationships in aerobic fermentation processes, *Biotechnol. Bioeng.* **23**:739–763.

Hennam, J. F., Cunningham, A. E., Sharp, G. S., and Atherton, K. T., 1982, Expression of eukaryotic coding sequences in *Methylophilus methylotrophus*, *Nature (Lond.)* **297**:80–82.

Higgins, I. J., Aston, W. J., Best, D. J., Turner, A. P. F., Jezequel, S. G., and Hill, H. A. O., 1984, Applied aspects of methylotrophy: bioelectrochemical applications, purification of methanol dehydrogenase and mechanism of methane mono-oxygenase, in: *Microbial Grown on C_1 Compounds*, Proceedings of 4th International Symposium (R. L. Crawford and R. S. Hanson, eds.), American Society for Microbiology, Washington, DC, pp. 297–305.

Hirose, Y., Enei, H., and Shibori, H., 1985, L-Glutamic acid fermentation, in: *Comprehensive Biotechnology*, Vol. 3 (M. Moo-Young, ed.), Pergamon Press, Oxford, pp. 593–600.

Hou, C. T., 1984, Other applied aspects of methylotrophs, in: *Methylotrophs: Microbiology, Biochemistry and Genetics* (C. T. Hou, ed.), CRC Press, Boca Raton, FL, pp. 145–166.

Hou, C. T., Patel, R. N., Laskin, A. I., Barnabe, N., and Marczak, I., 1979, Microbial oxidation of gaseous hydrocarbons: production of methyl ketones from their corresponding secondary alcohols by methane and methanol-grown microbes, *Appl. Env. Microbiol.* **38**:135–142.

Hsiao, H. Y., Wei, T., and Campbell, K., 1986, Enzymatic production of L-serine, *Biotechnol. Bioeng.* **28**:857–867.

Huang, T. L., Fang, B. S., and Fang, H. Y., 1985, Oxidation of secondary alcohols to methylketones by immobilized yeast cells, *J. Gen. Appl. Microbiol. (Tokyo)* **31**:125–134.

Izumi, Y., Asana, Y., Tani, Y., and Ogata, K., 1977, Mutants of an obligate methylotroph, formation of valine and leucine by analog-resistant *Methylomonas aminofaciens*, *J. Ferment. Technol.* **55**:452–458.

Izumi, Y., Takizawa, M., Tani, Y., and Yamada, H., 1982, L-Serine production by resting cells of a methanol-utilizing bacterium, *J. Ferment. Technol.* **60**:269–276.

Izumi, Y., Mishra, S. K., Ghosh, B. S., Tani, Y., and Yamada, H., 1983, NADH production from NAD^+ using a formate dehydrogenase system with cells of a methanol-utilizing bacterium, *J. Ferment. Technol.* **61**:135–142.

Jezequel, S. G., and Higgins, I. J., 1983, Mechanistic aspects of biotransformations by the monooxygenase system of *Methylosinus trichosporium* OB3b, *J. Chem. Tech. Biotechnol.* **33B**:139–144.

Karube, I., Okada, T., and Suzuki, S., 1982, A methane gas sensor based on methane-oxidising bacteria, *Anal. Chim. Acta* **135**:61–67.

Keune, H., Sahm, H., and Wagner, F., 1976, Production of L-serine by the methanol-utilizing bacterium *Pseudomonas* 3ab, *Eur. J. Appl. Microbiol. Biotechnol.* **2**:175–184.

Kierstan, M., 1982, The enzymatic conversion of ethanol to acetaldehyde as a model recovery system, *Biotechnol. Bioeng.* **24**:2275–2277.

Kieslich, K., and Sebek, O. K., 1979, Microbial transformations of steroids, in: *Annual Reports on Fermentation Processes*, Vol. 3 (D. Perlman, ed.), Academic Press, New York, pp. 275–304.

Kubota, K., Kageyama, K., Maeyashiki, I., Yamada, K., and Okumura, S., 1972, Fermentative production of L-serine, production of L-serine from glycine by *Corynebacterium glycinophilum* nov. sp., *J. Gen. Appl. Microbiol.* **18**:365–375.

Kuraishi, M., Tareo, H., Ohkouchi, N., Matsuda, N., and Nagai, I., 1979, SCP process development with methanol as substrate, in: *Microbiology Applied to Biotechnology*, Proceedings of 12th International Congress of Microbiology, Verlag Chemie, Weinheim, pp. 111–124.

Kvenvolden, K. A., 1988, Methane hydrate. A major reservoir of carbon in the shallow geosphere? *Chem. Geol.* **71**:41–51.

Leak, D. J., and Dalton, H., 1986, Growth yields of methanotrophs. 1. Effect of copper on the energetics of methane oxidation, *Appl. Microbiol. Biotechnol.* **23**:470–476.

Lim, W. J., and Tani, Y., 1988, Production of L-methionine-enriched cells of a mutant derived from a methylotrophic yeast, *Candida boidinii*, *J. Ferment. Technol.* **66**(6):643–647.

Linton, J. D., and Niekus, H. G. D., 1987, The potential of one-carbon compounds as fermentation feedstocks, in: *Microbial Growth on C_1 Compounds*, Proceedings of 5th International Symposium (H. W. van Verseveld and J. A. Duine, eds.), Martinus Nijhoff, Dordrecht, pp. 263–271.

Linton, J. D., Watts, P. D., Austin, R. M., Haugh, D. E., and Niekus, H. G. D., 1986, The energetics and kinetics of extracellular polysaccharide production from micro-organisms possessing different pathways of C1 assimilation, *J. Gen. Microbiol.* **132**:779–788.

Little, C. D., Palumbo, A. V., Herbes, S. E., Lidstrom, M. E., Tyndall, R. L., and Gialmer, P. J., 1988, Trichloroethylene biodegradation by a methane-oxidizing bacterium, *Appl. Environ. Microbiol.* **54**:951–956.

Ljungdahl, L. G., 1983, Formation of acetate using homo-acetate–fermenting anaerobic bacteria, in: *Organic Chemicals from Biomass* (D. L. Wise, ed.), Benjamin/Cummings, Menlo Park, CA, pp. 219–248.

Longin, R., Cooney, C. L., and Demain, A. L., 1982, Studies in the overproduction of indole-containing metabolites by a methanol-utilizing yeast, Hansenula polymorpha, *Appl. Biochem. Biotechnol.* **7**:281–293.

de Maeyer, E., Skup, D., Prasad, K. S. N., de Maeyer-Guignard, J., Williams, B., Meacock, P., Sharp, G., Pioli, D., Hennam, J., Schuch, W., and Atherton, K., 1982, Expression of a chemically synthesized human alpha-1-interferon gene, *Proc. Natl. Acad. Sci. USA* **79**:4256–4259.

Mayer, K, P., Grbic-Galic, D., Semprini, L., and McCarty, P. L., 1988, Degradation of trichloroethylene by methanotrophic bacteria in a laboratory column of saturated aquifer material, *Water Sci. Technol.* **20**:175–178.

Mazumder, T. K., Nishio, N., Hayaishi, M., and Nagai, S., 1986, Production of corrinoids including vitamin B_{12} by *Methanosarcina barkeri* growing on methanol, *Biotechnol. Lett.* **8**:843–848.

McNairney, J., 1984, Modification of a novel protein product, *J. Chem. Tech. Biotechnol.* **34B**:206–214.

Mizumo, S., and Imada, Y., 1986, Conversion of methanol to formic acid through the coupling of the enzyme reactions of alcohol oxidase, catalase and the formaldehyde dismutase, *Biotechnol. Lett.* **8:**79–84.

Mogren, H., 1979, SCP from methanol–the Norprotein process, *Process Biochem.* **14**(3):2–7.

Moore, A. T., Vira, A., and Fogel, S., 1989, Biodegradation of *trans*-1,2-dichloroethylene by methane-utilizing bacteria in a aquifer simulator, *Environ. Sci. Technol.* **23:**403–406.

Morinaga, Y., Yamanaka, S., and Takimani, K., 1981a, L-Serine production by methanol-utilizing bacterium *Pseudomonas* MS31, *Agric. Biol. Chem.* **45:**1419–1424.

Morinaga, Y., Yamanaka, S., and Takimani, K., 1981b, L-Serine production by temperature-sensitive mutants of methanol-utilizing bacterium *Pseudomonas* MS31, *Agric. Biol. Chem.* **45:**1425–1430.

Morinaga, Y., Tani, Y., and Yamada, H., 1982, L-Methionine production by ethionine-resistant mutants of facultative methylotroph, *Pseudomonas* FM18, *Agric. Biol. Chem.* **46:**473–480.

Morinaga, Y., Yamanaka, S., and Takimani, K., 1983, L-Serine production improved by analogue resistant mutants of a methanol-utilizing bacterium, *Agric. Biol. Chem.* **47:**2113–2114.

Morinaga, Y., Tani, Y., and Yamada, H., 1984, Homocysteine transmethylation in methanol-utilizing bacteria and its application to L-methionine production, *Agric. Biol. Chem.* **48:**143–148.

Nakayama, K., 1985, Lysine, in: *Comprehensive Biotechnology*, Vol. 3 (M. Moo-Young, ed.), Pergamon Press, Oxford, pp. 607–620.

Nakayama, K., Kobata, M., Tanaka, Y., Nomura, T., and Katsumata, R., 1975, Biological preparation of L-glutamic acid, *Ger. Offen.* **2:**458,206.

Natori, Y., and Nagasaki, T., 1981, Enhancement of coenzyme Q_{10} accumulation by mutation and effects of medium components on the formation of coenzyme Q homologs by *Pseudomonas* N842 and mutants, *Agric. Biol. Chem.* **45:**2175–2182.

Nishio, N., Tanaka, M., Matsuno, R., and Kamikubo, T., 1977, Production of vitamin B_{12} by methanol-utilizing bacteria, *Pseudomonas* AM1 and *Microcyclus eburneus*, *J. Ferment. Technol.* **55:**200–203.

Oldenhuis, R., Vink, R. L. J. M., Janssen, D. B., and Witholt, B., 1989, Degradation of chlorinated aliphatic hydrocarbons by *Methylosinus trichosporium* OB36 expressing soluble methane monooxygenase, *Appl. Environ. Microbiol.* **55:**2819–2826.

Oki, Y., Kitai, A., Kouno, K., and Ozaki, A., 1973, Production of L-glutamic acid by methanol-utilizing bacteria, *J. Gen. Appl. Microbiol. (Tokyo)* **19:**79–83.

Patel, R. N., Hou, C. T., Laskin, A. I., Derelanko, P., and Felix, A., 1979, Microbial production of methylketones: purification and properties of a secondary alcohol dehydrogenase from yeast, *Eur. J. Biochem.* **101:**401–406.

Patel, R. N., Hou, C. T., Laskin, A. I., Felix, A., and Derelanko, P., 1980, Microbial conversion of gaseous hydrocarbons: production of methyl ketones from corresponding *n*-alkanes by methane-utilizing bacteria, *Appl. Environ. Microbiol.* **39:**727–733.

Patel, R. N., Hou, C. T., Laskin, A. I., and Felix, A., 1982, Microbial oxidation of hydrocarbons: properties of a soluble methane monooxygenase from a facultative methane-utilizing organism, *Methylobacterium* sp. strain CRL-26, *Appl. Environ. Microbiol.* **44:**1130–1137.

Powell, K. A., and Rodgers, B. L. F., 1984, Single cell protein, in: *Methylotrophs: Microbiology, Biochemistry and Genetics* (C. T. Hou, ed.), CRC Press, Boca Raton, FL, pp. 119–144.

Roels, J. A., 1980, Application of macroscopic principles to microbial metabolism, *Biotechnol. Bioeng.* **22:**2457–2514.

Sahm, H., and Wagner, F., 1973, Microbial assimilation of methanol. The ethanol and methanol-oxidising enzymes of the yeast *Candida boidinii*, *Eur. J. Biochem.* **36:**250–256.

Sakai, Y., and Tani, Y., 1986, Formaldehyde production by cells of a mutant of *Candida boidinii* S2 grown in methanol-limited chemostat culture, *Agric. Biol. Chem.* **50:**2615–2620.

Senior, P. J., and Windass, J., 1980, The ICI single cell protein process, *Biotech. Lett.* **2:**205–210.

Shay, L. K., Hunt, H. R., and Wegner, G. H., 1987, High-productivity fermentation process for cultivating industrial microorganisms, *J. Indust. Microbiol.* **2:**79–85.

Shimizu, S., Ishida, M., Kata, N., Tani, Y., and Ogata, K., 1977a, Derepression of FAD pyrophosphorylase and flavin changes during growth of *Kloeckera* sp. no. 2201 on methanol, *Agric. Biol. Chem.* **41:**2215–2220.

Shimizu, S., Ishida, M., Tani, Y., and Ogata, K., 1977b, Production of flavin-adenine dinucleotide by methanol-utilizing yeasts, *J. Ferment. Technol.* **55:**630–632.

Silman, N. J., Carver, M. A., and Jones, C. W., 1989, Physiology of amidase production by *Methylophilus methylotrophus:* isolation of hyperactive strains using continuous culture, *J. Gen. Microbiol.* **135:**3153–3164.

Sirirote, P., Yamane, T., and Shimizu, S., 1986, Production of L-serine from methanol and glycine by resting cells of a methylotroph under automatically controlled conditions, *J. Ferment. Technol.* **64:**389–396.

Solomons, G. L., 1983, Single cell protein, *CRC Crit. Rev. Biotechnol.* **1:**21–58.

Solomons, G. L., 1985, Production of biomass by filamentous fungi, in: *Comprehensive Biotechnology*, Vol. 3 (M. Moo-Young, ed.), Pergamon Press, Oxford, pp. 483–505.

Suzuki, M., Berglund, A., Unden, A., and Heden, C. G., 1977, Aromatic amino acid production by analogue-resistant mutants of *Methylomonas methanolophila* 6R, *J. Ferment. Technol.* **56:**466–475.

Suzuki, T., Yamane, T., and Shimizu, S., 1986, Mass production of poly-beta-hydroxybutyric acid by fully automatic fed batch culture of methylotroph, *Appl. Microbiol. Biotechnol.* **23:**322–329.

Tanaka, Y., Araki, K., and Nakayama, K., 1980, Strain improvement of *Nocardia butanica* for microbial conversion of glycine into L-serine, *J. Ferment. Technol.* **58:**163–170.

Tani, Y., Kanagawa, T., Hanpongkittikun, A., Ogata, K., and Yamada, H., 1978, Production of L-serine by a methanol-utilizing bacterium *Arthrobacter globiformis* SK200, *Agric. Biol. Chem.* **42:**2275–2279.

Tani, Y., Mitani, Y., and Yamada, H., 1982, Utilization of C_1 compounds: phosphorylation of adenylate by oxidative phosphorylation in *Candida boidinii (Kloeckera* sp.) no. 2201, *Agric. Biol. Chem.* **46:**1097–1099.

Tani, Y., Yonehara, T., Mitani, Y., and Yamada, H., 1984, ATP production by sorbitol-treated cells of a methanol yeasts, *Candida boidinii (Kloeckera* sp.) no. 2201, *J. Biotechnol.* **1:**119–127.

Tani, Y., Yoon, B-D., and Yamada, H., 1985a, Production of cytochrome *c* by an obligate methylotroph, *Methylomonas* sp. YK1, *Agric. Biol. Chem.* **49:**2385–2391.

Tani, Y., Sakai, Y., and Yamada, H., 1985b, Isolation and characterization of a mutant of a methanol yeast *Candida boidinii* S2, with a higher formaldehyde productivity, *Agric. Biol. Chem.* **49:**2699–2706.

Terui, G., 1972, Tryptophan, in: *The Microbial Production of Amino Acids* (K. Yamada, S. Kinoshita, T. Tsunoda, and K. Aida, eds.), Kodansha, Tokyo, pp. 515–531.

Thill, G., Davis, G., Stillman, C., Tschopp, J. F., Craig, W. S., Velicelbi, G., Greff, J., Akong, M., Stroman, D., Torregrossa, R. and Siegel, R. S., 1987, The methylotrophic yeast *Pichia pastoris* as a host for heterologous protein production, in: *Microbial Growth on C_1 Compounds* H. W. van Verseveld and J. A. Duine, eds.), Nijhoff, Dordrecht, pp. 289–296.

Unilever, 1987, Process for preparing a catalase-free oxidase and a catalase-free oxidase-containing yeast, and the use thereof, Eur. Patent 242007.
Urakami, T., Terao, I., and Nagai, I., 1981, Process for producing bacterial single cell protein from methanol, in: *Microbial Growth on C_1 Compounds, Proceedings of the 3rd International Symposium* (H. Dalton, ed.), Heyden, London, pp. 349–359.
van Ravenswaay Claasen, J. C., and van der Linden, A. C., 1971, Substrate specificity of the paraffin hydroxylase of *Pseudomonas aeruginosa*, *Ant. van Leeuw.* **37**:339–352.
Verduyn, C., van Dijken, J. P., and Scheffers, W. A., 1983, A simple sensitive and accurate alcohol electrode, *Biotechnol. Bioeng.* **25**:1049–1055.
Vongsuvanlert, V., and Tani, Y., 1988, L-Iditol production from L-sorbose by a methanol yeast, *Candida boidinii (Kloeckera* sp.) no. 2201, *J. Ferment. Technol.* **66**:517–523.
Wegner, G. H., 1990, Emerging applications of the methylotrophic yeasts, in: *Microbial Growth on C_1 Compounds, Proceedings of the 6th International Symposium* (J. R. Andreesen and B. Bowien, eds.), *FEMS Microbiol. Rev.* **87**(Special Issue), pp. 279–283.
Weijers, C. A. G. M., van Ginkel, C. G., and de Bont, J. A. M., 1988, Enantiomeric composition of lower epoxyalkanes produced by methane, alkane, and alkene-utilizing bacteria, *Enz. Microb. Technol.* **10**:214–218.
Wilkinson, T. G., Topiwala, H. H., and Hamer, G., 1974, Interactions in a mixed bacterial population growing on methane in continuous culture, *Biotechnol. Bioeng.* **16**:41–59.
Windass, J. D., Worsey, M. J., Pioli, E. M., Pioli, D., Barth, P. T., Atherton, K. T., Dart, E. C., Byrom, D., Powell, K., and Senior, P. J., 1980, Improved conversion of methanol to single cell protein by *Methylophilus methylotrophus*, *Nature (Lond.)* **287**:396–401.
Yamada, H., Morinaga, Y., and Tani, Y., 1982, L-Methionine overproduction by ethionine-resistant mutants of obligate methylotroph strain OM33, *Agric. Biol. Chem.* **46**:47–55.
Yamada, H., Miyazaki, S. S., Shirae, H., and Izumi, Y., 1985, Threonine production from glycine and ethanol by a methanol utilizing bacterium, *J. Ferment. Technol.* **63**:507–513.
Yamada, H., Miyazaki, S. S., and Izumi, Y., 1986, L-Serine production by a glycine resistant mutant of methylotrophic *Hyphomicrobium methylovorum*, *Agric. Biol. Chem.* **50**:17–21.
Yonehara, T., and Tani, Y., 1988, ATP production by a methanol yeast, *Candida boidinii (Kloeckera* sp.) no 2201: effects of sorbitol treatment and zinc on cell structure as to ATP production, *Agric. Biol. Chem.* **52**:909–914.
Yoon, B. D., Uena, M., and Tani, Y., 1987, Improvement in cytochrome *c* production by glycine analog-resistant mutants of *Methylomonas* sp., *J. Ferment. Technol.* **65**:629–634.

Species Index

Acetobacter
  methanolicus, 158
  methanolicus MB58, 166, 171
Acidomonas, 71
Amycolatopsis methanolica, 72
Arthrobacter, 72
  B-175, 172
  globiformis, 258
P1, 72, 155, 172
Azotobacter, 127
Bacillus
  methanicus, 26
  methylicus, 3
  sp. PM6, 156
  sp. S2A1, 156
  stearothermophilus, 253
Blastobacter, 68, 73
Brevibacterium fuscum, 24, 166
Butyribacterium methylotrophicum, 74
Candida
  boidinii, 209
  boidinii KD1, 218
  boidinii no. 2201, 234, 265
  boidinii no. 2207, 259
Fusarium graminearum, 251
Hansenula
  anomola, 260
  polymorpha, 208, 209, 223
Hyphomicrobium, 71–72
  E.G., 2
  methylovorum KM146, 258
  X, 158
Klebsiella pneumoniae, 124, 127, 259
Kloeckera sp. 2201, 220
Methanohalophilus mahii, 74
Methanomonas methano-oxidans, 26
Methanosarcina barkeri, 74, 262
Methylobacillus
  flagellatum, 187
  flagellatum KT, 163
  glycogenes, 55

Methylobacillus (cont.)
  description of, 56–57
Methylobacter
  chroococcum, 51
  description of genus, 50
Methylobacterium, 192
  AM1, 131
  description of genus, 63
  ethanolicum, 120
  extorquens AM1, 158, 192
  organophilum XX, 192
  sp. strain CRL-26, 93
Methylococcus
  capsulatus (Bath), 93, 133
  capsulatus (strain M), 99
  description of genus, 49
  fulvus, 52
  gracilis, 34
  luteus, 34, 52
  minimus, 52
  mobilis, 52
  thermophilus, 34, 52
  ucrainicus, 52
Methylocystis
  description of genus, 51
  parvus OBBP, 116, 264
Methylomonas, 158
  aminofaciens, 261
  clara, 55, 186, 251
  description of genus, 49
  flagellata, 52, 266
  gracilis, 52
  margaritae, 52
  methanica, 151
  methanitrificans, 52
  methanooxidans, 52
  methanophila, 260
  methylovora, 186
  pelagica, 53
  rubra, 35
  rubrum, 99

SPECIES INDEX

Methylomonas (cont.)
   YK1, 261
Methylophaga
  marina, 60
    description of, 61
  thalassica, 61
    description of, 61
Methylophilus
  methylotrophus, 12, 55, 158
    description of, 58–59
  viscogenes, 263
Methylosinus
  description of genus, 50
  sporium, 93
  sp. strain 6, 124
    trichosporium OB3b, 93, 133
Methylovarius, 52, 59
Methylovibrio soehngenii, 52
Mycobacterium vaccae 10, 166
Nitrosomonas, 11
Nocardia sp. 239, 166, 172, 260
Paracoccus denitrificans, 5, 73, 162
Pichia
  pastoris, 230, 231, 235, 252
  pinus, 209, 224
Protaminobacter sp., 260

Pseudomonas
  3ab, 258
  aminovorans, 67–69
  C, 54, 163
  extorquens NR-1, 259
  FM518, 259
  methanica, 3, 26
  MS31, 258
  oleovorans, 69
  oxalaticus, 5, 162
  sp. K, 264
  sp. KM193, 258
  W6, 171
Rhizobium meliloti, 70
Rhodococcus erythropolis, 161
Rhodopseudomonas, 73
  acidophila, 267
Sporomusa acidoverans, 74
Thiobacillus, 73
Torulopsis sonorensis, 210
Xanthobacter
  autotrophicus, 73
  autotrophicus GJ10, 158
  H4-14, 161
  strain 25a, 169

Subject Index

Acetyl CoA oxidation pathway, 201
Adamantane, 102
Alcohol oxidase, 213, 214, 231, 233, 255, 266
  genes, 227
  mutants, 224
Alcoscan, 231
Amidase, 272
Amino acid production, 257–261
Ammonia assimilation genes, 127–130
Ammonia-oxidizing bacteria, 11
Aromatic amino acids, 260
  production, 260
Autotroph, 150
Auxotrophy markers, 189

Biocatalysis, 266
Biodegradation, 271–272
Biopol, 264
Biosensors, 266
Biotechnology, 246–247
Biotransformation, 267, 271
Broad-host-range vectors, 183
Bromomethane, 90

Catabolite repression, in yeast, 225
Catalase, 214
$^{13}C/^{14}C$-labelled metabolites, production of, 257
Chicken ovalbumin, 253
Chromosomal mapping, methylotrophs, 188–191
Chromosome-mobilizing plasmids, 13
Codon usage
  Methylococcus, 138–139
  Methylosinus, 142
Coenzyme Q, 44
Cofactor regeneration, 265
Cyclopropanol, 159
Cysts, 28
Cytochrome c, 261

Cytochrome P450, 100–103
Cytoplasmic NAD-linked formaldehyde dehydrogenase, 161

Dichloroethylene, 103
Dichloromethane, 5, 10, 71, 117
  mutagenesis, 117–118
Dihydroxyacetone, production, 232
Dihydroxyacetone kinase, 214
Dihydroxyacetone synthase, 214
Dimethyl sulfide, 3, 71
Dimethyl sulfoxide, 71
Dye-linked aldehyde dehydrogenase, 161

Economics, 246–247
Electroporation, 123
Enhanced oil recovery, 264
Epoxides, 10
Exospores, 27

FAD, 233
  biosynthesis, 220
Fermentation design, 255
Ferromagnetic coupling, 97
Formaldehyde
  metabolism, yeasts, 220–222
  oxidation, 160–162, 163
    in RuMP cycle methylotrophs, 162–166
Formaldehyde dehydrogenase, 214
  yeast, 214
Formate, oxidation, 160–162
Formate dehydrogenase, 214, 231, 266
  molybdenum cofactor, 162
  soluble NAD-dependent, 162
  yeast, 215
Fructose bis phosphate (FBP), 152
Fructose bis phosphate aldolase, 152, 157
Fructose bis phosphate aldolase/transaldolase, 157

SUBJECT INDEX

Gif system, 102
Global warming, 2
Glucose-6-phosphate dehydrogenase, 166
Glutamate, 260
Glutamate dehydrogenase, 12, 252
Glutamine synthetase, 127
  gene, 127–130
Glutamine synthetase/glutamate synthase
  pathway, 252
Glycerate kinase, 201
Glycerol production, 232
Glycine, 258
Glycosylation, 228, 235, 256
Guanosine monophosphate, 251

Hemerythrin, 95, 99
Hepatitis B surface antigen, 228, 235, 256
Heterologous gene expression, in yeasts,
  234–235
Hexulosephosphate isomerase, 166–168
Hexulosephosphate synthase, 152, 166–168
Hfr-like donors, 190–191
Human tumor necrosis factor, expression
  in yeast, 228

ICI process, 250
Inosine monophosphate, 251
γ-Interferon, 253
In vitro transcription/translation, 128
Iron ligands, 108
Isocitrate lyase, 14
Isoprenoid compounds, 44

2-Keto-3-deoxy-6-phosphogluconate
  (KDPG), 152
KDPG aldolase (KDPGA variant), 152
KDPG aldolase/transaldolase, 157

Leucine, 261
Lysine, 260

Malyl CoA lyase, 201
Marine methane-utilizing bacteria, 53
Marine methanol-utilizing bacteria, 59–60
Marker exchange, 184
  mutagenesis, 125
Mass spectrometry, 98
Methane, 2, 9

Methane monooxygenase, 85
  active site, 108–110
  amino acid sequence homologies, 108–110
  biotransformations, 268–270
  catalytic cycle, 99
  cooxidation, 101
  copper ions, 89
  electron transfer to, 93
  gene cluster, 135
  gene expression, 139
  gene probe, 139
  genes, 133–142
  hydroxylase component (Protein A), 93–98
  iron center, 95–97
  iron-sulfur center, 92, 137
  kinetic mechanism, 98–99
  mechanism of action, 103–108
  mutants, 117
  particulate, 88–90
  polypeptide sequence homologies, 143
  protein B, 92–93, 97
  protein C, 90–92, 137
  soluble, 90–93
  stereochemistry, 102
  substrates for, 101–103
  substrate specificity, 99
Methane thiol, 3
Methanogenic bacteria, 2
Methanol, 9
  regulation of metabolism in yeast, 218–220
  carbon flux in yeasts, 222
Methanol dehydrogenase, 158, 266
  Bacillus, 159
  in Gram-negative bacteria, 157–158
  in Gram-positive bacteria, 158–160
  NAD-linked, 159–160
  regulation, 168–169
Methanol dehydrogenase genes, in methanotrophs, 130–133
Methanol oxidase, 270–271
Methanol oxidation (Mox)
  genes, 191–199
    characterization, 196–199
    codon usage, 198
    functions, 195
    in *Methylobacterium*, 192–194
    signal sequence, 197

SUBJECT INDEX 285

Methanol oxidation (Mox) (*cont.*)
   transcriptional start site, 197
   mutants, 194–196
Methanol regulated genes, in yeast, 225–226
Methanol-utilizing bacteria, plasmids in, 186
Methanol-utilizing yeasts, 207–244
   application in industry, 229
   enzymology, 213–218
   life cycle, 211
   taxonomy, 210–213
Methanotrophs
   antibiotic sensitivites, 39
   bacteriophages, 120
   carbon assimilation pathways, 29
   classification scheme, 26
   conjugation systems, 120–123
   DNA:DNA homologies, 43
   gene expression, 129–130
   internal membranes, 29
   mutagenesis, 116–118
   pigmentation, 34
   plasmids in, 118–120
   transformation, 123
Methazotrophs, 150
Methionine production, 259–260
Methyl ketones, 271
Methylophilan, 263
Methylotrophs
   biotechnology, 9–11
   definition of, 150
   ecology of, 7–9
   expression of foreign genes, 191
   genetics of, 11–14, 183–206
   mutagenesis, 187
   phylogeny, 75–77
   vectors for, 185
Modifier protein, 158
Mouse dihydrofolate reductase, 253
Mutagenesis, yeasts, 222–228
Mycoprotein, 251

NAD(P)-linked glucose-6-phosphate dehydrogenase, 163
Naphthalene oxidation, 118
Nitrogen fixation, 126
Nitrogen fixation (*nif*) structural genes, 124
Norprotein, 250

*Ntr*A-dependent promoters, 128, 141
Nuclear magnetic resonance spectroscopy, 217

Oligonucleotide probe, 134
μ-Oxo bridge iron center, 95

Periplasmic space, 157, 159, 160, 261
Peroxisome, 214, 220
Phenylalanine, 260
Phosphatase, 214
Phosphofructokinase, 152, 157
6-Phosphogluconate dehydrogenase, 163, 166, 170
Phospho-3-hexuloisomerase, 152
Pink-pigmented facultative methylotrophs (PPFMs), 4, 63
Poly 54, 163
Poly-β-hydroxybutyrate (PHB), 263, 264–165
Polysaccharides, 263–264
Pressure cycle fermentor, 253
Probion, 250, 251
Propene oxidation, 97
Propene oxide, 269–270
Provesteen, 250, 252, 253–256
Pruteen, 9, 145, 250, 251, 252–253
Purple acid phosphatase, 95
Pyrroloquinoline quinone (PQQ), 131, 157, 196
   synthesis genes, 199–200

Quinoprotein alcohol dehydrogenase, 158

Reduced glutathione, 161
Restricted facultative methylotrophs, 61–62
Ribonucleotide reductase, 95, 136, 141
   B_2 proteins, 134
   B_2 subunit, 108, 109
16S Ribosomal RNA, 8, 75
Ribulose monophosphate
   cycle, 151–157
      distribution, 155
      energetics of, 154
      genes, 202
      mutants, 187
      regulation, 169–173
   pathway
      distribution of variants, 156–157
      occurrence of, 154–156

Ribulose-5-phosphate 3-epimerase, 173
Ribulose-5-phosphate isomerase, 173
Rosette formation, 38
R-Prime mapping, 189–190

Serine
  pathway genes, 200–202
  production, 257–259
Single cell protein, 9, 229, 246, 248–256
  from methanol, 249–256
Sorbitol, 233, 234
Synthesis gas, 87, 249

Taxonomy, 23
Tetrad analysis, 223
Tetrahydrofolate, 161
T7 expression vector, 130
Thermotolerant methylotrophic *Bacillus*, 159
Transaldolase, 173
Transketolase, 173
Transposon mutagenesis, methylotrophs, 187–188
Trichloroethylene, 10, 103, 272
Tryptophan, 260
Tumor necrosis factor, 256
Tyrosine, 260

Valine, 261
Versatile methylotrophs, 156
Vitamin B_{12}, 262
Vitamins and coenzymes, production of, 261–262

Xanthan gum, 264
Xylulose monophosphate (XuMP) cycle, 216–218
  key enzymes, 216

Yeast
  auxotrophic mutants, 222
  classical genetics, 223–224
  expression of genes in, 255
  expression of heterologous genes, 227–228
  fermentations, 255
  genetics and molecular biology, 222–228
  methanol pathway mutants, 224–225
  2-μm plasmid, 226
  plasmid vectors for, 226–227
  production of chemicals, 232–234
  single-cell protein, 253
  transformation systems, 226–227